Philosophies of the Sciences

Philosophies of the Sciences
A Guide

Edited by Fritz Allhoff

WILEY-BLACKWELL

A John Wiley & Sons, Ltd., Publication

This edition first published 2010
© 2010 Blackwell Publishing Ltd

Blackwell Publishing was acquired by John Wiley & Sons in February 2007. Blackwell's publishing program has been merged with Wiley's global Scientific, Technical, and Medical business to form Wiley-Blackwell.

Registered Office
John Wiley & Sons Ltd, The Atrium, Southern Gate, Chichester, West Sussex, PO19 8SQ, United Kingdom

Editorial Offices
350 Main Street, Malden, MA 02148-5020, USA

9600 Garsington Road, Oxford, OX4 2DQ, UK

The Atrium, Southern Gate, Chichester, West Sussex, PO19 8SQ, UK

For details of our global editorial offices, for customer services, and for information about how to apply for permission to reuse the copyright material in this book please see our website at www.wiley.com/wiley-blackwell.

The right of Fritz Allhoff to be identified as the author of the editorial material in this work has been asserted in accordance with the Copyright, Designs and Patents Act 1988.

Library of Congress Cataloging-in-Publication Data
Philosophies of the sciences: A guide / edited by Fritz Allhoff.
     p. cm.
   Includes bibliographical references and index.
   ISBN 978-1-4051-9995-7 (hardcover : alk. paper)   1. Science—Philosophy.
I. Allhoff, Fritz.
   Q175.G782 2010
   001.01–dc22
                                                        2009018244

A catalogue record for this book is available from the British Library.

Set in 10.5/13pt Minion by Graphicraft Limited, Hong Kong
Printed in Singapore by Ho Printing Singapore Pte Ltd

01   2010

Contents

Notes on Contributors

Fritz Allhoff is an Assistant Professor in the Department of Philosophy at Western Michigan University and a Research Fellow in the Centre for Applied Ethics and Public Philosophy at the Australian National University. At Western Michigan University, he is also an adjunct Assistant Professor in the Mallinson Institute for Science Education and the Director of the History and Philosophy of Science Working Group. He has held visiting fellowships at the Center for Philosophy of Science at the University of Pittsburgh and at the Brocher Foundation (Geneva, Switzerland). He has research and teaching interests in the history and philosophy of biology, from Darwin through contemporary debates.

William Bechtel is a Professor in the Department of Philosophy and the interdisciplinary programs in Science Studies and Cognitive Science at the University of California, San Diego. His research has sought to characterize mechanistic explanations, the processes by which they are developed and evaluated, and the ways they are represented. His current focus is on dynamical activity in self-organizing and sustaining mechanisms in biology, especially ones exhibiting oscillations. His recent books include *Discovering Cell Mechanisms* (2006) and *Mental Mechanisms* (2008).

Otávio Bueno is a Professor in the Department of Philosophy at the University of Miami. His research concentrates in philosophy of science, philosophy of mathematics, and philosophy of logic, and he is currently developing a fictionalist conception that is empiricist about science and nominalist about mathematics. He has held visiting professorships or fellowships at Princeton University, the University of York (UK), the University of Leeds, and the University of São Paulo. He has published papers in journals such as *Philosophy of Science, Synthese, Journal of Philosophical*

Logic, Studies in History and Philosophy of Science, British Journal for the Philosophy of Science, Analysis, Erkenntnis, and *History and Philosophy of Logic.*

Chris J.J. Buskes is a Lecturer in the Philosophy of Science at the Radboud University Nijmegen, The Netherlands. His work focuses on epistemology and general philosophy of science, with a special interest in the philosophy of biology. Among his publications is *The Genealogy of Knowledge* (1998), which offers a critical appraisal of evolutionary approaches to epistemology and philosophy of science. His latest book *Evolutionair Denken* (*Evolutionary Thinking,* 2006), aimed at a general public, was awarded the 2007 Socrates Goblet for the most stimulating Dutch book on philosophy and is now being translated into Spanish, German, and Polish.

Henk W. de Regt is a Lecturer in Philosophy of Science at the Faculty of Philosophy, VU University Amsterdam, The Netherlands. His present research focuses on the topics of scientific understanding and values in science. He has published in journals such as *Philosophy of Science, Synthese, British Journal for the Philosophy of Science,* and *Studies in History and Philosophy of Science.* De Regt is co-founder and member of the Steering Committee of the European Philosophy of Science Association, co-founder of the Society for the Philosophy of Science in Practice, and editor-in-chief of the Dutch philosophy journal *Algemeen Nederlands Tijdschrift voor Wijsbegeerte.*

Richard DeWitt is a Professor in the Department of Philosophy at Fairfield University. His areas of specialization include the history and philosophy of science, logic, and the philosophy of mind. His recent publications include work on many valued and fuzzy logics, especially the question of the axiomatizability of fuzzy logics; a publication (with R. James Long) involving medieval logic; and an introductory book on the history and philosophy of science.

Matthew H. Haber is an Assistant Professor in the Department of Philosophy at the University of Utah. His research in philosophy of biology has focused on conceptual issues in systematics, including phylogenetic inference, the nature of biological taxa and hierarchies, foundational issues in nomenclature and classification, and the structure of phylogenetic thinking. He also has research interests looking at the overlap of philosophy of biology and applied bioethics, most notably concerning the ethics of part-human embryonic research.

Andrew Hamilton is an Assistant Professor in the School of Life Sciences at Arizona State University, where he is active in the Center for Biology and Society, the Center for Social Dynamics and Complexity, and the International Institute for Species Exploration. His research focuses on the conceptual and theoretical foundations of the biological sciences, particularly evolutionary theory and systematics. Hamilton's work has appeared in philosophy, history, and science journals, including *Philosophy of Science, Biological Theory, Isis*, and *PLoS Biology*.

Daniel M. Hausman is the Herbert A. Simon Professor in the Department of Philosophy at the University of Wisconsin, Madison. His work lies mainly at the intersection between economics and philosophy and addresses issues concerning methodology, rationality, causality, and ethics; and he was the co-founder of the journal *Philosophy and Economics*. Among his many publications, *The Inexact and Separate Science of Economics* (1992), *Economic Analysis, Moral Philosophy, and Public Policy* (jointly with Michael McPherson, 2006), and *The Philosophy of Economics: An Anthology* (3rd edn., 2007) address general philosophical questions concerning economics.

Mitchell Herschbach is a PhD candidate in Philosophy and Cognitive Science at the University of California, San Diego. His dissertation integrates research from analytic philosophy of mind, phenomenology, cognitive science, developmental psychology, and neuroscience about human social understanding, particularly our folk psychological capacities and the cognitive mechanisms underlying them.

Maarten G. Kleinhans is a Lecturer in Earth Surface Dynamics at the Faculty of Geosciences, University of Utrecht, The Netherlands. His publications focus on the question of how the beautiful dynamical forms and patterns (and lack thereof) in rivers, deltas, and coastal seas are the result of smaller-scale transport and sediment size-sorting processes on Earth and on Mars. In addition, he works on philosophy of geosciences in practice in collaboration with philosophers of science. He is a member of the Young Academy of the Royal Netherlands Academy of Arts and Sciences and of the Descartes Centre for the History and Philosophy of the Sciences and the Humanities, University of Utrecht.

Daniel Little is Chancellor and Professor of Philosophy at the University of Michigan, Dearborn. His areas of research and teaching fall within the philosophy of the social sciences and the philosophy of history. His books include *Microfoundations, Method and Causation: Essays in the Philosophy of the Social Sciences* (1998) and *The Paradox of Wealth and Poverty: Mapping the Ethical Dilemmas of Global Development* (2003); his current book, *History's Pathways: Towards a New Philosophy of History*, will appear in 2010.

Aidan Lyon is a PhD candidate in Philosophy at the Research School of Social Sciences, at the Australian National University. His current research lies in the areas of philosophy of science, philosophy of physics, and philosophy of probability. His dissertation investigates how the concept of objective probability is used in branches of science such as quantum mechanics, statistical mechanics, and evolutionary theory. Some of his recent work has been on the indispensability of mathematical explanations in science, and on mathematical realism.

Edouard Machery is an Assistant Professor in the Department of History and Philosophy of Science at the University of Pittsburgh, as well as a resident fellow of the Center for Philosophy of Science (University of Pittsburgh) and a member of the Center for the Neural Basis of Cognition (Carnegie Mellon University and University of Pittsburgh). His research focuses on the philosophical issues raised by the cognitive sciences. He has published on a wide range of topics in the philosophy of psychology, including the nature of concepts, racial cognition, evolutionary psychology, innateness, and moral psychology. He is the author of *Doing without Concepts* (2009).

Jay Odenbaugh is an Assistant Professor in the Department of Philosophy and a faculty member of the Environmental Studies Program at Lewis and Clark College. His research is in the philosophy of biology, especially ecology and environmental ethics. He is currently working on a book entitled *On the Contrary: A Philosophical Examination of the Environmental Sciences and their Critics*, and editing a collection of essays entitled *After Environmentalism* (with Bill Chaloupka and Jim Proctor).

Samir Okasha is a Professor of Philosophy at the University of Bristol. Previously, he has worked at the University of York (UK), the London School of Economics, and the National University of Mexico. Much of his work focuses on conceptual and foundational problems in evolutionary biology, though he also works in general philosophy of science. His most recent book, *Evolution and the Levels of Selection* (2006), offers a comprehensive examination of the debate over units and levels of selection.

Joachim Schummer is Heisenberg Fellow at the University of Darmstadt. After double-graduation in chemistry and philosophy he has held teaching and research positions at the University of Karlsruhe, University of South Carolina, University of Darmstadt, Australian National University, and University of Sofia. His research interests focus on the history, philosophy, sociology, and ethics of science and technology, with emphasis on chemistry and, more recently, nanotechnology. Schummer has published the first monograph (*Realismus und Chemie/Realism and Chemistry*; 1996) and more than 30 papers on the philosophy of chemistry. His recent collaborative book publications include *Discovering the Nanoscale* (2005), *Nanotechnology Challenges* (2006), and *The Public Image of Chemistry* (2007). He is the founding editor of *Hyle: International Journal for Philosophy of Chemistry*.

Figures

Unit 1
Introduction

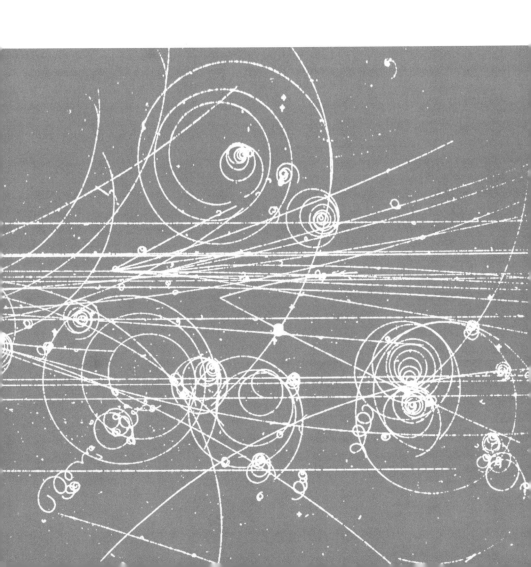

1 Philosophies of the Sciences[1]

Fritz Allhoff

Philosophy of science is changing. Or maybe it already has. During the twentieth century, there was a tremendous amount of brilliant work done in the philosophy of science by people such as Rudolf Carnap (1928), Karl Popper (1935),[2] Thomas Kuhn (1996), Imre Lakatos (1999),[3] and so on. But what defined much of that work was its generality: all of these people, as well as the traditions in which they worked, were concerned with general methodological issues in science. Whether talking about the logical structure of science, the demarcation of science from other endeavors, the nature of scientific change, or the features and merits of scientific research programs, these contributions almost completely transcended disciplinary boundaries within science. Certainly people were very impressed with physics, especially the experimental success of Einstein's theory of general relativity;[4] philosophy of physics was thoroughly launched in the early twentieth century and, like philosophy of logic and philosophy of mathematics, it has roots in ancient Greece.[5] And Popper evinced some hostility to natural selection – complaining that it was "almost tautological" (Popper 1935, §37) – though he eventually recanted (Popper 1978). But aside from a few exceptions, most of the work for which philosophy of science in the twentieth century will be remembered – and especially from the 1920s through the 1960s – is not endemic to any particular scientific discipline.

---

[1] I thank Zvi Biener and Marc-Alspector Kelly for helpful comments on this Introduction.

[2] See also Popper 1963.

[3] Lakatos's "Science and Pseudoscience" (1973) is a celebrated public lecture and quite accessible. It is reprinted in many places and available online at http://www.lse.ac.uk/collections/lakatos/scienceAndPseudoscience.htm (accessed October 3, 2008).

[4] See, for example, Dyson, Eddington, and Davidson 1920.

[5] See, for example, Sklar 1992, esp. 11–25. Biology also has some ancient roots as Aristotle wrote about it at length. See, especially, Aristotle 1984a, 774–993. See also Aristotle 1984b, 994–1086.

Shortly after the publication of *Structure*, and with all due respect to Kuhn, there started to be a paradigm shift away from this sort of approach to the philosophy of science or, if that is too strong, at least an increased attention given to individual sciences. Debates over reductionism, especially as catalyzed by Ernest Nagel's work,[6] brought individual sciences into discussions about the philosophy of science. However, the issue of whether biology, for example, reduces to physics is neither a question in the philosophy of biology nor a question in the philosophy of physics, but rather a general one in the philosophy of science, having to do with how to think about bridge laws, translations, and so on. In other words, merely talking about biology or physics – as relevant to some more general question – hardly means that one has started doing philosophy of biology or philosophy of physics. Some of those issues can be incorporated into the philosophy of biology – such as whether Mendelian genetics reduces to molecular biology – but these issues were not yet raised in the early 1960s, nor really until two or three decades later.[7]

In the 1960s and 1970s, philosophical attention began to be afforded to individual sciences in a new way. Again, this is not to imply that no philosophical attention had been given to those sciences before these decades, but rather that new emphases were born. While being careful not to overstate the case, the emergence of philosophy of biology played a large role in this transition. It is probably safe to say that those interested in biology had begun to tire of attempts to subsume biology under physics and were highly motivated to show that they had their own (irreducible) research program. People like Morton Beckner (1959), Marjorie Grene (1959), David Hull (1964, 1969, 1974), Kenneth Schaffner (1967a, 1967b), William Wimsatt (1972), Michael Ruse (1973), and others made substantial contributions to philosophy of biology, particularly in terms of launching it as a *sui generis* discipline and not one wholly subsumed by either physics (whether philosophically or scientifically) or the philosophy of science more generally.[8]

---

[6] See, especially, Nagel 1961.

[7] See, for example, Kitcher 1984 and Waters 1990. See Schaffner 1967a, 1967b, for earlier attempts.

[8] Or at least so goes the standard story, to which a challenge is issued in Byron forthcoming. Byron collects bibliometric data from four major philosophy of science journals during the time period, 1930–59, arguing that philosophy of biology was already underway. See also Grene and Depew 2004.

I actually think that Charles Darwin should be given more credit *vis-à-vis* philosophy of biology than he usually is. The levels of selection debate, for example, starts with Darwin, even if he was sloppy on the relevant question. Compare, for example, Darwin 1964, 61–3 and Darwin 1998a, 137; the former postulates selection on individuals and the latter selection on groups. Darwin also had some underappreciated thoughts on scientific explanation; see Darwin 1998b, 9–14.

Even some biologists – as opposed to academic philosophers – helped to bring more attention to philosophical problems in biology.[9]

As we close out the first decade of the twenty-first century, the once-fledgling field of philosophy of biology has now flourished, and other philosophies of sciences have emerged. Cognitive science, for example, hardly existed as a *scientific field* through most of a twentieth-century psychological landscape dominated by behaviorism.[10] Its emergence in the 1970s has catalyzed tremendous success in understanding the physical basis of mind and behavior, and in a very short time period it has spawned an entire philosophical discourse.[11] Judging from professional meetings and job listings in philosophy, the philosophy of cognitive science – perhaps including its ally, philosophy of psychology – has become one of the most active and popular sub-disciplines within the philosophy of science.

This book presents a collection of essays which celebrate the individual sciences and their attendant philosophical issues. The hope, though, is not just to study each of these in isolation, but rather to present them collectively, in the hope that there might be some sort of cross-fertilization of ideas. For example, maybe scientific laws are thought of differently in physics than in biology and, before this book, little attention has been given to how laws are thought of in the earth sciences at all. Reductionism is an issue in myriad scientific fields, yet many of them have different takes on why they (do not) reduce to physics. Scientific explanation is another area in which different sciences all owe some account, yet these projects are variously executed and rarely look like anything the deductive-nomological (or inductive-statistical) models would have required. By thinking about how individual sciences each carry out aspirations that all sciences have in common, we can gain a better understanding of the associative theoretical foundations.

Let me offer some remarks on the organization of the volume and the essays contained within. The original idea was to do "philosophies of the special sciences" which, roughly, includes all the sciences except for physics.[12]

---

[9]   See, for example, Gould and Lewontin 1979.

[10]   See Bechtel and Herschbach, "Philosophy of the Cognitive Sciences," §1 (Chapter 10 in this volume), for a more detailed discussion.

[11]   To be fair, there were some progenitors to the contemporary movement, including, in some respects, Aristotle and René Descartes. David Hume would certainly have been sympathetic to current projects, and his moral psychology, for example, was somewhat empirical, if not after underlying mechanisms.

[12]   See, for example, Fodor 1974.

But there were several reasons inveighing against this approach. First, it is important to have an essay on the philosophy of physics in a book like this, both for pedagogical reasons and to provide appropriate context for many of the other essays. Second, it is important to have an essay on general scientific methodology for the same reasons; including only philosophies of the special sciences would have precluded either of these essays. Third, it is important to have essays on the philosophies of mathematics, logic, and probability. These might seem like odd inclusions in the book since those three are not even widely regarded as (physical) sciences. Nevertheless, there are obvious reasons to include them insofar as nearly all other sciences make fundamental use of mathematics, logic, and/or probability and are therefore (philosophically) beholden to them. In recent years, philosophical issues in probability have been especially important to more general questions in the philosophy of science (e.g., confirmation theory), so the essay addressing this topic certainly deserves to be included. Lastly, unlike many of the other sciences considered, these have long and substantive histories; we might usefully compare these more mature disciplines to the more nascent ones.

The first unit contains this introductory essay, as well as a general introduction to the philosophy of science. The second unit, "Philosophy of the Exact Sciences," offers the aforementioned essays on logic, mathematics, and probability. In the third, we turn to the natural sciences, exploring the standard triumvirate of physics, chemistry, and biology, as well as the earth sciences. The fourth unit, "Philosophy of the Behavioral and Social Sciences," starts with "Philosophy of the Cognitive Sciences," which is complemented by the following essay, "Philosophy of Psychology." This fourth unit concludes with "Philosophy of Sociology" and "Philosophy of Economics."

Finally, the following essays are meant to be widely accessible; the contributors were told to conceive of their audiences as "advanced undergraduates." A primary aim for this book is to change the way that (at least some) philosophy of science courses are taught; rather than just focusing on general methodological issues, students can be exposed to specific philosophical discussions pertaining to individual sciences. This is not to say that some general grounding in philosophy of science is not essential for that project, whether in a previous course or in the first half of the term. Rather, the essays that follow should complement more traditional studies in the philosophy of science and, indeed, will not make sense without such background exposure. A strong secondary target for the book, though, is researchers in the fields: it is hoped that a philosopher of physics, for example, might gain perspective by learning about the philosophy of chemistry, or even any of the other fields. Again, cross-fertilization is a principal aim of the book and the fields represented should have a lot to say to each other.

The essayists were charged to present central debates in their fields, while not taking a stance on the associative resolutions; references are presented for tracking down further discussion. There are obvious challenges in presenting an entire discipline within even a long 8,000–10,000 word essay, so a premium was placed on breadth and, again, accessibility. I hope that the essays are valuable and important, and I thank the contributors for sharing their expertise and dealing admirably with all too much editorial feedback.

References

Aristotle. 1984a. "History of Animals," trans. D.W. Thompson. In Jonathan Barnes (ed.), *The Complete Works of Aristotle*. Princeton, NJ: Princeton University Press, 774–993.

Aristotle. 1984b. "Parts of Animals," trans. W. Ogle. In Jonathan Barnes (ed.), *The Complete Works of Aristotle*. Princeton, NJ: Princeton University Press, 994–1086.

Beckner, M. 1959. *The Biological Way of Thought*. New York: Columbia University Press.

Byron, J.M. Forthcoming. "Whence Philosophy of Biology?" *British Journal for the Philosophy of Science*.

Carnap, R. 1928. *Der Logische Aufbau der Welt*. Leipzig: Felix Meiner Verlag. English translation: R.A. George. 1967. *The Logical Structure of the World and Pseudoproblems in Philosophy*. Berkeley, CA: University of California Press.

Darwin, C. 1964. *On the Origin of Species*. Cambridge, MA: Harvard University Press.

Darwin, C. 1998a. *The Descent of Man*. New York: Prometheus Books.

Darwin, C. 1998b. *The Variation of Animals and Plants under Domestication*, vol. 1. Baltimore, MD: Johns Hopkins University Press.

Dyson, F.W., A.S. Eddington, and C.R. Davidson. 1920. "A Determination of the Deflection of Light by the Sun's Gravitational Field, from Observations Made at the Total Eclipse of May 29, 1919." *Philosophical Transactions of the Royal Society of London* 220: 291–333.

Fodor, J. 1974. "Special Sciences (Or: The Disunity of Science as a Working Hypothesis)." *Synthese* 28: 97–115.

Gould, S.J. and R.C. Lewontin. 1979. "The Spandrels of San Marco and the Panglossian Paradigm: A Critique of the Adaptationist Programme." *Proceedings of the Royal Society of London, Series B* 205.1161: 581–98.

Grene, M. 1959. "Two Evolutionary Theories, I–II." *British Journal for the Philosophy of Science* 9: 110–27, 185–93.

Grene, M.G. and D.J. Depew. 2004. *The Philosophy of Biology: An Episodic History.* New York: Cambridge University Press.

Hull, D.L. 1964. "The Metaphysics of Evolution." *British Journal for the Philosophy of Science* 3: 309–37.

Hull, D.L. 1969. "What Philosophy of Biology Is Not." *Synthese* 20: 157–84.

Hull, D.L. 1974. *Philosophy of Biological Science.* Englewood Cliffs, NJ: Prentice Hall.

Kitcher, P. 1984. "1953 and All That: A Tale of Two Sciences." *Philosophical Review* 93.3: 335–73.

Kuhn, T. 1996. *The Structure of Scientific Revolutions.* 3rd edn. Chicago: University of Chicago Press.

Lakatos, I. 1973. "Science and Pseudoscience." http://www.lse.ac.uk/collections/lakatos/scienceAndPseudoscience.htm (accessed October 3, 2008).

Lakatos, I. 1999. *The Methodology of Scientific Research Programmes.* J. Worrall and G. Currie (eds.). Cambridge: Cambridge University Press [1978].

Nagel, E. 1961. *The Structure of Science: Problems in the Logic of Scientific Explanation.* New York: Harcourt, Brace & World.

Popper, K. 1935. *Logik der Forschung.* Vienna: Julius Springer Verlag. English translation: *The Logic of Scientific Discovery.* 6th edn. London: Hutchinson, 1974 [1959].

Popper, K. 1963. *Conjectures and Refutations: The Growth of Scientific Knowledge.* London: Routledge.

Popper, K. 1978. "Natural Selection and the Emergence of Mind." *Dialectica* 32.3–4: 339–55.

Ruse, M. 1973. *The Philosophy of Biology.* London: Hutchinson.

Schaffner, K.F. 1967a. "Approaches to Reduction." *Philosophy of Science* 34: 137–47.

Schaffner, K.F. 1967b. "Antireductionism and Molecular Biology." *Science* 157: 644–7.

Sklar, L. 1992. *Philosophy of Physics.* Boulder, CO: Westview Press.

Waters, C.K. 1990. "Why the Antireductionist Consensus Won't Survive the Case of Classical Mendelian Genetics." *PSA 1990* 1: 125–39.

Wimsatt, W.C. 1972. "Teleology and the Logical Structure of Function Statements." *Studies in the History and Philosophy of Science* 3: 1–80.

2 Philosophy of Science

Richard DeWitt

 One new to the history and philosophy of science might reasonably wonder whether philosophy has anything substantive to do with science (§1). In this introductory essay, we will have a chance to explore a number of examples from science, and in each case we will find that it does not take long to uncover difficult and substantive philosophical issues. In many cases, these philosophical issues directly affect the day-to-day workings of science, such as confirmation, falsification, realism/instrumentalism, and underdetermination (§2). In other cases, the questions are more inherently philosophical in nature, such as scientific explanation, laws of nature, and inductive reasoning (§3). This essay provides an introduction to these topics, laying foundations for discussions that will recur throughout the volume.

1.   What Does Philosophy Have to Do with Science?

When people first hear the phrase "philosophy of science," a common reaction is a puzzled sort of look, often followed by a question along the lines of "what does philosophy have to do with science?" The question is perfectly reasonable. Philosophy and science are often viewed as quite separate enterprises, seemingly having little to do with one another. Philosophy, for example, is often viewed as pursuing the big, meaning-of-life type questions, with science pursuing more down-to-earth type questions.

One of my main goals in this introductory essay will be to illustrate that, no matter where in science one looks, one quickly uncovers substantive philosophical issues. In short, philosophy has a great deal to do with science. Some of the philosophical issues we will uncover are directly relevant to, and directly affect, the day-to-day workings of scientists, while others are philosophical questions and issues concerning some of the foundational concepts and tools used by scientists. But the bottom line is that philosophy is everywhere in science.

My strategy to illustrate these connections between science and philosophy is straightforward: we will look into a variety of aspects of science, and in each case, we will find that it does not take long to uncover substantive philosophical issues. Along the way we will also have a chance to get an overview of some basic and recurring topics in the philosophy of science.

Aside from the introduction and conclusion, this essay is divided into two general sections, each with several subsections. The first major section explores a number of philosophical issues that are relevant to the day-to-day workings of science. The second major section turns to a sampling of philosophical topics involving some of the fundamental concepts and tools of science, for example, scientific explanation, scientific laws, and inductive reasoning.

I should stress that this is an introductory essay, in which we will be quickly passing through a variety of issues and topics. In later essays in this book, you will have a chance to look into some of these issues (and many others) more carefully. You might want to think of this introductory essay as a sort of philosophy of science sampler, with which we can get a taste of some of the ways philosophy ties in with science and a flavor of some of the recurring topics and issues in the philosophy of science.

2.   Philosophical Issues in Scientific Practice

In this first major section, we look briefly at some ways philosophical issues arise in science, with particular emphasis on some issues relevant to the workings of science. There are a large number of such issues, of which we will see just a few. Let us begin with a look at a common characterization of one of the main processes involved in science.

2.1   *Confirmation and disconfirmation*

It is a fairly common belief that science proceeds by observing some class of phenomena – for example, the observed movement of the stars and planets across the night sky. And then one generates a theory, perhaps the theory that the earth and other planets move about the sun in elliptical orbits, with the sun occupying one focus of the ellipse. Then one finds ways to test that theory, generally by making predictions about what we would expect to observe assuming the theory is on the right track – for example, that the point of light we call 'Mars' should be in such and such a position in the night sky on a particular day at a particular time. Then one checks to see if the predictions are correct, and if they are, this provides confirming evidence for the theory, and if not, this disconfirms (or falsifies) the theory.

There is not much question that this process – observation, theory generation, designing tests, and confirmation/disconfirmation – plays an important role in science. But what is not widely appreciated is just how subtle this simple-sounding process is. In fact, the process just outlined is often presented, and I think often (wrongly) viewed, as relatively straightforward, almost a cookbook approach to science. Sure, the details and ingredients might change from one recipe to the next, but overall, a common view is that the scientific process is not that much different from the relatively straightforward recipe-following process most of us employ in our kitchens.

This science-as-recipe-following view is a substantial misconception about the workings of science. When we probe into this process just a bit, we quickly find some difficult, and largely philosophical, issues lurking just beneath the surface. For example, let us look into just the confirmation/disconfirmation aspect of the process; that is, the stage where one checks to see if predictions made by a theory are accurate. As a concrete example, consider again the theory mentioned above, that the earth and other planets move about the sun in elliptical orbits, with the sun occupying one focus of an ellipse. This theory (or, at least, the part involving the motion of Mars) was first put forward by Johannes Kepler (1571–1630) in 1609.[1] In the centuries following the publication of Kepler's theory, predictions based on the theory – such as when and where a planet will appear in the night sky – were quite accurate, and certainly far more accurate than any competing astronomical theory. This large body of accurate predictions provided a vast amount of confirming evidence for Kepler's theory.

On the recipe-following view of science, one might think that centuries' worth of accurate predictions would show that a theory is correct beyond any reasonable doubt. But is this right? Do all these accurate predictions show that Kepler's theory is definitely (or, at least, almost definitely) correct? The answer is clearly no. One general reason (not the only reason, but an important one) is because this sort of reasoning – that is, inferring that a theory is correct based on the theory making accurate predictions – is a

---

[1]   For a recent translation of Kepler's 1609 work, see Gingerich and Donahue 1993.

species of inductive reasoning. And the inductive nature of this reasoning has important implications for theory confirmation.

In general, what characterizes inductive reasoning is that, even in the strongest sort of such reasoning, and when all the premises (that is, the reasons given in support of the conclusion) given are correct, it is still possible for the conclusion to be wrong. So by its very nature, inductive reasoning can at best provide support for a theory but can never show that a theory is definitely correct. (Consider a simple example to illustrate the fallibility of inductive reasoning: I am writing this on a Friday morning, and yesterday morning, as I do every Thursday morning, I put the trash out, because the trash has always been picked up late on Thursday afternoon. So I inferred, inductively and based on years of confirming instances of the trash being picked up on Thursday afternoons, that the trash would again be picked up this Thursday afternoon. But not so. It is Friday morning, and the trash is still out there. I am now inferring, inductively again, that the company is behind because of recent holidays and that the trash will be picked up today. We will see.)

So again, inductive reasoning can at best provide support for a theory but can never show that a theory is definitely correct. And it is worth emphasizing that in the case of Kepler's theory, after over 200 years of accurate predictions, increasingly accurate measuring devices suggested that predictions about planetary motion, based on Kepler's theory, were not entirely accurate (the amount of error was small, but measurable). So in spite of 200 plus years of confirming evidence for Kepler's theory, it turns out that the theory is not quite right after all.

The example above is typical of what happens in science (and in everyday life, for that matter). We might have years or even centuries of confirming evidence for a theory, in the form of accurate predictions made from the theory. Yet it might (and often does) turn out that eventually new evidence arises (often as a result of new technology and new inventions) that is no longer compatible with the original theory.[2]

---

[2] Considerations related to those above lead some to endorse what is often referred to as the "pessimistic induction." This is the inductive reasoning that, since most of our scientific theories in the past have turned out to be mistaken, we can inductively infer that most of our current best theories will likewise eventually turn out to be mistaken. Note that there are a sizable number of examples, in the form of theories that seemed well confirmed but later turned out to be mistaken, that help support the conclusion. On the other hand, also note that this is a bit of inductive reasoning, and as with all inductive reasoning, the conclusion may turn out to be mistaken regardless of how much confirming evidence there is. Such is the nature of inductive reasoning.

What about situations in which a prediction made by a theory does not turn out to be accurate, that is, cases of disconfirmation? If we cannot, by inductive confirmation reasoning, show that a theory is definitely correct, can we at least, with disconfirming evidence, show that a theory is definitely (or at least almost definitely) not correct?

At first glance, it might appear so. But as is almost always the case in science, disconfirmation turns out to be more complex than it appears. A simple example will suffice to see the complexities. Suppose it is Valentine's Day, and you reason that if your sweetheart loves you, then you will receive a gift from him or her. But no gift arrives. This seems a case of disconfirmation, in which you made a prediction (the arrival of a gift) based on a theory (your sweetheart loves you), and no gift arrived. And the conclusion suggested by the failed prediction, unfortunately, is that your sweetheart does not love you.

But wait. There are any number of other explanations for the lack of the gift. Perhaps your sweetheart did send a gift, but the delivery van crashed. Or perhaps the van was hijacked. Or had a flat tire. Or maybe the gift was delivered to the wrong address. Or perhaps a rival for your sweetheart's attention stole the gift before you could find it, or your roommate is hiding it as a joke, or any of an indefinite number of other possibilities.

In this context, such other possibilities are often referred to as "auxiliary hypotheses."[3] More specifically, auxiliary hypotheses are all the additional factors (in many cases there are a large number of such factors) needed to get the prediction from the theory. That is, one can expect to observe the predicted observation only if the theory, and all the auxiliary hypotheses, are correct. So if no gift arrives for you on Valentine's Day, the only conclusion you can reasonably draw is that either your sweetheart does not love you, or the van broke down, or the van was hijacked, or had a flat tire, or a rival stole the gift, or your roommate is hiding it as a joke, and so on for all the other possible auxiliary hypotheses. In short, in most situations in which a predicted observation does not occur, there are countless explanations, other than the falsity of the main theory, for why the prediction was not observed. So when faced with an incorrect prediction, one typically cannot have great confidence that the theory in question is incorrect.

---

[3] The terms used to describe this idea have changed over the years, with, for example, Duhem speaking of "various hypotheses" or "imaginable assumptions" (Duhem 1954), Hempel using terms such as "subsidiary hypotheses" (see, for example, Hempel 1965b), Popper speaking of "auxiliary assumptions" (Popper 1963), and so on. In recent decades the term "auxiliary hypotheses" has become more or less standard, though in all the discussions mentioned the key ideas were largely the same.

For an instance of this in a scientific context, consider again the example above of Kepler's theory. At about the same time as Kepler was publishing his views, the telescope was invented and was being used (mainly by Galileo) for astronomical observations.[4] As a new instrument, the telescope provided new data relevant to the issue of whether the sun moves around the earth (as in the traditional system), or whether the earth moves around the sun.

Some of this new evidence seemed incompatible with the theory that the sun moves around the earth. For example, through a telescope Venus can be seen to go through a range of phases much like the moon, which for reasons we need not go into here initially seems problematic for the earth-centered view. That is, it seems that if the earth-centered view is correct, then Venus should not be observed to go through phases.

But as usual, auxiliary hypotheses are lurking under the surface. It turns out that the phases of Venus (again for reasons we will skip over here) are only a problem if both the sun and Venus revolve around the earth. So, in this case, the real reasoning is "if the sun moves around the earth, and Venus does also, then Venus should not be observed to go through phases." And as a matter of fact, even before the telescope was invented one of the most prominent astronomers of the time had proposed a theory in which the earth was the center of the universe, the moon and sun orbited the earth, and all the other planets orbited the sun. The system is now referred to as the *Tychonic system* (after its proposer, Tycho Brahe, 1546–1601).[5] Moreover, the phases of Venus are exactly what would be expected if the earth-centered Tychonic system is correct (and so the phases of Venus actually provide confirming evidence for the earth-centered Tychonic system).

In short, the auxiliary hypotheses lurking under the surface mean that the discovery of the phases of Venus did not in any way rule out an earth-centered system. As usual, one could hold to the earth-centered system by rejecting the auxiliary hypothesis that Venus moves around the earth. And as noted, this sort of system, with the earth at the center of the universe

---

[4]   Galileo published the first set of observations with the telescope in 1609, under the title *Sidereus Nuncius* (usually translated as "The Starry Messenger" or "The Starry Message"). In the next few years Galileo would publish additional observations from the telescope, including the discovery of the phases of Venus. For a translation of the 1609 work, see van Galilei 1989.

[5]   A good description of the Tychonic system can be found in Dreyer 1953.

but with Venus (and the other planets) orbiting the sun, had already been proposed ten years before the discovery of the phases of Venus.[6]

In general, it is difficult to rule out, definitively, any theory. There may be no better way to reinforce this point than by noting that, even 400 years later, there is no shortage of advocates of the Tychonic system, that is, advocates of the view that the earth is the center of the universe. In fact, if one takes the original Tychonic system, and modifies it to include Kepler's contribution of elliptical orbits (and a few other modifications from twentieth-century physics), the resulting system (call it the modernized Tychonic system) is an earth-centered system that makes the same major predictions of the vastly more widely accepted sun-centered view of our solar system. So the bottom line is that, even with a seemingly outdated theory such as the earth-centered theory, it is difficult to find straightforward observational evidence that definitively shows the theory to be incorrect.

2.2   Karl Popper and falsificationism

A topic closely related to the confirmation/disconfirmation issues discussed above concerns a common view on what distinguishes scientific from non-scientific theories. For example, there is widespread agreement that, say, Kepler's theory on planetary orbits is a scientific view, whereas, say, astrology is not. (One can make a case that astrology at one time could have been considered properly scientific, but the general consensus is that astrology is no longer a genuinely scientific theory.)

What distinguishes scientific theories from non-scientific theories? There is no widespread consensus on the answer to this question, but one of the most popular answers, especially among working scientists (as well as a good number of philosophers of science), has its roots in the work of Karl Popper (1902–94).[7]

In outline, Popper's approach is to emphasize the disconfirmation (or at least attempted disconfirmation) of theories, rather than focusing on confirmation. Part of the reason for this is because confirmation often seems

---

[6]   I should note that there are good reasons for preferring the Kepler-style sun-centered view of our solar system over earth-centered views such as the Tychonic system. But the reasons are surprisingly complex, and far more subtle, than are generally realized. And none of the reasons are straightforwardly or easily observable.

[7]   A good starting source for Popper's views is Popper 1959.

too easy. For example, if I check my astrological forecast for this week, I find that I should not expect straight answers at work, leading to frustration with some work projects. And indeed, as I reflect on this past week, I have been a bit frustrated with some committee work, partly because of a sense of not receiving information as complete as I would like.

If we look again at the pattern of confirmation reasoning from the previous section, we can note that this astrology example fits the pattern. But there is broad consensus that this sort of example provides no real confirming evidence for astrology. The reason seems clear: the astrological prediction is much too vague and broad, so there is almost no chance of not observing something this week that would fit the prediction.

Situations similar to those illustrated by the astrology example arise in mainstream science as well. Consider again the examples discussed in the previous section, namely Kepler's theories on planetary orbits versus the (modernized) Tychonic view of planetary orbits. As noted, the vast majority of predictions generated by each of these theories are identical. So even though these are competing theories, we can easily generate thousands of predictions for each of these theories (for example, predictions about where all the planets, moon, sun, stars, etc. should appear in the sky), and these predictions, for both theories, will all be accurate.

Partly because of the relative ease with which one can generate confirming evidence for a theory, Popper argued that we should de-emphasize confirmation and instead look to disconfirmation as a key criterion for whether a theory should be considered scientific. Moreover, according to Popper, the riskier a theory, the better it is. That is, the best examples of scientific theories will be those that make specific, straightforwardly testable predictions, such that if those predictions turn out to be incorrect, this will strongly suggest that the theory is incorrect. In short, the best theories are those that expose themselves to being falsified. (By way of contrast, note that because of the vagueness of its predictions, astrology would not meet this criterion, and thus, on Popper's view, astrology should not count as a legitimate scientific theory.)

As a corollary to this, our best theories, according to Popper, are not those that have piled up tons of confirming predictions; rather, our best theories are those that have passed these sorts of tests in which the theory was exposed to the possibility of being falsified. An example (one Popper himself used) might help clarify this.

In the decades and century following the publication of Kepler's theory of planetary orbits, that theory was fleshed out by various advances, most notably, Newton's physics.[8] We need not go into the details here, but, broadly speaking, Newton's physics provided a broader theoretical framework, of which Kepler's earlier theory was a sort of subset. So, for the remainder of this example, I will focus on Newton's theory, with the understanding that it contains Kepler's account of planetary motion.

Newton's theory was capable of making quite specific predictions – for example, predictions as to the position of planets. And coupled with the increasingly accurate means of measuring planetary positions, Newton's theory seems a good example of a theory at risk.

And indeed, in the 1800s astronomers were able to determine that the orbit of Uranus did not match the predictions of Newton's theory. But, as it turns out, an additional planet (Neptune) was discovered, and the influence of Neptune on Uranus nicely explains, using Newton's theory, the earlier observed perturbations in the orbit of Uranus. Moreover, Newton's theory had been used to predict the possible existence of an unknown planet, exactly because such a planet would explain the oddities with the orbit of Uranus.

So, Newton's theory is one that makes specific predictions that are reasonably straightforward to test, and, as such, Newton's theory is, according to Popper, a good example of a theory at risk. And moreover, Newton's theory is a good example of a theory that has survived such tests, including initially problematic situations such as illustrated by the case involving the perturbations of the orbit of Uranus.

In summary, on Popper's view, our best theories are not those that have endless cases of confirming predictions, but rather, our best theories are those that have survived specific cases where they could have been shown to be wrong. And in addition, it is primarily this characteristic – that is, being at risk in the sense of being able to be shown to be incorrect – that distinguishes theories that are genuinely scientific (for example, Newton's physics) from those that are not (for example, astrology).

The above discussion is a fairly rough sketch of Popper's views on these matters. But suffice it to say that this approach of Popper's – primarily, viewing attempts at falsifying theories as more important to science than confirming evidence, and viewing potential falsification as the key characteristic of a scientific theory – is a fairly common view among working scientists.

As noted, Popper's views have been quite influential among working scientists and philosophers of science. But such views have also played

---

[8]   For a recent translation, with commentary, of Newton's principal work, see Newton 1999.

important roles in key court cases. For example, in an important case in Arkansas in 1982 (*McClean v. Arkansas Board of Education* 1982), the judge ruled that creation science was not science, and Popper's views (especially the importance of a theory being capable of being falsified) played an important role in that decision. Similar comments hold for a more recent ruling in a Dover, Pennsylvania case (*Kitzmiller v. Dover Area Sch. Dist.* 2005), in which intelligent design was found not to be a legitimate scientific theory, again largely on Popperian grounds.

But in spite of the popularity of Popper's views, the views outlined above are a bit more problematic than they might at first seem. The basic complicating factors are largely those discussed in the preceding section. Recall that in any case where a theory appears to be falsified (that is, any case in which an expected prediction is not observed), there are always auxiliary hypotheses involved. For example, consider again the issues described above involving the orbit of Uranus. When observations indicated that the orbit was not what Newton's theory predicted, the reaction was not to jump to the conclusion that Newton's theory was incorrect. Instead, researchers did the reasonable thing when faced with problematic evidence from a key and well established theory such as Newton's – they looked to auxiliary hypotheses as an alternative explanation for the failed predictions. And in this case, they found an alternative explanation, namely, a new planet.

Moreover, as argued most notably by another prominent figure in twentieth-century philosophy of science (as well as physics and mathematics), Imre Lakatos (1922–74), even if a new planet had not been discovered to account for the failed predictions, there is essentially no chance we would have abandoned Newton's theory (Lakatos 1980). Newton's theory had become so key to science, and so well established, that scientists would almost certainly have continued to look into alternative auxiliary hypotheses rather than rejecting the Newtonian theory. Part of the view defended by Lakatos, now generally accepted, is that long-term, well entrenched theories are quite resistant to any sort of straightforward falsification, and that theory change in such cases is a much more complex process (involving, for example, complex interactions within and between competing research programs).[9]

Issues such as those raised by Lakatos provide problems that go beyond what we can thoroughly look into here. There is no doubt that falsification

---

[9] This view of theory change as involving entire approaches and research programs in science has ties to the work of one of the more prominent names in history and philosophy of science, namely, Thomas Kuhn (see Kuhn 1962).

is an important tool in science, but there is likewise not much doubt that the process of falsification, and of theory acceptance and rejection, is a more complex process than that outlined above. Suffice it to say that, as usual, the situations involving scientific theories, and issues in the philosophy of science, tend to be much more complex and difficult than they often appear to be at first glance.

2.3   The underdetermination of theories

Another topic that ties in closely with those discussed above is the notion of the *underdetermination* of theories. Underdetermination is a common topic in the philosophy of science, and so worth taking a moment to explore. However, issues surrounding underdetermination are, as usual, complex, and in this brief section we will just take a brief look at this subject.

The key ideas behind underdetermination are most closely associated with the work of Pierre Duhem (1861–1916; Duhem was a physicist, though one with a lifelong interest in philosophical issues related to science), and also with W.V.O. Quine (1908–2000).[10] As usual, an example will help illustrate this notion. Consider again the Newton/Kepler sun-centered view, and the modernized Tychonic (earth-centered) view. As mentioned above, even though these are competing theories (one being an earth-centered theory; the other a sun-centered theory), both make the same predictions about observations such as the phases of Venus, where and when planets should be observed in the night sky, the positions and movements of the stars, times of solar and lunar eclipses, times of sunrise and moonrise, and so on.

In short, most (arguably all) of the straightforward observational evidence is compatible with both the earth-centered, modernized Tychonic system, and the more widely accepted Kepler-style sun-centered view of the solar system. In general, this is what is meant by underdetermination. That is, the basic idea behind underdetermination is that often (maybe typically, or even always, depending on your views on the subject), the available evidence will be compatible with two or more competing theories. Or, as it is often phrased, theories are underdetermined by the available data.

---

[10]   See, for example, Duhem 1954. See also Quine 1951, 1964, 1969. The underdetermination of theories is one aspect of what is often referred to as the "Quine-Duhem Thesis," though that thesis includes views beyond those described in this section.

I should note again that the issues involved in underdetermination are more complex than the quick sketch above might suggest. For example, there are a variety of ways, some very controversial and some not controversial at all, of construing underdetermination. As an example of an uncontroversial construal, there is little question that, at times, there have in fact been two or more competing theories equally supported by the available evidence. At the much more controversial end of the spectrum, some argue that such considerations support the view that scientific theories are entirely social constructs, such that there are no objective reasons to support one theory over another any more than there are objective reasons to support one ice-cream preference over another.

To review quickly §§2.1–2.3: we began by noting that the process of observation, theory generation, testing, and confirmation/disconfirmation was an important process in science. And moreover, it seems to be a common conception that this process is a more or less recipe-following process, with few or no philosophical or conceptual issues involved. Thus far we have discussed primarily the confirmation/disconfirmation part of this process, but as we have seen, just a little probing under the surface reveals some difficult, and largely philosophical, issues involved in this process.

Though we do not have time to consider them in detail here, similar issues arise when one looks into the other aspects of the process. For example, consider the observation part of the process. Presumably, one observes data before generating possible theories to explain that data. But it is far from clear that it is possible to gather data in a theory-neutral way. (For example, can we be theory-neutral about any subject, and can we gather data without already having some theoretical framework to help make sense of that data?) So it is far from clear that one can objectively gather data prior to having theories about that data.

Theory generation is likewise a complex process, and no one has ever developed anything remotely resembling a recipe for producing theories. When one looks at historical examples and tries to reconstruct the process by which a scientist arrived at a theory, the range of processes we find are astounding. In short, there does not appear to be any sort of easily articulated process or procedure by which one generates theories.

And it is likewise for the testing aspect of the process. Designing a test of a theory is almost never simple or straightforward. Some of the reasons for this involve the role of auxiliary hypotheses, discussed above, and the sorts of issues discussed above involving inductive reasoning. But also there is simply the raw complexity of many tests of theories. Over the past few centuries, theories in science have tended to become increasingly complex, and, not surprisingly, many of the tests of such theories have likewise become increasingly complex. For example, when Einstein proposed his theory of relativity early in the 1900s, one of the unusual predictions one could make from the theory was that starlight should be bent when passing near a large object such as the sun. But putting this prediction to the test was complex. To note just one complexity, the mathematics involved were complex enough that, in order to solve the equations, a good number of simplifying assumptions, known to be incorrect, had to be made (for example, that the sun is perfectly spherical, uninfluenced by outside forces, and non-rotating – none of which is correct).[11]

In summary, none of the features of the processes outlined – observation, theory generation, testing, or confirmation/disconfirmation – are in any way straightforward or any sort of recipe-following procedure. As we are seeing, it does not take long to uncover, in science, substantive philosophical issues.

2.4   Instrumentalism and realism

The discussions above largely concerned issues involved in theory confirmation and disconfirmation. In this section we will look at a slightly different topic, one involved, roughly, with the question of what we want from scientific theories.

One answer to this question, which almost everyone agrees on, is that we want scientific theories to be able to handle the relevant data. In particular, we want theories to be able to make accurate predictions, and we evaluate theories in large part based on how accurate their predictions are.

So, for example, there is general agreement that Newton's physics (developed mainly in the late 1600s) makes more accurate predictions than does Aristotle's physics, and that modern physics (for example, Einstein's special and general theories of relativity, developed early in the 1900s) makes more accurate predictions than does Newton's physics (for a brief overview of Aristotle, Newton, and Einstein, see DeWitt 2004).

But suppose we ask the question of whether Newton's theories provided a better model, or picture, of reality, than did Aristotle's theories, and whether Einstein's theories in turn provide a better model or picture of reality than

---

[11]   For an interesting discussion of some of these complexities, see Laymon 1984.

do Newton's theories. That is, in general, is one theory better than another not just because it gives more accurate predictions, but, in addition, because it provides a better model, or picture, of reality? Is reality the business of science; that is, should we expect, and even require, a good theory to provide an accurate picture of the way things really are?

With respect to this question, we find a substantial amount of disagreement. On the one hand are those who maintain that science ought to attempt to provide theories that not merely handle the data, but that reflect the way things really are. Stephen Weinberg, a leading physicist of recent decades and winner of the 1979 Nobel Prize in physics, is not hesitant to express sentiments such as that the job of science "is to bring us closer and closer to objective truth" (Weinberg 1998), and that there is "a necessity built into nature itself" (Weinberg 1992), and that in large part the job of a scientist is to find theories that capture the necessity inherent in nature. Einstein likewise often seemed to embrace such sentiments. For example, Einstein objected all his life to one of the most predictively successful theories we have ever had, namely quantum theory. Einstein fully recognized that quantum theory was enormously successful in terms of making accurate predictions, but he objected to the theory because he was convinced the theory could not possibly reflect the way the universe really is.

The view expressed by Weinberg and Einstein (and plenty of others, both scientists and philosophers of science), that theories in science ought to reflect the way the universe really is, is generally referred to as *realism*. And again, the basic idea behind a realist approach to science is that a good theory ought not only to make accurate predictions, but should in addition be required to model accurately, or picture, the way the universe really is.

In contrast to realism is a view usually referred to as *instrumentalism* (also sometimes referred to as *operationalism*). Those in this camp believe the main job of a theory is to handle the data (by, for example, making accurate predictions), and whether a theory accurately models or pictures reality is of no importance. This view is also found prominently among both working scientists and philosophers of science. For example, an instrumentalist attitude toward science is advocated in almost every standard college-level text on quantum physics, in which the student of quantum physics is generally encouraged to focus on making accurate predictions, and to leave worries about reality to the philosophers.

Notice a key aspect of realism and instrumentalism: namely, instrumentalism and realism are clearly philosophical views. They certainly are not themselves scientific theories – they are not, for example, designed to handle any body of observed data, nor can they be used (not in any straightforward sense) to generate predictions about what might be observed. Nor does it make much sense to speak of putting instrumentalism and realism to any sort of experimental test. That is, some of the key hallmarks of scientific theories – the ability to use them to make predictions, being able to put them to experimental tests, and so on – do not apply to instrumentalism and realism.

Nor are instrumentalism and realism parts of scientific theories. This is easy enough to see. One and the same theory (e.g., quantum theory) might well be taken with an instrumentalist approach by some scientists and a realist approach by others. In each case the theory is the same, so the instrumentalism and realism must not be part of the theory itself.

So instrumentalism and realism are neither scientific theories, nor parts of scientific theories. Instead, instrumentalism and realism are attitudes toward scientific theories, and philosophical attitudes at that.

Notice also that whether a particular scientist is more of an instrumentalist or a realist might well affect the approach he or she takes toward science. To give one example: in the beginning of a classic 1905 paper by Einstein (the paper in which he introduces what is now called the special theory of relativity), Einstein indicates that part of his motivation for pursuing this project stemmed from what he viewed as oddities in the application of the usual equations in physics used to handle cases where an electrical current is induced in a wire coil, either by moving the coil past a stationary magnet or by moving the magnet past a stationary coil (Einstein 1905). Notably, there were no problems with the predictions one got from the existing theory. So from an instrumentalist view (that is, if one primarily wants a theory to provide accurate predictions), there were no problems in this case for the existing theory.

The only puzzle (and this is part of what Einstein was interested in) stemmed from taking a realist perspective. Roughly, the worry was that the existing theory treated the two cases (magnet moving past a stationary coil versus coil moving past a stationary magnet) differently. And although Einstein did not phrase it this way, one way to put the problem is by asking how could the magnet and coil "know" which was stationary and which was moving? They obviously cannot, and so it seems strange to have different parts of the existing theory apply depending on which is stationary and which is moving.

Again, note that this worry is only a worry from a realist perspective – again, the predictions from the existing theory were accurate, so there are no worries from an instrumentalist perspective. It is not difficult to find other examples of this sort, in which whether a working scientist takes more of a realist or an instrumentalist approach affects the way he or she approaches science. And it is worth noting that in the example above, Einstein's realist leanings were part of what led to one of the most important theories of the twentieth century.

So once again we see it does not take long to find philosophical issues – in this case, instrumentalist and realist attitudes – cropping up in science. And as with the issues involved in the previous section, these philosophical issues are not of interest only to philosophers, but rather, they often influence the actual doing of science.

3.   Philosophical Issues in Scientific Foundations

The issues discussed above are related not only in that they are philosophical issues, but also in that they are philosophical issues that influence the workings of science. In this section, we consider a sampling of issues related to science that are philosophically interesting, but are such that they (at least typically) do not directly influence the actual workings of science. So these are not so much philosophical issues *in* science, but more philosophical issues *about* science (or, more precisely, about some basic concepts and tools of science).

3.1   Scientific explanation

Recall that in §2.1, we looked at the basic confirmation and disconfirmation aspect of science; that is, the process of generating predictions and then checking to see if the predictions are correct. And we saw that even this basic aspect of science had intriguing (and largely philosophical) issues just beneath the surface. In this section we probe into a topic with close ties to issues we found in that section; namely, issues related to scientific explanation.

Let us begin by returning to an earlier example. Consider again the theory that planets move about the sun in elliptical orbits, with the sun at one focus of the ellipse. Again, this idea was first proposed by Kepler in the early 1600s and is now often referred to as Kepler's first law of planetary motion. Also important is what we now usually refer to as Kepler's second law of planetary motion. Let us take a moment to understand this second law.

Consider the elliptical orbit of a planet, say Mars, with the sun at one focus of the ellipse. Suppose we think of this as a sort of elliptical pie. And imagine we cut this elliptical pie as follows. Along the crust (that is, Mars' orbit) we will mark the points where Mars is at, say, 30-day intervals over the course of a year. (This will mean we will have 12 marks along the crust, one for Mars' position at the beginning of each 30-day interval. Incidentally, the time interval we use does not matter, so long as we mark the crust at equal time intervals.) Now we will cut the pie into 12 pieces, and in particular we will cut the pie from the position of the sun (recall this will be at one focus of the ellipse) to the marks on the crust. The pieces of this pie will have different shapes; for example, some will be shorter than others (because the sun is closer to one edge of the ellipse than the other), some will be narrower than others, some wider, and so on. But Kepler's second law is, roughly, this: each piece will contain the same amount of pie. That is, although each piece will have a different shape, each piece will have the same area. (This is why the law is commonly referred to as the "equal areas" law. Another way of describing Kepler's second law is that a line drawn from the sun to the planet will sweep out equal areas in equal time.)

In addition to having substantial predictive power (for example, Kepler's first and second laws provided substantially more accurate predictions than any competing theory), Kepler's laws also seem to have substantial explanatory power. For example, suppose we are observing the night sky, and we ask why Mars appears where it does. We could answer this question using Kepler's laws. The full explanation would not be trivial, but neither is it terribly complex. In providing an explanation, we might start with the position of the earth and Mars, in their respective orbits, at the beginning of the year. Then, using Kepler's first and second laws, we could deduce where the earth and Mars will be at this moment. That would then show us where, from our perspective on earth, Mars should appear in the night sky. And that explains why Mars appears where it is tonight – Mars' position in the night sky is a deductive consequence of Kepler's laws together with some initial conditions (in this case, the positions of the earth and Mars at the beginning of the year).

If we look more closely at this style of explanation, the pattern is as follows: from a scientific law or laws (in this case, Kepler's laws of planetary motion), together with some set of initial conditions (here, the initial positions of the earth and Mars at the beginning of the year), we deduce that such and such should be observed (here, the current position of Mars in the night sky). And such a deduction seems to provide a nice explanation of why we observed (or are observing) some phenomenon.

The style of explanation just described has come to be called the "covering law" model of explanation. Although many scholars have contributed to this model of explanation, this style of explanation is most closely associated with Carl Hempel (1905–94) (for an early account, see Hempel and Oppenheim 1948, Hempel 1965a). And indeed, this account of explanation seems, intuitively, to be on the right track.

As a brief aside, the model of explanation illustrated above is often termed the "deductive-nomological" model (or D-N model; "deductive" because of the deductive nature of the reasoning involved and "nomological" because of the involvement of laws). Also, there is a variety of related models of explanation which, because of time and space constraints, we will not explore here. Most notable among these (and usually viewed as another type of covering law style explanation) is the "inductive-statistical" (or I-S) model, which is similar to the D-N model except that the inferences involved are inductive inferences from statistical laws (for an early investigation, see Hempel 1962). It is worth noting that a variety of alternative approaches to the D-N and I-S style of explanation have also been defended over the years.[12] For this section, we will focus primarily on the covering law model, as illustrated by the example above involving Kepler's laws, and attempt to convey a flavor of the difficulties this model has encountered.

One basic problem with the covering law model is that it seems to miss a certain asymmetry that is present in explanation. This "asymmetry" issue requires some explanation, and, as usual, an example might help. Consider again the case above, where we used Kepler's laws, together with the position of the earth and Mars at the beginning of the year, to explain the current position of Mars. Notice that the covering law model of explanation allows us to run this example in the other direction. In particular, from Kepler's laws of planetary motion, together with the current position of Mars, as viewed from earth in the night sky, we can deduce the position of Mars and earth at the beginning of this year. So on

---

[12]   For an overview of such accounts, see Salmon 1989. In addition to Salmon's paper, Kitcher and Salmon 1989 provides a good sampling of a variety of approaches to explanation.

the covering law model, Kepler's laws, together with the current position of Mars as viewed from earth, explain the positions of Mars and the earth at the beginning of the year.[13]

But this just seems wrong. It seems we should be able to explain future events from past events, but not vice versa. That is, although past events seem able to explain current events, it does not seem right to try to explain past events from current events. And that is the sense in which explanation is asymmetric. And again, the covering law model seems simply to miss this characteristic of explanation.

It seems we need to supplement the covering law model, or look for alternative accounts of scientific explanation, or both. In the remainder of this section, I will try to sketch (very) briefly some of these approaches.

One initially appealing way of supplementing the covering law account is by appealing to the notion of causation. We do seem to have strong intuitions that past events can cause current events, but not vice versa. That is, causation seems to have the sort of symmetry lacking in the covering law model. So perhaps an account of explanation along the general lines of the covering law model, supplemented with an account of causation, might do the trick.

However, causation is a notoriously slippery subject. One of the earliest substantive analyses of the concept of causation dates back to the Scottish philosopher David Hume (1711–76) (Hume 1975). In a nutshell, Hume argued that causation is not something we observe in nature; rather, what we literally observe are events followed reliably by other events. These "constant conjunctions" between events (for example, events such as two pool balls colliding are reliably followed by events such as the pool balls moving away from each other) are all we actually observe. We do not observe any causation. And, Hume continues, if causation is not something we observe "out there" in nature, then the concept of causation must come from us, that is, it must be contributed by our minds.

I think you can see some of the issues raised by these sorts of Humean concerns about causation. For example, if scientific explanation depends on the notion of causation, and causation is not "out there" in nature but is instead contributed by our minds, then we have to suspect that explanations in science are not the sort of objective, independent accounts I think we initially tend to think of them as.

---

[13] For an early example of this sort of objection to the covering law model, see Bromberger 1966.

This is, of course, just a brief sketch of some difficult issues. Suffice it to say that, although much interesting work has been done in recent decades on the notion of causation, and its potential role in helping to understand scientific explanation, nothing close to any sort of a consensus view on these matters has evolved (Sosa and Tooley 1993, Woodward 2003). Causation is, as noted, an incredibly slippery issue.

As mentioned earlier, some alternative accounts of explanation, ones not based on the covering law model, have been (and are being) explored. For example, some have proposed that a sort of explanatory unification is central to explanation (Friedman 1974, Kitcher 1981). To take one quick case, in Newtonian physics certain similarities among phenomena (say the similarity in the forces you experience as a result of acceleration or deceleration, and the gravitational forces you experience from being near a large body such as the earth) have to be taken as a sort of curious and unrelated coincidence. In contrast, on Einstein's general relativity (Einstein 1916), these two phenomena are seen as resulting from the same set of basic laws. In this sense, Einstein's theory unifies these previously disparate phenomena, and as noted, this sort of unification is seen by some as key to explanation.

These are, of course, not the only options and avenues being explored concerning explanation in science, and needless to say, even a cursory survey of all the alternatives is beyond the scope of this essay. I think one can, though, sum up the current situation by saying that there is, at present, nothing remotely resembling a consensus regarding scientific explanation. This is an intriguing situation, in that scientific explanation seems remarkably fundamental to science, and yet it is remarkably difficult to provide an account of what scientific explanation is.

3.2   Scientific laws

At various points in our discussion so far, we have appealed to the notion of a scientific law. And indeed, scientific laws seem to play a central role in science. Kepler's laws of planetary motion, for example, are an indispensable part of Kepler's account of the motion of planets. Laws also seem to be central to more philosophical approaches to science we have discussed – for example, the covering law model of scientific explanation.

But what is a scientific law? For the sake of a concrete example, let us again think about one of Kepler's laws of planetary motion, say the equal areas law. Again, this law says that a line from the sun to a planet will sweep out equal areas in equal time. Our initial inclination is to view this law as capturing some sort of fundamental, and presumably exceptionless, regularity about the universe (or at least a fundamental regularity about planetary motion).

We also, at least initially, tend to view this as an objective regularity, that is, as a regularity that exists independently of us. So, for example, our initial inclinations are to think that even if humans had never evolved on earth, planetary orbits still would have worked in accordance with Kepler's law, that is, a line drawn from the sun to a planet would still have swept out equal areas in equal time.

So, in this section, let us focus on just these two features we tend to associate with scientific laws, namely, that laws such as Kepler's equal areas law capture an exceptionless regularity about the universe, and that such laws are objective, that is, independent of us.

Let us begin with some considerations about regularities. Notice that regularities, even exceptionless regularities, are everywhere. But we are not inclined to count most of these regularities as scientific laws. For example, as far as I can recall, I have always put my pants on left-leg first, and probably always will. Suppose this in fact turns out to be the case, that is, I always have and always will put my pants on left-leg first. Then this would be an exceptionless regularity, but it certainly is not a scientific law. Likewise, it is an exceptionless regularity that there has never been a single English sentence of a million words (and almost certainly never will be), but this regularity is also not a scientific law. And so on for countless other regularities.

Intuitively, there seems to be a substantial difference between accidental regularities (for example, my dressing routine) and regularities that we tend to think of as scientific laws (for example, Kepler's equal areas law of planetary motion). And in fact, there is a fairly standard account of this difference.

The difference between accidental and law-like regularities is usually accounted for by appealing to what are called "counterfactual conditionals." Counterfactual conditionals (or just "counterfactuals") are something everyone is familiar with, though you may not be familiar with the term. The idea is straightforward. Suppose you say, "If I had studied harder for that test, then I would have gotten an A on it," or "If the battery in my alarm had not died, then I would have been on time for my appointment." Note that these are conditional sentences (that is, "if/then" sorts of sentences), in which the antecedent of the conditional (the "if" part) is false; hence the name "counterfactual conditional" (that is, contrary to fact). And again, we use counterfactual conditionals all the time to express what would have happened if antecedent conditions had been other than they actually were.

Now consider again Kepler's equal areas law, and suppose we focus on a particular planet, say Jupiter. Notice that Kepler's law would have held for Jupiter even if a wide variety of facts about Jupiter had been different. For example, if Jupiter had been more massive than it is, or less massive, or had a rocky composition rather than being a gas giant, or was located at a different distance from the sun, and so on. In all these cases, it still would have held true that a line drawn from the sun to Jupiter would have swept out equal areas in equal time. In short, Kepler's law remains correct even under a wide variety of counterfactual conditions.

In contrast, we can easily imagine any number of conditions under which the regularity about my dressing behavior would no longer have held. For example, my pants regularity probably would not have held had my parents taught me to put on pants in a different manner, or perhaps if I had been left-handed, or if as a child I had sustained an ankle sprain that made it difficult to put my pants on left-leg first, or any number of other perfectly trivial counterfactual conditions. So in contrast with the regularity expressed by Kepler's law, my pants-putting-on behavior is not true under a variety of counterfactual conditions.

In short, there seems to be a sort of necessity about the regularities captured by scientific laws, where that necessity is lacking in accidental regularities. And that necessity seems to be nicely captured by counterfactual statements. In particular, law-like regularities remain true under a variety of counterfactual conditions, whereas accidental regularities do not. As such, counterfactuals seem key in distinguishing law-like regularities from accidental regularities.

However, it does not take long to uncover surprisingly difficult issues involving counterfactuals (see Quine 1964, Goodman 1979). To look into one issue among many, counterfactuals seem to be inescapably context- and interest-dependent. For example, consider the counterfactual above, when we imagined you saying to yourself, "If I had studied harder for that test, then I would have gotten an A on it." Whether this counterfactual is true or not depends substantially on the context surrounding the counter-factual, as well as on other background information. For example, suppose you wound up missing the test entirely, perhaps because the battery in your alarm clock was dead. In that case, the counterfactual would be false. Likewise, the counterfactual would have been false if you had the wrong notes for the exam (in which case it is likely no amount of studying would earn you an A). And so on for countless other contextual issues.

These sorts of considerations raise a variety of issues. For example, if we need counterfactuals to distinguish law-like regularities from accidental regularities, and the truth of counterfactuals is context-dependent, and what the relevant context is depends on our interests in various situations, then it begins to look like which regularities count as laws are likewise dependent on our interests. And in this case, scientific laws begin to look as if they are dependent on us, rather than being objective and independent.[14]

Or another issue: largely because of the context-dependence of counterfactuals, it has been difficult to provide an account of under what conditions counterfactuals are true and false.[15] So if the distinction between accidental and law-like regularities depends upon the truth of the relevant counterfactuals associated with those regularities, and it is difficult to give a principled account of the truth and falsity of those counterfactuals, then it seems we will have difficulty giving a principled account by which we can distinguish accidental from law-like regularities.

Other issues concerning law-like regularities arise as well. To look at one of these issues, although quite briefly, consider again Kepler's laws of planetary motion. As it turns out, there is an important sense in which these are not exceptionless regularities. We have already seen (in §2.2) that in the 1800s it was recognized that there were some peculiarities with the orbit of Uranus, and in particular, the motion of Uranus was found not to be in strict accordance with Kepler's laws. As another example, in July of 1994 a quite substantial comet (Comet Shoemaker-Levy 9) collided with Jupiter. The collision's effect on Jupiter's orbit was not large, but it was enough of an impact to slightly affect Jupiter's orbit, such that Kepler's laws did not quite hold for that time period near the comet's collision.

Although the dramatic collision of the comet and Jupiter was an unusual event, less dramatic events influence planetary orbits all the time. The gravitational effects of Neptune on Uranus, as well as the gravitational effects of various bodies (other planets, comets, asteroids, and so on), will

---

[14] We might want to simply abandon the common notion that science is an objective enterprise, and adopt the view that what we tend to think of as scientific discoveries are more like scientific inventions. Largely for reasons of space, we cannot go into this avenue of thought, though we should note that no small number of scholars have defended this idea.

[15] For one influential, though far from entirely agreed-upon, attempt to give such an account, see Lewis 1973.

likewise affect the orbit of a planet, and in ways so as to make Kepler's laws not strictly apply.

The usual view is that the regularities expressed by scientific laws (and indeed, almost all regularities, whether accidental or law-like) apply with the understanding that certain "all else equal" clauses apply. For example, if Uranus were not affected by the gravitational influences of Neptune, or had comet Shoemaker-Levy 9 not smashed into Jupiter in July of 1994, and were these planets not influenced by the gravitational influences of other planets, and not influenced by the gravitational influences of the millions of asteroids in the asteroid belt, and not influenced by the presence of the Galileo spacecraft, and not influenced by other similar factors, then Kepler's laws would hold.

These sorts of "all else equal" clauses are typically referred to as "*ceteris paribus*" clauses, and such *ceteris paribus* clauses seem an indispensable part of scientific laws. *Ceteris paribus* clauses seem to have a number of problematic features, of which we will only briefly note a couple. First, do not overlook that such clauses have close ties to counterfactual conditionals, and as such bring with them the issues discussed above concerning counterfactuals. In addition, *ceteris paribus* clauses are notoriously vague. For example, in the example above involving Jupiter, what exactly counts as influences "like" the ones described? Whether one thing is like another again seems very much context- and interest-dependent. So once again, if scientific laws require interest-dependent criteria, it would appear difficult to maintain that scientific laws are objective, that is, independent of us, not to mention difficult to give a principled account of when such conditions hold.

There are far more issues surrounding the role played by *ceteris paribus* clauses in science than just those outlined above.[16] But as with previous discussions, the discussion above will, I hope, at least convey a flavor of some of these issues.

3.3 Hume's problem of induction

As noted earlier in this essay, an important component of science is the making of predictions about what we would expect to observe. Nor is this sort of reasoning unique to science – we all use this sort of reasoning

---

[16] For more on issues involving *ceteris paribus* clauses in science, see Cartwright 1983, Lange 2000, 2002.

all the time, probably every day. Also notice – and this is of particular relevance for the main topic of this section – that this sort of reasoning is reasoning about what we can expect to observe in the future. Again, there is nothing seemingly unusual about this – we reason about the future all the time in everyday life, and reasoning about the future is certainly a critical aspect of science.

Concerning reasoning about the future, Hume (whose views on causation we have already encountered) seems to have been the first to notice some important points about such reasoning (Hume 1975). The first point Hume noticed was this: if we are to trust our reasoning about the future, we have to assume that what happens in the future will be similar to what has happened in the past.

To see this, consider what the world would be like if what happened in the future was not like what has happened in the past. For example, suppose we lived in a world where each day was very different from the day before. Maybe the force of gravity changes from day to day; maybe the sun rises in the east one day and from some other random location the next day; maybe the weather changes randomly and unpredictably, being below freezing one day and 100 degrees the next; and so on. In such a world, we could not make the sorts of predictions about the future that we take for granted.

So, to repeat the first key point of Hume, to trust our reasoning about the future, we have to assume that what happens in the future will be similar to what has happened in the past. But now comes Hume's second key point: there seems to be no way to justify this assumption, that is, there seems to be no way to justify the assumption that what happens in the future will be like what has happened in the past.

Note that one way we could *not* justify this assumption is by using circular reasoning, that is, by using the assumption to justify itself. To see a simple case of circular reasoning (and why it does not work to justify a claim), consider this example. My new students sometimes tell me they have heard I sometimes goof around with the class, and, reasonably enough, would like to know when they can trust what I am saying, and distinguish it from when I am just goofing. Suppose I answer by going to the board at the front of the class, and writing:

What DeWitt writes on the board is trustworthy.

And I tell the class that that is how they can tell: if I write something on the board, it is trustworthy.

But now a particularly inquisitive student raises a perfectly reasonable question, namely, why should they trust what I just wrote on the board? And suppose I reply, "Because I wrote it on the board, and as you can see by what's written on the board, what DeWitt writes on the board is trustworthy."

That is blatantly circular reasoning. I am using the claim "what DeWitt writes on the board is trustworthy" to justify the claim "what DeWitt writes on the board is trustworthy." And that sort of circular reasoning – using a claim to justify itself – is no justification at all.

But notice – and this is really at the heart of Hume's second key point – that exactly this sort of circularity is lurking with the key claim that what happens in the future will be like what has happened in the past. Why? Because the only way we seem to have to justify this claim is noting that what happened today is similar to what happened yesterday, and what happened yesterday was similar to what happened the day before, and so on. That is, we are justifying the claim "what happens in the future will be similar to what has happened in the past" by appealing to past experience. But as Hume noted, any such reasoning – that is, any reasoning about the future based on past experience – must already *assume* that what happens in the future will be similar to what has happened in the past. That is, in trying to justify the claim "what happens in the future will be similar to what has happened in the past," we have to *assume* the truth of that exact claim, that is, that what happens in the future will be similar to what has happened in the past. And as with the "what DeWitt writes on the board" case, this is circular and thus no justification at all.

So Hume noticed a very general, and very curious, point about any reasoning concerning the future; namely all such reasoning rests on an assumption that cannot be logically justified. And the more general implication is difficult to avoid: if all reasoning about the future rests on a logically unjustified assumption, then our conclusions based on such reasoning, that is, our conclusions about what is likely to happen in the future, are equally logically unjustified.

4.   Conclusion

As noted in the introduction to this essay, this has been a sampler, with a bit of tasting and testing of various philosophical issues related to science.

As we have seen, it does not take long to find substantial and difficult philosophical issues in science. Some of these philosophical issues have an impact on the actual doing of science, while others we surveyed involve interesting and difficult philosophical issues, for example on foundational concepts and tools of science, though ones not directly impacting the workings of science.

What is included in the sections above is by no means a thorough survey of the range of philosophical issues in science – there are far more issues and puzzles than we have had time to survey here. Nor have any of the issues been covered in any depth. Again, the main metaphor was one of a sampler, in which we could try various flavors of issues and problems found in the philosophy of science.

Later essays in this collection will allow you to go into issues in substantially greater depth. In some cases the essays will take you further into issues touched upon above. In other cases, you will have a chance to explore new issues. But in all cases, I think you will find what we found above, namely, that philosophy is everywhere in science.

References

Bromberger, S. 1966. "Why Questions." In R. Colodny (ed.), *Mind and Cosmos: Essays in Contemporary Science and Philosophy*. Pittsburgh, PA: University of Pittsburgh Press.

Cartwright, N. 1983. "The Truth Doesn't Explain Much." In N. Cartwright (ed.), *How the Laws of Physics Lie*. Oxford: Oxford University Press.

DeWitt, R. 2004. *Worldviews: An Introduction to the History and Philosophy of Science*. Malden, MA: Blackwell Publishing.

Dreyer, J. 1953. *A History of Astronomy from Thales to Kepler*. New York: Dover Publications.

Duhem, P. 1954. *The Aim and Structure of Physical Theory*, trans. P. Wierner. Princeton, NJ: Princeton University Press, [1906].

Einstein, A. 1905. "On the Electrodynamics of Moving Bodies." *Annalen der Physik* 17.

Einstein, A. 1916. "The Foundations of the General Theory of Relativity." *Annalen der Physik* 49.

Friedman, M. 1974. "Explanation and Scientific Understanding." *Journal of Philosophy* 71.1: 5–19.

Gingerich, O. and W. Donahue. 1993. *Johannes Kepler New Astronomy*. Cambridge: Cambridge University Press.

Goodman, N. 1979. *Fact, Fiction, and Forecast.* 4th edn. Cambridge, MA: Harvard University Press, [1955].

Hempel, C. 1962. "Deductive-Nomological vs. Statistical Explanation." In H. Feigle and G. Maxwell (eds.), *Minnesota Studies in the Philosophy of Science*, vol. 3. Minneapolis: University of Minnesota Press.

Hempel, C. 1965a. *Aspects of Scientific Explanation and Other Essays in the Philosophy of Science.* New York: Free Press.

Hempel, C. 1965b. "Empiricist Criteria of Cognitive Significance: Problems and Changes." In C. Hempel, *Aspects of Scientific Explanation and Other Essays in the Philosophy of Science.* New York: Free Press.

Hempel, C. and P. Oppenheim. 1948. "Studies in the Logic of Explanation." *Philosophy of Science* 15: 135–75.

Hume, D. 1975. *Enquiries Concerning Human Understanding and Concerning the Principles of Morals*, L.A. Selby-Bigge and P.H. Nidditch (eds.). Oxford: Clarendon Press.

Kitcher, P. 1981. "Explanatory Unification." *Philosophy of Science* 48.4: 507–31.

Kitcher, P. and W. Salmon (eds.). 1989. *Scientific Explanation.* Minneapolis: University of Minnesota Press.

Kitzmiller v. Dover Area Sch. Dist. 2005. 400 F. Supp. 2d 707, 727 (M.D. Pa.). Available online at http://www.pamd.uscourts.gov/kitzmiller/kitzmiller_342.pdf (accessed March 10, 2008).

Kuhn, T. 1962. *The Structure of Scientific Revolutions.* Chicago: University of Chicago Press.

Lakatos, I. 1980. *The Methodology of Scientific Research Programmes: Vol. I*, J. Worrall and G. Currie (eds.). Cambridge: Cambridge University Press.

Lange, M. 2000. *Natural Laws in Scientific Practice.* New York: Oxford University Press.

Lange, M. 2002. "Who's Afraid of *Ceteris-Paribus* Laws? Or: How I Learned to Stop Worrying and Love Them." *Erkenntnis* 57: 407–23.

Laymon, R. 1984. "The Path from Data to Theory." Reprinted in J. Leplin, *Scientific Realism.* Berkeley, CA: University of California Press.

Lewis, D. 1973. *Counterfactuals.* Cambridge, MA: Harvard University Press.

McLean v. Arkansas Board of Education. 1982. 529 F. Supp. 1255, 1258–1264 (ED Ark.).

Newton, I. 1999. *The Principia: Mathematical Principles of Natural Philosophy*, trans. B.I. Cohen and A. Whitman. Berkeley, CA: University of California Press.

Popper, K. 1959. *The Logic of Scientific Discovery.* London: Hutchinson.

Popper, K. 1963. *Conjectures and Refutations.* London: Routledge & Kegan Paul.

Quine, W.V.O. 1951. "Two Dogmas of Empiricism." *Philosophical Review* 60: 20–43.

Quine, W.V.O. 1964. *Word and Object.* Cambridge, MA: MIT Press.

Quine, W.V.O. 1969. *Ontological Relativity and Other Essays.* New York: Columbia University Press.

Salmon, W. 1989. "Four Decades of Scientific Explanation." In P. Kitcher and W. Salmon (eds.), *Scientific Explanation*. Minneapolis: University of Minnesota Press.

Sosa, E. and M. Tooley (eds.). 1993. *Causation*. Oxford: Oxford University Press.

van Galilei, G. 1989. *Sidereus Nuncius*, trans. A. van Helden. Chicago: University of Chicago Press, [1609].

Weinberg, S. 1992. *Dreams of a Final Theory*. New York: Pantheon Books.

Weinberg, S. 1998. "The Revolution That Didn't Happen." *New York Review of Books* 45.15.

Woodward, J. 2003. *Making Things Happen: A Theory of Causal Explanation*. Oxford: Oxford University Press.

Unit 2

Philosophy of
the Exact Sciences

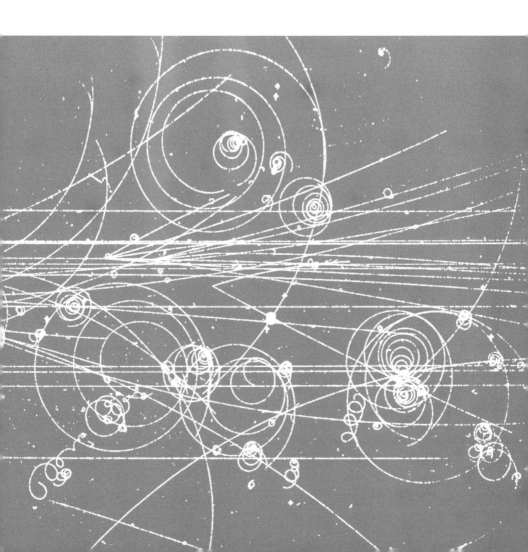

3   Philosophy of Logic

Otávio Bueno

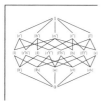

In this essay, I explore three central issues in the philosophy of logic. First, I examine the issue of the conceptual analysis of logical consequence. I focus the discussion on Alfred Tarski's model-theoretic account and how it embodies some of the central features of logic: universality, formality, truth-preservation, and *a priori*. Second, I discuss the issue of logical pluralism: whether, given a language, there is only one admissible specification of the logical consequence relation (as the logical monist insists), or whether there are several such specifications (as the logical pluralist contends). Third, I consider the issue of the applications of logic, and I explore the connections between the application of logic and of mathematics. Along the way, the interconnections among these issues are explored, in order to indicate how different philosophical conceptions of logic emerge.

1.   Introduction

The central concept of logic is that of logical consequence: whether, in an argument, a given statement (the conclusion) logically follows from certain other statements (the premises). This is a simple point, but it raises a number of intriguing philosophical issues. In this essay, I will focus on three of these issues which have shaped much of the recent philosophy of logic and are thus central to the philosophical reflection about this field. Depending on how these issues are addressed, and the connections that are drawn between them, distinctive conceptions of logic emerge.

The first issue is concerned with conceptual analysis. Can the concept of logical consequence be *analyzed*? And if so, how? That is, are there necessary and sufficient conditions that specify the extension of this concept; that determine the conditions under which a statement is a logical consequence of another (or of a group of others)? Given that, in natural language, we do have an intuitive, informal notion of what follows from what, at stake here is a process of uncovering the conditions that characterize the concept of logical consequence. (This is the issue of the *conceptual analysis of logical consequence*.)

Second, after this concept is analyzed, the issue arises as to whether there is only one correct specification of the concept, or whether the latter admits multiple specifications. The *logical monist* insists that there is only one such specification; the *logical pluralist*, in turn, argues that there are several correct specifications of logical consequence. (This is the *logical pluralism issue*.)

Third, once the plurality of logic is resolved, the issue arises as to how the application of logic can be accommodated. Is the application of logic similar to the way in which mathematical theories are applied? Or is logic applied in some other way? (This is the issue of the *applications of logic*.)

The articulation of a philosophy of logic involves careful reflection about these issues. All of them emerge, in one way or another, from the conceptual analysis of logical consequence. In this essay, I will examine each of these issues in turn, exploring, along the way, the interconnections among them.

2. Logical Consequence

Conceived as a field of inquiry, logic deals primarily with the concept of logical consequence. The aim is to determine what follows from what, and to justify why this is the case. Intuitively, we have some informal ideas about which conclusions seem to follow (or not) from certain premises. For instance, from the statement "Olivia was playing with the books," we correctly conclude that, "Someone was playing with the books." But we do not generally conclude that, "The little girl was playing with the books" – unless we have some additional information about Olivia.

Part of logic's concern is to explain why such judgments are correct – when they are indeed correct – and to provide an account of other, often more controversial, cases in which it may not be clear whether the conclusion does follow from the premises. As a result, it is no surprise that *logical consequence* becomes such a central concept. But logic is not any old study of logical consequence; it is a particular sort of study of this relation.

2.1   Formality and universality

From its early stages, logical studies has addressed the concept of logical consequence in a particular way. Since its inception in Aristotle's work, and continuing through the mathematically developed work of Gottlob Frege (1967), logic has been conceived as a study of the relation of consequence having certain features. Central among them is the fact that logic is both universal and formal. It is formal in the sense that the validity of an argument is a matter of the form of the argument. It is universal in the sense that once a valid argument form is uncovered, it does not matter how that form is instantiated: every particular instance of the argument form will be valid as well.

But under what conditions is an argument valid – or, equivalently, under what conditions is the conclusion of an argument a logical consequence of its premises? The usual formulation, which is cashed out in modal terms, is: an argument is valid if, and only if, it is impossible for the premises to be true and the conclusion false. What is the source of this impossibility? The conjunction of the premises and the negation of the conclusion of a valid argument is inconsistent; that is, such a conjunction leads to a contradiction of the form *A and not-A*. Consider a simple example:

(P1)   Julia and Olivia are having dinner.
(C)   Therefore, Julia is having dinner.

Clearly, if we have the conjunction of the premise ("Julia and Olivia are having dinner") and the negation of the conclusion ("Julia is not having dinner"), we obtain a contradiction, namely: Julia is having dinner and Julia is not having dinner. Therefore, the argument is valid.

What is the logical form of the argument above? Note that the first premise of the argument, (P1), is equivalent to:

(P1′)   Julia is having dinner and Olivia is having dinner.

Thus, (P1′) is a statement of the form: *P and Q*. The form of the argument is then:

(P1′)   *P and Q*
(C)   Therefore, *P*

Now, we can clearly see why any argument with this form is bound to be valid. The conjunction of the premise (*P and Q*) and the negation of the conclusion (*not-P*) leads immediately to the contradiction: *P and not-P*.

This is, of course, an extremely simple example, but it already illustrates the two significant features of logic: universality and formality. The validity of the argument ultimately depends on its logical form: the interplay between the logical form of the premise and the conclusion. And once the logical form of the argument is displayed, it is easy to see the universality of logic at work. It does not matter which sentences are replaced by the variables *P* and *Q*. Any instance will display the same feature: we obtain a contradiction if we conjoin the premise and the negation of the conclusion. The argument is indeed valid.

These two features of logic go hand in hand. By being formal, logic can be universal: the logical form of a valid argument ultimately grounds the fact that every instance of the argument will also be valid. Suppose, in turn, that all instances of a given argument form are valid. This suggests that there is something underlying the validity of all such instances: a valid argument form.

2.2   The model-theoretic account: truth preservation *and* a priority

But there is another way of thinking about what grounds the validity of an argument. Why is it that in a valid argument it is impossible to conjoin the premises and the negation of the conclusion? On the truth-preservation account, this is because valid arguments are *truth-preserving*. If the premises are true, the conclusion *must* be true. Thus, if we were to make the conjunction of the premises of a valid argument with the negation of the conclusion, we would obtain a contradiction. But even on the truth-preservation account, the formal aspect of logic is crucial. After all, it is ultimately due to its logical form that an argument is truth-preserving.

All of these features – formality, universality, and truth preservation – are clearly articulated in the most celebrated account of logical consequence in the literature. It is the proposal advanced by Alfred Tarski in terms of mathematical models (see Tarski 1936b). In order to formulate his account, Tarski first introduces what he takes to be a necessary condition for logical consequence, something that can be called the "substitution requirement." This requirement connects a given class *K* of sentences and a particular sentence *X* as follows:

If, in the sentences of the class K and in the sentence X, the constants – apart from purely logical constants – are replaced by any other constants (like signs being everywhere replaced by like signs), and if we denote the class of sentences thus obtained from K by K', and the sentence obtained from X by X', then the sentence X' must be true provided only that all sentences of the class K' are true. (Tarski 1936b, 415; emphasis omitted)

What is required here is a systematic replacement – or interpretation – of the expressions in the sentences of K and in X, so that the truth of the new sentence X' is guaranteed by the truth of the new sentences of K'. This is accomplished by a series of steps. First, the occurrences of the same expressions are replaced by the same substitution expressions. That is, if on a given interpretation, "cat" is replaced by "mat," this replacement takes place in *every* occurrence of "cat," and the latter is *always* replaced by "mat" in that interpretation. Second, expressions of the same logical type are replaced by expressions of the corresponding type. That is, predicates are replaced by predicates; individual constants are replaced by individual constants; and so on. Finally, the logical constants that occur in the sentences of K or in X are never replaced or interpreted.[1] They are held fixed over any interpretation. As long as these steps are taken, if the sentences of K' are true, sentence X' must be true as well.

We have in place here the core ideas of truth preservation, formality, and universality. The sort of replacement that Tarski envisions with the substitution requirement is one in which the logical form of the sentences in question is preserved, even though everything else changes. After all, the logical form of a sentence is ultimately given by the connections among the logical constants that occur in that sentence, and these constants are never replaced when the interpretation of the sentences of K and X is given. What emerges from this is truth preservation: the truth of the sentences in K' guarantees the truth of X'. And this result holds universally – as long as the conditions in the substitution requirement are met.

As an illustration of the substitution requirement, suppose that the class K contains two sentences:

(K_1)   Either Olivia is sleeping or she is playing with her toys.
(K_2)   Olivia is not sleeping.

---

[1]   The logical constants include the logical connectives – items such as negation ("not"), conjunction ("and"), disjunction ("or"), and the conditional ("if-then") – and the quantifiers: the universal quantifier ("for all"), and the existential quantifier ("for some").

And suppose that X is the sentence:

(X)   Olivia is playing with her toys.

We can now form the class K' by replacing all the non-logical terms in the sentences of K; that is, all of the expressions in K's sentences except for "either ... or ..." in K_1, and "not" in K_2. We would then obtain, in a particular interpretation, two new sentences, such as:

(K_1')   Either Julia is dancing or she is resting.
(K_2')   Julia is not dancing.

In this case, sentence X' will be:

(X')   Julia is resting.

Now, it is easy to see that if all the sentences of the class K' – that is, sentences (K_1') and (K_2') – are true, then sentence X' must be true as well. And clearly exactly the same result will be obtained for any other interpretation – any other replacement of the expressions in K and X. The substitution requirement is therefore met.

While developing his account, Tarski was particularly concerned with the formal aspect of logical consequence. In fact, for him, logical consequence and formal consequence amount, basically, to the same concept. The logical consequence relation is one that holds in virtue of the form of the sentences in question. As Tarski points out:

> Since we are concerned here with the concept of logical, i.e., *formal*, consequence, and thus with a relation which is to be uniquely determined by the form of the sentences between which it holds, this relation cannot be influenced in any way by empirical knowledge, and in particular by knowledge of the objects to which the sentence X or the sentences of the class K refer. The consequence relation cannot be affected by replacing the designations of the objects referred to in these sentences by the designations of any other objects. (Tarski 1936b, 414–15; emphasis in original)

By highlighting the formal component of the consequence relation, Tarski is in a position to articulate, in this passage, a significant additional feature of logic. Since logical consequence depends on considerations of logical form, it becomes entirely independent of empirical matters. In fact, each particular interpretation of the non-logical vocabulary could be thought of as one way in which empirical (non-logical) considerations could impinge on the consequence relation. Given the substitution requirement, however, none of the particular interpretations actually changes the consequence relation. As a result, the *a priori* character of the logical consequence relation – and of logic more generally – emerges. By requiring the invariance of the consequence relation over different interpretations of the non-logical expressions, Tarski is in fact detaching the relation of logical consequence from the vicissitudes of empirical domains. Thus, the logical consequence relation is not open to revision on empirical grounds. The relation does not change when the designation of objects and predicates shifts from one interpretation to another.

With these considerations in place, we now have all of the resources to present Tarski's analysis of the concept of logical consequence. As noted above, Tarski's proposal is formulated in terms of *models*. A model is an interpretation – a systematic replacement of the non-logical expressions satisfying the substitution requirement – of a class of sentences in which all of these sentences are true. In this case, Tarski notes: "The sentence X follows logically from the sentences of the class K if and only if every model of the class K is a model of the sentence X" (Tarski 1936b, 417). Clearly, the truth-preservation component is in place here. After all, the truth of the sentences of K, which is exemplified in each model of K, guarantees the truth of X. Moreover, as we saw, the substitution requirement underlies the formulation of the models under consideration, since each model presupposes the satisfaction of that requirement. And given that, as we also saw, the universality, formality, and *a priori* character of logic emerges from this requirement, it is no surprise that the concept of logical consequence, as characterized by Tarski, also inherits these features.

2.3   *Modal matters*

Tarski's analysis of logical consequence also has an intriguing feature. Recall that when we first introduced the concept of logical consequence above, we did it in modal terms: an argument is valid if, and only if, it is *impossible* for the premises to be true and the conclusion false. This means that if the premises of a valid argument are true, the conclusion *must* be true as well. There is a *necessity* involved in the connection between premises and conclusion. However, *prima facie*, we do not seem to find such a necessity explicitly stated in Tarski's analysis. We simply have a universal claim about models: in a valid argument – that is, one in which the conclusion follows logically from the premises – every model of the premises is a model of the conclusion. Where does the necessity come from?

There are two ways of answering this question. At the time Tarski presented his analysis, there was considerable skepticism, at least in some quarters, about modal notions. These notions were perceived as being poorly understood and genuinely confusing, and it would be better to dispense with them entirely. Much of the early work of logical positivists, particularly Rudolph Carnap, can be seen as an attempt to avoid the use of modal notions. In fact, even W.V. Quine's well known distrust of modality can be interpreted as the outcome of this positivist hangover. In light of this historical background, it is not surprising that Tarski would try to avoid using modal terms explicitly in his analysis of logical consequence.

But there is a second, stronger response. Tarski did not think he needed to use modal concepts in his characterization of logical consequence given that the models he was invoking already had the relevant modal character. A mathematical model represents certain possibilities. If certain statements can be true together, that is, if there is no inconsistency in their conjunction, and the latter is, hence, possible, then there will be a model – a particular interpretation of the non-logical expressions involved – in which these sentences are all true. If the sentences in question are inconsistent and, thus, their conjunction impossible, no such model will be available. Thus, models are already modal in the relevant way. That is why mathematicians can replace talk of what is possible (or impossible) with talk of what is true in a model (or in none).

Of course, this response presupposes the adequacy of models to represent possibilities, so that to every way something can be, there is a model that is that way. Perhaps this is not an unreasonable assumption when we are dealing only with mathematical objects. But since the scope of logic goes beyond the domain of mathematics – after all, we use logic to reason about all sorts of things – one may wonder whether the Tarskian is really entitled to this assumption.[2]

---

[2] These are worries of a modalist sort, that is, of someone who thinks that logic is inherently modal in character, and the modality involved cannot be eliminated in terms of models or other replacements (see Shalkowski 2004). And the adequacy of Tarski's account of logical consequence has indeed been challenged (see, in particular, Etchemendy 1990, McGee 1992). As one would expect, given the influence of Tarski's approach, it did not take too long for defenses of Tarski to emerge (see, e.g., Ray 1996, Sher 1996, Shapiro 1998).

2.4   *Logical consequence and logical truth*

But whether we adopt an explicitly modal characterization of logical consequence, or a model-theoretic one (in terms of models), once the concept of logical consequence is formulated, the notion of logical truth can be characterized as well. A sentence X is logically true if, and only if, X is a logical consequence from the empty set. In other words, any interpretation of a logical truth X is a model of X; that is, any such interpretation makes X true. So, it does not matter which set of sentences K we consider, X will be true in any interpretation. This includes, of course, the case in which K is empty.

There are those who have insisted, or at least have implied, that logical truth, rather than logical consequence, is the fundamental concept of logic. In the very first sentence of his elementary textbook, *Methods of Logic*, W.V. Quine notes: "Logic, like any science, has as its business the pursuit of truth. What are true are certain statements; and the pursuit of truth is the endeavor to sort out the true statements from the others, which are false" (Quine 1950, 1). This suggests a certain conception of logic according to which its aim is to determine the logical truths and to distinguish them from the logical falsities. In fact, Quine continues: "Truths are as plentiful as falsehoods, since each falsehood admits of a negation which is true. But scientific activity is not the indiscriminate amassing of truths; science is selective and seeks the truths that count for most, either in point of intrinsic interest or as instruments for coping with the world" (Quine 1950, 1). So, it seems that we are after *interesting* logical truths, even though it is not entirely clear how such truths can be characterized. What exactly would be an *un*interesting logical truth? Quine, of course, does not pause to consider the issue. And he does not have to. After all, as he seems to acknowledge a few paragraphs after the quotation just presented,[3] the central concept of logic is not logical truth, but logical consequence (see also Hacking 1979, Read 1995, Beall and Restall 2006).

---

[3]   Later in the introduction to *Methods of Logic*, Quine notes: "The chief importance of logic lies in implication, which, therefore, will be the main theme of this book. Techniques are wanted for showing, given two statements, that the one implies the other; herein lies logical deduction" (Quine 1950, 4).

3. Logical Pluralism

Given the centrality of logical consequence, the question immediately arises: is it possible that there are multiple specifications of the concept of logical consequence for the same language? Or when we are given a language, are we also given one, and only one, logic that goes with that language? What is at stake here is the issue of *logical pluralism*.

This issue could have been presented, in a somewhat more dramatic way, by asking the question: is there just one right logic, or are there many such logics? The logical monist insists that there is only one (see, e.g., Quine 1970, Priest 2006b); the logical pluralist contends that there are many (see, e.g., da Costa 1997, Bueno 2002, Beall and Restall 2006, Bueno and Shalkowski 2009). Of course, the notion of a "right logic" is not exactly transparent. But one way of making it a bit clearer is to say that the right logic is the one that provides, to borrow Stephen Read's useful phrase, "the correct description of the basis of inference" (Read 1988, 2). When we are given a logic, we are given certain inferences that are sanctioned as valid. A correct description of these inferences is one that sanctions as valid those inferences that intuitively seem to be so and does not sanction as valid those inferences that intuitively do not seem to be so. The logical pluralist insists that there are many correct descriptions of the basis of inference, whereas the monist denies that this is the case.

3.1 Logical disagreement

But is there really a disagreement between the logical monist and the logical pluralist? If the pluralist and the monist were talking about *different* notions of inference – in particular, different concepts of validity – there may not be any conflict between their views. Both can be perfectly right in what they assert about logic if they are not talking about the same concept.

As an illustration, suppose, for the sake of argument, that the monist has a certain conception of inference, for example, according to which the meaning of the logical constants is fixed *a priori* and is completely determined. On this conception, there is only one correct description of the basis of inference: the description that faithfully captures and preserves the meaning of the logical constants. Suppose, again for the sake of argument, that the pluralist understands inference in a *different* way: inferences are a matter of specifying certain rules that govern the behavior of the logical vocabulary, and questions of meaning do not even arise. On this understanding of inference, there is more than one correct account of the basis of inference such that any specification of the rules for the logical vocabulary will do.

In this hypothetical case, the pluralist and the monist do not seem to be talking about the same concept of validity. The monist can agree with the pluralist that given different specifications of the rules for the behavior of the "logical vocabulary," there would be various correct accounts of the basis of inference. It is just that, on the monist's view, these rules are not really about what she takes the logical vocabulary to be, since for her the meaning of the latter is specified independently of any such rules. So, given the pluralist's rules, it becomes clear that the pluralist is not talking about the logical constants that the monist is. Thus, no disagreement emerges.

There is, however, something strange with this assessment. After all, *prima facie*, there does seem to be a genuine disagreement between the logical monist and the logical pluralist. One of them sees a plurality of specifications of the concept of logical consequence where the other sees none. One identifies several correct descriptions of the basis of inference; the other insists that there is only one. Moreover, they do seem to be talking about the *same* issue: the proper characterization of the concept of logical consequence. To make sense of this debate as a *genuine* debate, both parties need to be talking about the *same* thing – logical consequence (or validity) – despite possible conceptual differences.

3.2 Logical pluralism by cases

Recently, JC Beall and Greg Restall have offered an interesting framework in terms of which logical pluralism can be articulated, and in this framework, the debate between the pluralist and the monist becomes authentic (Beall and Restall 2006). On their view, there is only one concept of logical consequence – or validity – and both the pluralist and the monist share that concept. The disagreement between them emerges from the fact that, for the logical pluralist as opposed to the monist, the concept of validity admits different specifications, different instances, depending on the *cases* one considers. In fact, Beall and Restall formulate the concept of validity explicitly in terms of cases (2006, 29–31): an argument is valid if, and only if, in every *case* in which the premises are true, the conclusion is true as well.

But what are the *cases*? Cases are the particular situations in terms of which the truth of statements is assessed. For example, in classical logic, as we saw above, cases are Tarskian models. These models specify *consistent* and *complete* situations. That is, in a Tarskian model, no contradictory statement is true, and for every sentence in the language, either that sentence is true (in the model) or it is false (in the model). So, if every Tarskian model of the premises of an argument is also a Tarskian model of the argument's conclusion, then the argument is valid. In this way, by focusing on Tarskian models, we obtain classical logic.

But if we broaden the kinds of cases we consider, we will make room for non-classical logics as well (Beall and Restall 2006). For example, if we consider *inconsistent* and complete situations, we allow for cases in which some contradictory statements of the form *A and not-A* hold. In this way, as we will see below, we obtain what are called paraconsistent logics. If we consider consistent and *incomplete* situations, we allow for cases in which *A or not-A* may *not* hold, given that, at a certain stage, perhaps not every property of a given object has been specified, and thus, it is not determined whether that object has or lacks a given property.[4] In this way, we obtain what are called constructive logics. If we consider consistent, complete but *non-distributive* situations, we allow for cases in which conjunction ("and") does *not* distribute over disjunction ("or").[5] In this way, as we will see, we obtain what are called quantum logics.

For the logical pluralist, each of this group of cases provides a perfectly acceptable specification of the concept of validity. We thus have perfectly acceptable logical consequence relations. Distinct logics then emerge. As an illustration, I will consider two examples: the cases of paraconsistent logic and of quantum logic.

---

[4] For example, suppose that mathematical objects are not taken to exist independently of us but are the result of certain mathematical constructions that we perform. On this view, a mathematical object is constructed in stages, and at each stage, the object will have only those properties that have been explicitly specified up to that point. It is then perfectly possible that at a certain stage it is not determined whether a mathematical object has or lacks a given property, since the relevant fact of the matter that settles the issue – namely, the explicit specification of the property in question – is not in place yet.

[5] This means, as we will see below, that the distributive law from classical logic fails. According to this law, conjunction ("$\wedge$") distributes over disjunction ("$\vee$") as follows: $P \wedge (Q \vee R) \leftrightarrow (P \wedge Q) \vee (P \wedge R)$. This principle, as will become clear shortly, is violated in so-called quantum logics.

3.3   Inconsistent cases and the Liar paradox

Suppose that in describing a certain situation, we stumble into inconsistency. For example, we may be trying to determine whether the following sentence is true or false:

(1)   Sentence (1) is not true.

Let us assume that sentence (1) is not true. But this is precisely what is stated by the sentence. So, the latter *is* true. Suppose, then, that sentence (1) is true. Thus, what is stated by that sentence is the case. But what is stated is that sentence (1) is not true. So, the sentence is *not* true. In other words, if sentence (1) is true, then it is not true; if sentence (1) is not true, then it is true. This entails that sentence (1) is both true and not true – a contradiction. Sentence (1) is called the Liar sentence, and what we have just obtained is the Liar paradox.

How can we make sense of a situation such as this? If we use classical logic, our immediate response would be to deny that the situation described by the Liar sentence could ever arise. After all, classical logic embodies a puzzling way of dealing with inconsistency: it is trivialized by any contradiction of the form *P and not-P*. That is, in classical logic, everything follows from a contradiction. Here is why:

(1)   *P and not-P*   Assumption.
(2)   *P*   From (1) by conjunction elimination.
(3)   *P or Q*   From (2) by disjunction introduction.
(4)   *not-P*   From (1) by conjunction elimination.
(5)   *Q*   From (3) and (4) by disjunctive syllogism.

Given that Q is arbitrary, and that all of the inferential steps above are valid in classical logic, we conclude that, according to those logics, everything follows from a contradiction. This means that classical logic is *explosive*: it identifies contradiction and triviality, by allowing us to derive any sentence whatsoever from a contradiction. (This latter inference is called, rather dramatically, explosion.)

In fact, this bewildering feature of classical logic can be obtained directly from the concept of validity. As we saw, every Tarskian model of the premises of a valid argument is a Tarskian model of the argument's conclusion. Suppose now that the premise in question is a contradiction of the form *P and not-P*. As we also saw, Tarskian models are consistent. So, there is no model that makes *P and not-P* simultaneously true. In other words, there are no models of *P and not-P*. Thus, there are no models of *P and not-P* in which the negation of the conclusion, *not-Q* (for an arbitrary sentence *Q*), is true. As a result, we conclude that (vacuously) every model of *P and not-P* is also a model of *Q*. Therefore, everything follows from a contradiction.

Given this result, it is not surprising that classical logicians avoid contradictions at all costs. If they ever stumbled into a contradiction, such as the one expressed in the Liar sentence, they would immediately run into triviality. Thus, understandably, these logicians look for ways to block the Liar paradox.[6] The trouble is that these strategies typically involve some restriction in the expressive power of the language that is used to express the Liar sentence. For example, Tarski (1936a) introduced an influential strategy in terms of a hierarchy of formal languages, which, in turn, was grounded on the distinction between object language and meta-language. The object language is the language we are interested in studying. This language has an important limitation: it does not allow us to express claims about the truth of sentences of the object language. It is only in the meta-language that we can formulate the concept of truth for sentences of the object language. (The meta-language, in turn, cannot express the concept of truth for its own sentences. We define the concept of truth for sentences of the meta-language in a meta-meta-language, and so on.) Since the object language does not have the resources to express the concept of truth, the Liar sentence, which depends crucially on that concept, cannot be expressed in the object language. As a result, the Liar paradox is blocked.

Tarski's strategy, however, has significant costs (see Priest 2006a). In order to avoid the inconsistency that emerges from the Liar, the strategy rules out something that takes place all the time in a natural language such as English: the possibility of expressing the Liar sentence. We thus have here a trade-off between inconsistency and expressive limitation. In order to ensure the consistency of his formal languages, Tarski imposes a significant limitation on their expressive power. Moreover, Tarski's strategy also introduces a rather artificial distinction between object language and meta-language. No such distinction seems to be found in natural languages.

---

[6] Various attempts to accommodate the Liar paradox can be found in Martin 1984, and in Beall 2007. For a discussion of why attempts to block the paradox based on classical logic ultimately do not succeed, see Priest 2006a. I will return to this point shortly.

3.4   Inconsistent cases and paraconsistent logic

Is there a different strategy to deal with the Liar paradox? Suppose that we claim that, in the case of the Liar sentence, appearances are not deceptive after all. The argument presented above, showing that we obtain a contradiction from the Liar sentence, is valid. It does establish that the Liar sentence is both true and false. In this case, as opposed to what we would expect given classical logic, we do have a case of a true contradiction. This is precisely what the dialetheist – someone who believes in the existence of true contradictions – argues. And the Liar sentence is one of the most important examples of the creatures in whose existence the dialetheist believes (see Priest 2006a, 2006b).

Clearly, for dialetheism to get off the ground, the underlying logic cannot be classical. Otherwise, for the reasons discussed above, we would obtain a trivial system – one in which everything would be derivable. And since we cannot use a trivial system to discriminate truths from falsities, such a system is entirely useless. It is, therefore, crucial that the logic used by the dialetheist be one in which not everything follows from a contradiction. So, the argument for explosion above needs to be blocked. Luckily for the dialetheist, there are such logics, called paraconsistent logics. Minimally, a logic is paraconsistent if the inference from a contradiction to an arbitrary sentence is *invalid* – in other words, explosion fails.[7]

---

[7]   Paraconsistent logics have a very interesting history (see da Costa, Béziau, and Bueno 1995). After some forerunners in the 1940s, particularly in Poland, Newton da Costa developed these logics in the late 1950s and in the 1960s in Brazil. From the 1970s until today, they have been articulated further in Australia, Brazil, and in the US, among many other countries.

It is interesting to note that Aristotelian logic is paraconsistent. After all, a syllogism with contradictory premises does not entail everything. Consider, for example, the following case. From the premises "Every man is mortal" and "Some man is immortal," we cannot validly infer that "Every tomato is blue." (For surveys of many paraconsistent systems, including paraconsistent syllogistic theories, see da Costa, Krause, and Bueno 2007, da Costa and Bueno 2001; see also Priest 2006a.)

Note also that one can be a paraconsistent logician without being a dialetheist, that is, without believing in the existence of true contradictions. In fact, most paraconsistent logicians are not dialetheists. For example, you may think that truth is not the relevant norm to assess theories, but something weaker, such as empirical adequacy, is (see van Fraassen 1980). Thus, despite your use of a paraconsistent logic, you are not committed to true theories – or to true contradictions!

But how can explosion fail? Suppose that our paraconsistent logic admits *three* truth-values: true (T), false (F) and both true-and-false (T-F).[8] And consider the following truth-conditions for the logical connectives (where *P* and *Q* range over sentences):

(Neg₁)   "*not-P*" is true if, and only if, "*P*" is false.
(Neg₂)   "*not-P*" is false if, and only if, "*P*" is true.
(Dis₁)   "*P or Q*" is true if, and only if, "*P*" is true or "*Q*" is true.
(Dis₂)   "*P or Q*" is false if, and only if, "*P*" is false and "*Q*" is false.
(Con₁)   "*P and Q*" is true if, and only if, "*P*" is true and "*Q*" is true.
(Con₂)   "*P and Q*" is false if, and only if, "*P*" is false or "*Q*" is false.

These are, of course, exactly the truth-conditions for the logical connectives of classical logic. But given that we have three truth-values, we need the two clauses for each connective. (These clauses are equivalent in classical logic, but not in Priest's paraconsistent logic LP – the Logic of Paradox.)

With these truth-conditions, it is not difficult to see how explosion is blocked. Suppose that the sentence *P* is both true-and-false (e.g., it is a sentence such as the Liar), and suppose that the sentence *Q* is false. Consider now disjunctive syllogism (the inference used in the last step of the argument for explosion discussed in the previous section):

(1)   *P or Q*   Assumption.
(2)   *not-P*   Assumption.
(3)   *Q*   From (1) and (2) by disjunctive syllogism.

Now, clearly "*P or Q*" is true. After all, "*P*", by hypothesis, is both true-and-false – and thus, in particular, true; hence, by condition (Dis₁), "*P or Q*" is true. Moreover, "*not-P*" is also true, given that "*P*" is both true-and-false, and hence, in particular, false; thus, by condition (Neg₁), "*not-P*" is true. As a result, the sentences in lines (1) and (2) above are true. However, "*Q*", by hypothesis, is false. This means that disjunctive syllogism is invalid, since we moved from true premises to a false conclusion. And given that this inference is required in the derivation of

---

[8]   Graham Priest developed in the 1970s a paraconsistent logic that has this feature (for details, see Priest 2006a, 2006b, and the remarks in the text below). It is called the Logic of Paradox (LP).

explosion, we can use the same counter-example to show that the latter is invalid as well. With a paraconsistent logic, contradictions no longer entail everything. They can be tamed.

Let us now recall the concept of validity offered by the logical pluralist: in every *case* in which the premises are true, the conclusion is true as well. As we saw, the classical logician considers as cases consistent and complete situations (Tarskian models). The paraconsistent logician, however, countenances a broader range of cases, which now include *inconsistent* and complete situations, such as those invoked in the Liar sentence. When cases of this sort are invoked, some classically valid inferences, such as explosion, are violated. Paraconsistent logic then emerges.

3.5   Quantum cases and quantum logic

By broadening the range of cases, however, we have room for additional non-classical logics. Suppose that we are now reasoning about quantum phenomena. By exploring certain features of the quantum domain, we obtain a rather unexpected result. A well known principle in classical logic is the so-called distributive law, according to which:

(Distr) $P \wedge (Q \vee R) \leftrightarrow (P \wedge Q) \vee (P \wedge R)$

However, under certain conditions, this law seems to be violated in the quantum domain. Which conditions are these?

First, according to standard quantum mechanics, each electron e has an angular momentum, or spin, in the x direction, which has only two possible values: $+\frac{1}{2} \vee -\frac{1}{2}$. In other words, if we denote by e_x the momentum of e in the x direction, then we have from quantum mechanics that: $e_x = +\frac{1}{2} \vee e_x = -\frac{1}{2}$. (Let us call this the *spin principle*.) Second, given the so-called Heinsenberg's principle, it is not possible to measure the spin of e in two distinct directions simultaneously.[9]

Suppose now that x and y are two distinct directions. We have just measured the momentum of e in the direction x, and obtained that: $e_x = +\frac{1}{2}$. Thus, the sentence "$e_x = +\frac{1}{2}$" is true. However, from the spin principle,

---

[9]   The measurement of the spin of an electron in one direction disturbs the spin of that electron in any other direction. Hence, we have the impossibility of measuring the spin in distinct directions at the same time.

we have that "$e_y = +\frac{1}{2} \lor e_y = -\frac{1}{2}$" is always true (at any instant). Thus, we conclude that the conjunction

(1) $e_x = +\frac{1}{2} \land (e_y = +\frac{1}{2} \lor e_y = -\frac{1}{2})$

is also true. If we apply to (1) the distributive law (Distr), it follows that:

(2) $e_x = +\frac{1}{2} \land (e_y = +\frac{1}{2} \lor e_y = -\frac{1}{2}) \leftrightarrow$
       $(e_x = +\frac{1}{2} \land e_y = +\frac{1}{2}) \lor (e_x = +\frac{1}{2} \land e_y = -\frac{1}{2})$

As we just saw, the left-hand side of this biconditional is true. However, given Heisenberg's principle, it is not possible to measure simultaneously the momentum of *e* in two distinct directions *x* and *y*. Thus, the right-hand side of the biconditional (2) is not true. Since one side of the biconditional is true and the other false, the biconditional fails. But this biconditional is an instance of the distributive law, which therefore fails in the quantum domain. We conclude that the application of the distributive law runs into difficulties in the quantum case. This result has motivated the development of various quantum logics, which are logics in which the distributive law does not hold (see, e.g., Birkhoff and von Neumann 1936, Hughes 1989, Rédei 1998).

Of course, there are ways to try to maintain classical logic given the problem presented by (2). However, up to this moment, none of the proposals has received unanimous acceptance. For example, someone could try to argue that it is the measurement that "creates" the value of the spin. Hence, the statement "$e_y = +\frac{1}{2} \lor e_y = -\frac{1}{2}$" is not true or false. In this way, we could try to restore the equivalence in (2), given that, on this intepretation, none of the sides is true. The problem here is that to accommodate this move, we need to introduce truth-value gaps, and recognize statements that lack truth-value. But this goes against classical logic with its standard, two-valued semantics.

This simple example also motivates conditions in which we may revise certain logical principles on *empirical grounds*. This suggests that, as opposed to the way in which Tarski has articulated the model-theoretic account, we may have good reason to revise a logical principle based on empirical considerations. In fact, the *a priori* character of logic has been contested in some quarters for this kind of reason (see, e.g., Putnam 1968, Bueno and Colyvan 2004). The example above indicates that a particular logical principle – the distributive law – seems inadequate to reason about quantum phenomena. After all, it leads us to infer results that violate conditions specified by quantum mechanics. As a result, Tarski's substitution requirement that was discussed above should be restricted. Depending on the domain we consider, such as the one dealing with quantum phenomena, we may conclude that certain replacements of the non-logical vocabulary are not truth-preserving, and thus are inadequate on empirical grounds. This does not mean that the substitution requirement does not hold at all. The idea is to restrict it: once a given domain is fixed, the inferences in *that* context are truth-preserving and do satisfy the condition of the substitution requirement. The latter simply does not hold in general.

We can now return to the concept of validity considered by the logical pluralist. What are the cases that the quantum logician entertains? Clearly, these are situations in which Heisenberg's principle and the spin principle hold. They are consistent and complete situations, in which conjunction does not distribute over disjunction. This is, of course, an unexpected result. It is formulated based on the capricious behavior of the phenomena in the quantum domain. Given that the distributive law does not seem to hold in this domain, a quantum logic can be adopted.

3.6 *Pluralism about logical pluralism*

Interestingly enough, there are distinct ways of understanding the logical pluralist picture. On the *pluralism by cases* approach just presented, there is only one concept of validity of an argument (an argument is valid if, and only if, in every *case* in which the premises are true, the conclusion is true as well). As we saw, what changes are the cases we consider: different cases lead to the formulation of different logics. If for classical logicians consistent and complete situations are the relevant cases, for quantum logicians, non-distributive situations need to be included, whereas para-consistent logicians emphasize the need for taking into account inconsistent situations as well.

But there is a second way of understanding logical pluralism. We can call this the *pluralism by variance* approach – or *pluralism by domains* (see, e.g., da Costa 1997, Bueno 2002, Bueno and Shalkowski 2009). On this view, there are different notions of validity, depending on the particular logical constants – that is, logical connectives and quantifiers – that are used in the specification of the concept of validity. On this understanding of logical pluralism, an argument is valid if, and only if, the conjunction of the premises and the negation of the conclusion is necessarily false.[10] If the conjunction and the negation invoked in this characterization are those from classical logic, we obtain the notion of validity from classical logic. Alternatively, if the conjunction and negation are those from paraconsistent (or quantum or some other) logic, we obtain the notion of validity from paraconsistent (or quantum or some other) logic (see Read 1988).[11]

According to the logical pluralist by variance, the debate between, for example, the classical and the paraconsistent (or quantum) logicians is a debate about what connectives need to be used in the characterization of the concept of validity. The logical monist thinks that there is only one right answer to the question about the connectives, whereas the logical pluralist thinks that there are many: different logics offer different, but also perfectly adequate, answers to this question.

Does that mean that the logical pluralist is a logical relativist? In other words, is the logical pluralist committed to the claim that all logics are equally adequate, that one is just as good as the other? Definitely not! After all, depending on the domain we consider, different logics would be appropriate – and other logics would be entirely inappropriate for such domains. As we saw above, the motivation for the development of quantum logics emerged from the inadequacy of some classically valid principles to accommodate inferences in the quantum domain. We also saw that, as opposed to paraconsistent logics, classical logic is not adequate to reason about inconsistent domains without triviality. Thus, we have here a clear emphasis on domains as the primary factor in the selection of a logic: inconsistent domains (paraconsistent logic), non-distributive domains (quantum logics), consistent and complete domains (classical logic). Hence, it is easy to see why we may call pluralism by variance *pluralism by domains*.

---

[10]   Note the explicitly modal character of the concept of logical consequence as formulated by the logical pluralist by variance. This point is not made explicit in the *formulation* of logical pluralism by cases, despite the fact that Beall and Restall recognize the significance of the modal issue (see Beall and Restall 2006, 14–16).

[11]   It should be pointed out that, in his 1988 book, Stephen Read seems to be a monist about logic – at least he is defending relevant logic, as opposed to classical logic, as the correct account of validity. (Relevant logic is a form of paraconsistent logic in which the premises need to be relevant for the derivation of the conclusion; see Read 1988.) But for Read, the issue among logicians who adopt different logics is the identification of the right group of connectives that have to be used to formulate the concept of validity. This is how the pluralist by variance thinks of the logical pluralism issue.

Can pluralists by domains make sense of logical disagreements? They can. What is at issue in debates about logic is the determination of the right connectives in the formulation of the concept of logical consequence. The format for that concept is the same: the impossibility of conjoining the premises and the negation of the conclusion in a valid argument. However, different instances of the concept of logical consequence emerge, depending on the connectives that are used in each particular context.

But what is the difference between cases (as used by the pluralist by cases) and domains (as invoked by the pluralist by domains)? Cases are the basic semantic structures (such as various kinds of models) that are employed to assess the truth-value of certain statements and thus to characterize various logics. Domains, in turn, are not semantic structures. They are the various kinds of things we reason about: inconsistent situations, quantum phenomena, constructive features of mathematics, and so on. These are very different things.

Not surprisingly perhaps, according to the logical pluralist by domains, logic does not have two of the traditional features that have been often associated with it. First, logic does not seem to be universal but is much more context-sensitive than is often thought. Changes in domains (e.g., from consistent to inconsistent domains, from distributive to non-distributive domains) are often associated with changes in logic (e.g., from classical to paraconsistent, from paraconsistent to quantum). Second, some of these changes in logic are motivated by empirical considerations. Again, the quantum case illustrates this situation beautifully. This suggests that logical principles need not be *a priori*: they can be motivated and revised on empirical grounds (see Bueno and Colyvan 2004). By not having these features, logic becomes much closer to empirical forms of inquiry, which typically are neither universal nor *a priori*.

These different formulations of logical pluralism highlight different ways of conceptualizing the issue of the plurality of logic. Whether we formulate the view in terms of cases or domains, logical pluralists agree that we need to make room for the plurality of logics as a crucial feature of a philosophical understanding of logic. The logical monist takes issue with this assessment, insisting that there is only one logic, which has all the elegant features that have been traditionally associated with the field (i.e., universality, formality, *a priori* character, and truth-preservation). Of course, the logical monist need not, but often does, identify logic with classical logic (Quine 1970 is a clear example). The challenge is then to provide an account that makes room for the perceived plurality of logics, and for the many applications of logics.

4. Applications of Logic

It is an undeniable fact that there are many logics: classical logic, modal logic, relevant logic, quantum logic, temporal logic, deontic logic, paraconsistent logic, free logic, and many more (for a description of many of these logics, see Gabbay and Guenthner 2001). Can a logical monist deny this fact? It would be very implausible, of course, if the monist were forced to do that. But the monist has resources to avoid being painted into such a corner.

Consider the distinction between pure and applied logic (da Costa 1997). We can think of a *pure logic* as a formal calculus that is studied primarily for its mathematical and formal features. We introduce a formal language, including connectives and quantifiers, specify rules for the behavior of the logical vocabulary, and define a concept of logical consequence for the language. We then study syntactic and semantic properties of the resulting language, and, whenever possible, prove a completeness theorem (to the effect that every logically valid statement is provable in the system). When we develop a logic in this way, we are not concerned with any applications it may have. We are simply developing a formal system.

This is entirely analogous to the notion of *pure mathematics*. In geometry, we may introduce various principles that characterize a domain of geometrical objects (such as points and curves), and we can then study these objects for their mathematical interest. In this way, we can introduce Euclidean systems – systems that satisfy the traditional assumptions about geometrical objects (in particular, in these systems the sum of the internal angles of a triangle is 180°). But we can also introduce non-Euclidean systems – systems in which some of the traditional assumptions about geometrical objects no longer hold (for example, the sum of the internal angles of a triangle is more than 180°). These two kinds of systems, although incompatible, can be studied side by side as pure mathematical theories.

Suppose, now, that we want to find out whether any of these geometrical systems, properly interpreted, apply to the physical world. We want to know what is the geometry of the universe. At this point, we leave the realm of pure mathematics and shift to *applied mathematics*. It is no longer enough to find out what follows (or not) from certain principles about geometrical objects. We need to develop suitable empirical interpretations of these principles that let us apply the pure geometrical systems to a messy, motley, often incongruous world. We can start with simple identifications, such as a light ray with a geometrical line, and try to work our way from there (see Putnam 1979).

Exactly the same point goes through in the case of *applied logic*. Logic can be applied to a huge variety of things: from electronic circuits and pieces of formal syntax through air traffic control, robotics, and computer programming to natural languages. In fact, paraconsistent logics have been applied to many of these items (see da Costa, Krause, and Bueno 2007). It is not difficult to see the motivation. For example, suppose that you are an air traffic controller operating two radars. According to one of them, there is an airplane approaching the airport from the southeast; according to the other, there is none. You have here contradictory reports, and no obvious way to decide which of the radars is correct. In situations such as this, it can be very useful to adopt a paraconsistent logic. The logic can be used to help you reason about these pieces of information without triviality; that is, without logically implying everything. This may help you determine which bits of information should be eventually rejected.

Of course, crucial among these applications of logic is the application to the vernacular (see Priest 2006b). Ultimately, what we want to find out when we apply a pure logic is the consequence relation (or the various consequence relations if logical pluralism turns out to be correct) that is (or are) used in the language we speak – although the other applications are important as well.

Now, reflection on what holds in a certain domain may give us reason to reconsider the adequacy of certain patterns of inference. For instance, as we saw, since the distributive law can be violated in quantum domains, we may adopt inferential procedures in which this law is not invoked when reasoning about quantum phenomena. This offers a certain conception of how new relations of logical consequence can be formulated, and subsequently applied.

In this respect, there are significant connections between the application of logic and the application of mathematics. In some cases, pure mathematical theories that were initially developed quite independently of concerns with applications, turned out to be particularly useful in applied contexts. A case such as the development of non-Euclidean geometries illustrates this point. In other cases, the need for the development of suitable mathematical theories emerged from the lack of appropriate mathematical tools to describe certain empirical situations. This was the case, for instance, with the early development of the calculus by Isaac Newton and Gottfried Leibniz. The calculus was a mathematical theory that emerged, basically, from the need to articulate certain explanations in physics.

The same points go through in the case of applied logic. In certain cases, pure logics have been developed in advance of particular applications. The use of paraconsistent logic in air traffic control or in robotics was certainly not part of the original motivation to formulate this logic. In other cases, certain needs in applied contexts demanded certain relations of logical consequence. Suitable logics were then subsequently developed. As we saw, quantum logics emerged from the difficulty that the distributive law generated in the domain of quantum mechanics. In retrospect, the connections between applied logic and applied mathematics should not be surprising, given that ultimately both deal with inferences and with suitable descriptions of particular domains.

At this point, the monist can invoke the distinction between pure and applied logic to make sense of the perceived plurality of logics. That there are many *pure* logics is an undeniable fact. But the central question is to determine what is the correct logic *applied* to the vernacular (see Priest 2006b). When we try to answer *this* question, the logical monist insists, we find that there is only one. The pluralist disagrees – and the debate goes on.

5. Conclusion

In this essay, we saw that the central concept of logic – that of logical consequence – can be formulated in different ways. Following a long-standing tradition, Tarski's model-theoretic formulation of the concept embodied some of the central features of logic: universality, formality, truth-preservation, and *a priori*. We also saw that, by using models in the characterization of logical consequence, Tarski's formulation ended up favoring one particular, but definitely important, logic: the classical one.

Following the logical pluralist framework offered by Beall and Restall, we then explored what happens to the characterization of logical consequence if we change the emphasis from models to other cases, such as inconsistent and complete situations, or consistent, complete and non-distributive situations. We saw that different logics would then emerge – in particular, paraconsistent and quantum logics. But the logical pluralist framework in terms of cases is not the only one, and we saw that logical pluralism can also be characterized in terms of domains, with the advantage that the explicitly modal force of the notion of logical consequence is highlighted. In turn, the logical pluralist by domains challenges some of the received features of logic; in particular, the latter's alleged *a priori* and universal character. For some, this is a *reductio ad absurdum* of this form of logical pluralism; for others, this shows the close connection that logic has with all our forms of inquiry, particularly in science and mathematics.

This point naturally led to the issue of the application of logic. And we explored some of the connections between the application of logic and of mathematics. Both have pure and applied counterparts. Both are sometimes motivated by, and developed from, the needs that emerge in empirical contexts. Both have been successfully applied to areas for which they were not originally devised. Given the close connections that logic has with our forms of inquiry, particularly on the pluralist by domains conception, it is not surprising to find such connections between these fields. This is as it should be. There is no reason to leave logic hanging in the clouds.

References

Beall, JC (ed.). 2007. *Revenge of the Liar: New Essays on the Paradox*. Oxford: Oxford University Press.

Beall, JC and G. Restall. 2006. *Logical Pluralism*. Oxford: Clarendon Press.

Birkhoff, G. and J. von Neumann. 1936. "The Logic of Quantum Mechanics." *Annals of Mathematics* 37: 823–43.

Bueno, O. 2002. "Can a Paraconsistent Theorist be a Logical Monist?" In W. Carnielli, M. Coniglio, and I. D'Ottaviano (eds.), *Paraconsistency: The Logical Way to the Inconsistent*. New York: Marcel Dekker, 535–52.

Bueno, O. and M. Colyvan. 2004. "Logical Non-Apriorism and the 'Law' of Non-Contradiction." In G. Priest, JC Beall, and B. Armour-Garb (eds.), *The Law of Non-Contradiction: New Philosophical Essays*. Oxford: Clarendon Press, 156–75.

Bueno, O. and S. Shalkowski. 2009. "Modalism and Logical Pluralism." *Mind* 118: 295–321.

Carnielli, W., M. Coniglio, and I. D'Ottaviano (eds.). 2002. *Paraconsistency: The Logical Way to the Inconsistent*. New York: Marcel Dekker.

da Costa, N.C.A. 1997. *Logiques classiques et non classiques*. Paris: Masson.

da Costa, N.C.A. and O. Bueno. 2001. "Paraconsistency: Towards a Tentative Interpretation." *Theoria* 16: 119–45.

da Costa, N.C.A., J.-Y. Béziau, and O. Bueno. 1995. "Paraconsistent Logic in a Historical Perspective." *Logique et Analyse* 150-151-152: 111–25.

da Costa, N.C.A., D. Krause, and O. Bueno. 2007. "Paraconsistent Logics and Paraconsistency." In D. Jacquette (ed.), *Philosophy of Logic*. Amsterdam: North-Holland, 791–911.

Etchemendy, J. 1990. *The Concept of Logical Consequence*. Cambridge, MA: Harvard University Press.

Frege, G. 1967. "Concept Script: A Formal Language of Pure Thought Modeled upon That of Arithmetic," trans S. Bauer-Mengelberg. In J. van Heijenoort (ed.), *From Frege to Gödel: A Source Book in Mathematical Logic, 1879–1931*. Cambridge, MA: Harvard University Press.

Gabbay, D.M. and F. Guenthner (eds.). 2001. *Handbook of Philosophical Logic*. 2nd edn. Dordrecht: Kluwer Academic Publishers.

Hacking, I. 1979. "What is Logic?" *Journal of Philosophy* 76: 285–318. Reprinted in R.I.G. Hughes (ed.), *A Philosophical Companion to First-Order Logic*. Indianapolis, IN: Hackett Publishing Company, 1993, 225–58.

Hughes, R.I.G. 1989. *The Structure and Interpretation of Quantum Mechanics*. Cambridge, MA: Harvard University Press.

Hughes, R.I.G. (ed.). 1993. *A Philosophical Companion to First-Order Logic*. Indianapolis, IN: Hackett Publishing Company.

Jacquette, D. (ed.). 2007. *Philosophy of Logic*. Amsterdam: North-Holland.

Martin, R. (ed.). 1984. *Recent Essays on Truth and the Liar Paradox*. Oxford: Clarendon Press.

McGee, V. 1992. "Two Problems with Tarski's Theory of Consequence." *Proceedings of the Aristotelian Society* 92: 273–92.

Priest, G. 2006a. *In Contradiction*. 2nd edn. Oxford: Oxford University Press.

Priest, G. 2006b. *Doubt Truth to Be a Liar*. Oxford: Oxford University Press.

Priest, G., JC Beall, and B. Armour-Garb (eds.). 2004. *The Law of Non-Contradiction: New Philosophical Essays*. Oxford: Clarendon Press.

Putnam, H. 1968. "The Logic of Quantum Mechanics." In *Mathematics, Matter and Method*. Philosophical Papers, vol. 1. 2nd edn. Cambridge: Cambridge University Press, 1979, 174–97.

Putnam, H. 1979. *Mathematics, Matter and Method*. Philosophical Papers, vol. 1. 2nd edn. Cambridge: Cambridge University Press.

Quine, W.V. 1950. *Methods of Logic*. Cambridge, MA: Harvard University Press. Introduction reprinted in R.I.G. Hughes (ed.), *A Philosophical Companion to First-Order Logic*. Indianapolis, IN: Hackett Publishing Company, 1993, 1–5; references are to the latter edition.

Quine, W.V. 1970. *Philosophy of Logic*. Englewood Cliffs, NJ: Prentice Hall.

Ray, G. 1996. "Logical Consequence: A Defense of Tarski." *Journal of Philosophical Logic* 25: 617–77.

Read, S. 1988. *Relevant Logic*. Oxford: Blackwell.

Read, S. 1995. *Thinking About Logic*. Oxford: Oxford University Press.

Rédei, M. 1998. *Quantum Logic in Algebraic Approach*. Dordrecht: Kluwer Academic Publishers.

Schirn, M. (ed.). 1998. *Philosophy of Mathematics Today*. Oxford: Oxford University Press.

Shalkowski, S. 2004. "Logic and Absolute Necessity." *Journal of Philosophy* 101: 1–28.

Shapiro, S. 1998. "Logical Consequence: Models and Modality." In M. Schirn (ed.), *Philosophy of Mathematics Today*. Oxford: Oxford University Press, 131–56.

Sher, G. 1996. "Did Tarski Commit 'Tarski's Fallacy'?" *Journal of Symbolic Logic* 61: 653–86.

Tarski, A. 1936a. "The Concept of Truth in Formalized Languages." In *Logic, Semantics, Metamathematics: Papers from 1923 to 1938*. 2nd edn. John Corcoran (ed.). Indianapolis, IN: Hackett Publishing Company, 1983, 152–278.

Tarski, A. 1936b. "On the Concept of Logical Consequence." In *Logic, Semantics, Metamathematics: Papers from 1923 to 1938*. 2nd edn. John Corcoran (ed.). Indianapolis, IN: Hackett Publishing Company, 1983, 409–20.

Tarski, A. 1983. *Logic, Semantics, Metamathematics: Papers from 1923 to 1938*. 2nd edn. John Corcoran (ed.). Indianapolis, IN: Hackett Publishing Company.

van Fraassen, B. 1980. *The Scientific Image*. Oxford: Clarendon Press.

van Heijenoort, J. (ed.). 1967. *From Frege to Gödel: A Source Book in Mathematical Logic, 1879–1931*. Cambridge, MA: Harvard University Press.

4 Philosophy of Mathematics

Otávio Bueno

In this essay, I examine three interrelated issues in the philosophy of mathematics. The first is an issue in ontology: do mathematical objects exist and, if so, what kind of objects are they? The second is an issue in epistemology: how do we have mathematical knowledge? The third is an issue about the application of mathematics: how can mathematics be applied to the physical world? These issues are, of course, interconnected, and I explore the relations between them by examining different philosophical conceptions about mathematics. I start with four versions of Platonism (Fregean Platonism, Gödelian Platonism, Quinean Platonism, and structuralist Platonism), and I focus on the different epistemological strategies that have been developed to explain the possibility of mathematical knowledge. Given that, on these views, mathematical objects exist and are not physically extended, accounts of how we can have knowledge of these objects are in order. I then consider three alternative nominalist approaches (mathematical fictionalism, modal structuralism, and deflationary nominalism), which deny the existence of mathematical objects. With this denial, nominalists owe us, in particular, an explanation of the application of mathematics. I then examine their strategies to make sense of this issue, despite the non-existence of mathematical entities. I conclude the essay by bringing together the various issues and the philosophical conceptions.

1. Introduction

Understood as a philosophical reflection about mathematics, the philosophy of mathematics has a long history that is intertwined, in intriguing and complex ways, with the development of philosophy itself. Some of the central figures in western philosophy – from Plato and Aristotle through

René Descartes and Gottfried Leibniz to Immanuel Kant and Rudolf Carnap – had more than a simple acquaintance with the mathematics developed in their own times. In fact, much of their best work was inspired by a careful consideration of the mathematics available to them, and some of them (such as Descartes and Leibniz) made lasting contributions to mathematics itself.

In this essay, rather than providing a historical overview of the philosophy of mathematics, I will focus on some of the central issues in the field, giving the reader a sense of where some of the current debates are. In particular, I will examine three interrelated issues. First, I will consider an *ontological* issue: the issue of the existence and nature of mathematical objects. Do they exist? And what kinds of things are mathematical objects? Are they mind-independent abstract entities (that is, entities that exist independently of us, but which are not located in space-time), or are mathematical objects particular mental constructions (the result of certain operations performed in the mathematicians' minds)? Moreover, should mathematical objects be thought of as individual entities (such as particular numbers), or as overall structures (such as various relations among numbers)? There are significant disagreements about how these questions should be answered, and distinctive proposals have been advanced to support the corresponding answers.

Second, I will examine an *epistemological* issue pertaining to the nature of our mathematical knowledge. How do we know the mathematics that we know? Do we have some form of access to mathematical objects akin to the perception of physical objects? And how is mathematical knowledge possible if mathematical objects are taken to be abstract entities that exist independently of us? Once again, distinctive views have been developed to address these problems.

Finally, I will discuss the problem of the *application of mathematics*. How can mathematics be applied to the physical world? Should we take mathematical objects to be indispensable to our best theories of the physical world? That is, can such theories be formulated in a way that they are without reference to mathematical objects? And if mathematical objects are indeed indispensable, does that require us to be committed to their existence?

As will become clear, these three issues are closely connected, and answers to one of them will constrain and, in part, shape answers to the others. Typically, to develop a particular philosophy of mathematics involves, at a minimum, the development of views about these issues. At least those who have approached the philosophy of mathematics systematically have attempted to do that. Thus, instead of examining these issues separately, I will consider them together, as part of the development of different conceptions of mathematics. I will start by describing four different versions of Platonism in mathematics – according to which mathematical objects exist and they are abstract – and how these various proposals have addressed the issues. I will then consider three versions of nominalism in mathematics – according to which mathematical objects do not exist – and discuss how these issues have been re-conceptualized by nominalists.

Like most philosophical proposals, those discussed in this essay face challenges. Nevertheless, I have decided to focus on the positive contributions they have made, leaving the assessment of their drawbacks for another occasion. Even if none of these proposals has settled the issues under consideration, each has made significant contributions to our understanding of the philosophical issues about mathematics. For this reason alone, they deserve close study.

2. Platonism in Mathematics

The basic ontological question in the philosophy of mathematics deals with the issue of the existence of mathematical objects. According to the Platonist, mathematical objects *exist* independently of us. They are *not mind-dependent* in the sense that their existence is not the result of our mental processes and linguistic practices. Furthermore, they are *abstract*, which is to say that they do not have physical extension (they are not located in space and time). So, mathematical objects are not the kind of thing with which we would expect to have causal interaction, given that we can interact causally only with physical objects. For example, the pitcher throws one ball at a time. When the batter hits that ball, he does not hit the number one; he could not hit a number that has no physical extension. He hits a physical object, the baseball.

For the Platonist, not only mathematical objects exist, they exist necessarily. That is, those objects that do exist necessarily do so. Mathematical objects, on this conception, could not have failed to exist. What is the source of this necessity? In other words, is there something that makes mathematical objects necessarily existing entities? Clearly, there is no causal process involved here, given the causal inertness of mathematical objects. So, the necessity has some other origin. An answer that some Platonists, such as Gottlob Frege (1974), have offered emerges from the concepts involved. Concepts are here understood not as mental, psychological states, such as beliefs or mental images. Such states are subjective, and they depend on whoever has them. Your mental image of a watermelon may, but need not, have something in common with mine. Concepts, on the Platonist view, are not subjective images or experiences on our minds. Rather, they are objective, mind-independent objects whose existence does not depend on us. And once certain concepts are in place, objects that satisfy these concepts need to exist. In brief, certain concepts require the existence of the corresponding objects. For example, consider the concept of a natural number. Ask yourself: how many objects fall under the concept *not identical to itself*? The answer, of course, is zero. That characterizes the number zero (that is, this condition specifies the object that falls under the concept *zero*). How many objects fall under the concept *identical to zero*? The answer, once again, is clear: only one object (namely, zero). We have now characterized the number one. How many objects fall under the concept *identical to one or zero*? Precisely two objects: zero and one. That characterizes the number two. And so on. This is, very briefly, Frege's characterization of natural numbers (Frege 1974).[1]

There are different forms of Platonism. Here I will consider four of them: Fregean Platonism (Frege 1974, Hale and Wright 2001), Gödelian Platonism (Gödel 1964, Maddy 1990), Quinean Platonism (Quine 1960, Colyvan 2001), and structuralist Platonism (Resnik 1997, Shapiro 1997). Given that all of these views are particular forms of Platonism, they have in common the contention that mathematical objects (broadly understood here to include mathematical structures): exist; are mind-independent; and are abstract (that is, they are not physically extended). Despite these common features, there are significant differences among these views. I will explore some of them below.

2.1 Fregean Platonism

For the Fregean Platonist, the abstract character of mathematical objects emerges from the kind of thing they are: objects that fall under certain

---

[1] Operations over natural numbers, such as addition and subtraction, can also be characterized, and arithmetic can be developed perfectly in this way (see Frege 1974; see also Boolos 1998, Hale and Wright 2001).

concepts. As we saw, Fregean concepts are abstract, mind-independent things; numbers and other mathematical objects inherit the same abstract character from the concepts under which they fall. The Fregean Platonist has no difficulty making sense of the objectivity of mathematics, given the mind-independence of the concepts involved in their characterization. Frege's own motivation to develop his proposal, which was initially developed to make sense of arithmetic, emerged from the need to provide a formulation of arithmetic in terms of logic. Frege had developed the first formal system of symbolic logic, and his goal was to show that arithmetical concepts could be reduced to logical concepts – such as identity, predication, negation, and conjunction – plus some definitions. Frege was, thus, what we now call a *logicist* about arithmetic. In the example discussed above, the number zero was characterized in terms of the concepts of negation, identity, and predication (i.e., the number of objects that fall under the concept *not identical to itself*).

Part of the motivation for Frege's logicism was epistemological. The logicist approach could offer a suitable account of our knowledge of arithmetic. (I will return to this point below.) But probably Frege's main motivation was of an ontological nature. He wanted to provide the right answer about the nature of mathematical objects, and, in particular, provide the proper characterization of the concept of number. In fact, the bulk of *Foundations of Arithmetic* (Frege 1974) is dedicated to a thorough and extremely critical discussion of the accounts of the concept of number in Frege's time. Given what Frege (correctly) perceived to be a morass of confusion, unclearness, and incoherence that prevailed in the discussions of the foundations of arithmetic from the seventeenth through the nineteenth centuries, he worked very hard to offer a clear, well grounded, and coherent logicist alternative. Due to the precise nature of the logical notions, Frege's reduction of arithmetic concepts to logical ones generated a clear-cut formulation of the former.

Central to Frege's strategy was the use of what is now called Hume's Principle: two concepts are equinumerous if and only if there is a one-to-one correspondence between them. Hume's Principle was used at various crucial points; for instance, to show that the number zero is different from the number one. Recall that the concept *zero* is characterized in terms of the number of objects that fall under the concept *not identical to itself*, and that the concept *one*, in turn, is characterized in terms of the concept *identical to zero*. Now, given that nothing falls under the concept *not identical to itself*, and only one object falls under the concept *identical to zero*, by Hume's Principle, these two concepts are not equinumerous. As a result, zero is distinct from one.

But how can one establish that Hume's Principle is true? Frege thought he could derive Hume's Principle from a basic logical law, which is called Basic Law V. According to this law, the extension of the concept *F* is the same as the extension of the concept *G* if and only if the same objects fall under the concepts *F* and *G*. Basic Law V seemed to be a fundamental logical law, dealing with concepts, their extensions, and their identity. It had the right sort of generality and analytic character that was needed for a logicist foundation of arithmetic.

There is only one problem: Basic Law V turns out to be inconsistent. It immediately raises Russell's paradox if we consider the concept *is not a member of itself*. To see why this is the case, suppose that there is such a thing as the set composed by all the sets that are not members of themselves. Let us call this set *R* (for Russell). Now let us consider whether *R* is a member of *R*. Suppose that it is. In this case, we conclude that *R* is not a member of *R*, given that, by definition of *R*, *R*'s members are those sets that are *not* members of themselves. Suppose, in turn, that *R* is *not* a member of *R*. In this case, we conclude that *R is* a member of *R* – since this is precisely what it takes for the set *R* to be a member of *R*. Thus, *R* is a member of *R* if, and only if, *R* is not a member of *R*. It then immediately follows that *R* is and is not a member of *R* – a contradiction.

Someone may say that this argument just shows that there is not such a thing as the Russell set *R* after all.[2] So, what is the big deal? The problem is that, as Russell also found out, it follows from Frege's system using suitable definitions that there *is* a set of all sets that are not members of themselves. Given the argument above establishing that there *is not* such a set, we have a contradiction. Frege's original reconstruction of arithmetic in terms of logic was in trouble.

But not everything was lost. Although Frege acknowledged the problem, and tried to fix it by introducing a new, consistent principle, his solution

---

[2] I am assuming here, with Frege and Bertrand Russell, that the underlying logic is classical. In particular, classical logic has the feature that everything follows from a contradiction; this principle is often called *explosion*. However, there are non-classical logics in which this is not the case; that is, on these logics, not everything follows from a contradiction. These logics are called paraconsistent (see, e.g., Priest 2006, da Costa, Krause, and Bueno 2007). If we adopt a paraconsistent logic, we can then study the properties of the Russell set in a suitable paraconsistent set theory (see, again, da Costa, Krause, and Bueno 2007).

ultimately did not work.[3] However, there was a solution available to Frege. He could have jettisoned the inconsistent Basic Law V and adopted Hume's Principle as his basic principle instead. Given that the only use that Frege made of Basic Law V was to derive Hume's Principle, if the latter were assumed as basic, one could then run, in a perfectly consistent manner, Frege's reconstruction of arithmetic. In fact, we could then credit Frege with the theorem to the effect that arithmetic can be derived in a system like Frege's from Hume's Principle alone. Frege's approach could then be extended to other branches of mathematics.[4]

2.2   Gödelian Platonism

If, for the Fregean, arithmetic is ultimately derivable in second-order logic plus Hume's Principle, for someone like Kurt Gödel, the truth of basic mathematical axioms can be obtained directly by intuition (see Gödel 1964, Maddy 1990). Frege thought that intuition played a role in how we come to know the truth of geometrical principles (which, for him, following Kant, were synthetic *a priori*);[5] but arithmetic, being derivable from logic, was analytic. For Gödel, however, not only the principles of arithmetic, but also the axioms of set theory can be apprehended directly by intuition. We have, Gödel claims, "something like a perception of the objects of set theory" (1964, 485). That is, we are able to "perceive" these objects as having certain properties and lacking others, in a similar way to that in which we perceive physical objects around us. That we have such a perception of set-theoretic objects is supposed to be "seen from the fact that the axioms [of set theory] force themselves upon us as being true" (Gödel 1964, 485).

But how exactly does the fact that the axioms of set theory "force themselves upon us as being true" support the claim that we "perceive" the objects of set theory? Gödel seemed to have a broad conception of perception, and

---

[3]   The principle that Frege introduced as a replacement for Basic Law V, although logically consistent, was inconsistent with the claim that there are at least two distinct numbers. Since the latter claim was true in Frege's system, the proposed principle was unacceptable. For an illuminating discussion, see Boolos 1998.

[4]   To implement a program along these lines is one of the central features of the neo-Fregean approach to the philosophy of mathematics. See, e.g., Hale and Wright 2001, and, for a discussion, Boolos 1998.

[5]   Synthetic *a priori* claims are claims with content (i.e., they are, as it were, about "the world"), but which are known independently of experience.

when he referred to the objects of set theory, he thought that we "perceived" the concepts involved in the characterization of these objects as well. The point may seem to be strange at first. With further reflection, however, it is not unreasonable. In fact, an analogous move can be made in the case of the perception of physical objects. For example, in order for me to perceive a tennis ball, and recognize it *as* a tennis ball, I need to have the concept of *tennis ball*. Without the latter concept, at best I will perceive a round yellow thing – assuming I have *these* concepts. Similarly, I would not be able to recognize certain mathematical objects *as* objects of set theory unless I had the relevant concepts. The objects could not be "perceived" to be set-theoretic except if the relevant concepts were in place.

Now, in order to justify the "perception" of set-theoretic objects from the fact that the axioms of set theory are forced upon us as being true, Gödel needs to articulate a certain conception of rational evidence (see Parsons 2008, 146–8). On Gödel's view, in order for us to have rational evidence for a proposition – such as an axiom of set theory – we need to make sense of the concepts that occur in that proposition. In making sense of these concepts, we are "perceiving" them. Mathematical concepts are robust in their characterization, in the sense that what they stand for is not of our own making. Our perception of physical objects is similarly robust. If there is no pink elephant in front of me right now, I cannot perceive one.[6] And you cannot fail to perceive the letters of this sentence as you read it, even though you might not be thinking about the letters as you read the sentence, but what the latter stands for. The analogy between sense perception and the "perception" of concepts is grounded on the robustness of both. The robustness requires that I perceive what is the case, although I can, of course, be mistaken in my perception. For instance, as I walk along the street, I see a bird by a tree. I find it initially strange that the bird does not move as I get closer to the tree; only to find out, when I get close enough, that there was no bird there, but a colorful piece of paper. I thought I had perceived a bird, when in fact I perceived something else. The perception, although robust – something was perceived, after all – is fallible. But I still perceived what was the case: a piece of paper in the shape of a bird. I just mistook that for a bird, something I corrected later. Similarly, the robustness of our "perception" of the concepts involved in an axiom of set theory is

---

[6] I can *imagine* one, but that is an entirely different story, since imagination and perception have very different functions, and each has its own phenomenology. I can perhaps *hallucinate* that there is a pink elephant in front of me, but again that would not be to *perceive* an elephant.

part of the account of how that axiom can force itself upon us as being true. By making sense of the relevant set-theoretic concepts, we "perceive" the latter and the connections among them. In this way, we "perceive" what is the case among the sets involved. Of course, similarly to the case of sense perception, we may be mistaken about what we think we perceive – that is part of the fallibility of the proposal. But the "perception" is, nevertheless, robust. We "perceive" something that is true.

This account of "perception" of mathematical concepts and objects is, in fact, an account of mathematical intuition. Following Charles Parsons (2008, 138–43), we should note that we have intuition of two sorts of things. We have intuition of objects (e.g., the intuition of the objects of arithmetic), and we have intuition that some proposition is true (e.g., the intuition that "the successor of a natural number is also a natural number" is true). The former can be called "intuition *of*," and the latter "intuition *that*." In the passage quoted in the first paragraph of this section, Gödel seems to be using the intuition *that* the axioms of set theory force themselves upon us to support the corresponding intuition *of* the objects of set theory. The robustness of both intuitions involved here is a central feature of the account.

2.3   Quinean Platonism

The Gödelian Platonist explores some connections between the "perception" of mathematical objects and perception of physical entities. The Quinean Platonist draws still closer connections between mathematics and the empirical sciences. If you have a strong nominalistic tendency, but find out that, in the end, you cannot avoid being committed to the existence of mathematical objects when you try to make sense of the best theories of the world, you probably are a Quinean Platonist. W.V. Quine himself was such a Platonist (see Quine 1960). Even when he acknowledges the indispensable role that reference to mathematical objects plays in the formulation of our best theories of the world, Quine insists that he is committed to only one kind of mathematical object: classes. All the other mathematical objects that he needs, such as numbers, functions, and geometrical spaces, can be obtained from them.

On the Quinean conception, Platonism is a matter of ontological honesty. Suppose you are a scientific realist about science; that is, you take scientific theories to be true (or, at least, approximately true), and you think that the terms in these theories refer. So, for example, in quantum mechanics, you acknowledge that it is crucial to refer to things like electrons, protons, photons, and quarks. These posits are an integral part of the theory, and positing their existence is central in the explanation of the behavior of the observable phenomena. These explanations include making successful predictions and applying quantum theory in various contexts.

Now, in the formulation of quantum mechanics, it is indispensable to refer not only to electrons and other quantum particles, but also to mathematical objects. After all, it is in terms of the latter that we can characterize the former. For instance, there is no way to characterize an electron but in terms of a certain group of invariants. These invariants are particular mathematical functions (particular mappings). So, to acknowledge commitment to the existence of electrons but deny the existence of the corresponding mathematical functions is to take back what needs to be assumed in order to express what such a physical object is. It is to indulge in double thinking, in ontological dishonesty. Quine's indispensability argument is an attempt to get us straight – particularly the scientific realists among us. According to this argument:

(P1)   We ought to be ontologically committed to all and only those entities that are indispensable to our best theories of the world.

(P2)   Mathematical entities are indispensable to our best theories of the world.

(C)   Therefore, we ought to be ontologically committed to mathematical entities.[7]

The point of the indispensability argument is to ensure that mathematical and physical posits be treated in the same way. In fact, on the Quinean picture, there is no sharp divide between mathematics and empirical science. Both are part of the same continuum and ultimately depend on experience. There is only a difference in degree between them. Certain mathematical theories are presupposed and used in various branches of science, but, typically, the results of scientific theories are not presupposed in the construction of mathematical theories. In this sense, the theories in mathematics are taken to be more general than the theories in science. Mathematics also plays a fundamentally instrumental role in science,

[7]   The central ideas of the indispensability argument can be found in Quine 1960. Hilary Putnam developed these ideas further in an explicitly scientific realist framework; see Putnam 1971. For a thorough and insightful examination and defense of the indispensability argument, see Colyvan 2001. The formulation of the argument presented here is due to Mark Colyvan.

helping the formulation of scientific theories and the expression of suit-able relations among physical entities.

Of course, not every mathematical theory has indispensable applications in science. For example, theories about inaccessible cardinals in set theory – that is, roughly, sets that cannot be reached, even in principle, from other sets below in the hierarchy of sets – do not seem to have found indispensable applications. For Quine, such theories constitute mathematical recreations, and do not demand ontological commitment, that is, commitment to the existence of the entities they posit. The situation is entirely different, however, when we consider those mathematical theories that are used in science. Given that we cannot even formulate the corresponding scientific theories without using (and quantifying over) mathematical objects – such as functions, numbers, and geometrical spaces – the existence of the latter is required as much as the existence of physical posits – such as neutrons, protons, and positrons. Platonism is the outcome of an honest understanding of the ontology of science.

2.4   Structuralist Platonism

The crucial feature of mathematical structuralism is to conceptualize math-ematics as the study of structures, rather than objects. Different forms of structuralism provide different accounts of structure (see, e.g., Resnik 1997, Shapiro 1997). But, crucially for the structuralist, it does not matter which mathematical objects one considers; as long as they satisfy the relevant struc-ture it will be sufficient to explain the possibility of mathematical knowledge.

We find this move in Michael Resnik's defense of structuralism. To explain the possibility of mathematical knowledge, Resnik introduces the notion of a *template*, which is a concrete entity – including things such as draw-ings, models, blueprints – and is meant to link the concrete aspects of our experience with abstract *patterns* (Resnik's term for structure). The crucial idea is that there are structural relations (such as an isomorphism[8]) between templates and patterns that allow us to represent the latter via the former. In particular, it is because there are such structural relations between

---

[8]   An isomorphism between two structures is a function that preserves the relations in these structures. For example, the structure of the even numbers (and the relation of addition between them) is isomorphic to the structure of the odd numbers (and the addition relation), since there is a function that maps each even number into one, and only one, odd number and that preserves the relation of addition between such numbers.

patterns and templates that mathematicians can use proofs – the process of creating and manipulating concrete templates via certain operations – to generate information about abstract patterns (Resnik 1997, 229–35). And given that mathematicians only have access to templates, no direct access to positions in patterns – that is, no direct access to mathematical objects – is presupposed in Resnik's picture.

A significant feature of patterns, on Resnik's view, is the fact that the positions in such patterns are incomplete. This means that there is no fact of the matter as to whether these positions have certain properties or not. Consider, for example, the second position in the natural number pattern (for simplicity, call that position "the number two"). It is not clear that there is a fact of the matter as to whether this position – the number two – is the same as the corresponding position in the real number pattern. In other words, it is not clear that there is a fact of the matter as to whether the number two in the natural number pattern is the same as the number two in the real number structure. After all, the properties that a position in a pattern has depend on the pattern to which it belongs. In the natural number pattern, the number two has the third position in the pattern – that is, the number three – as its immediate successor. But this is not the case in the context of the real number pattern. Of course, in the real number pattern, the immediate successor of the number two *that is also a natural number* is the number three. But to say that this is the same property as the one in the natural number pattern is already to assume that the corresponding numbers are the same, which is the point in question. As a result, it is not clear how one could decide issues such as these.[9]

For Resnik, rather than a problem, this incompleteness of the positions in a pattern (or structure) is a significant feature of mathematics. Ultimately, we will not be able to decide several issues about the identity of the positions in structure. We can decide, however, issues about the mathematical structures themselves, where such incompleteness typically does not emerge. That is expected on a structuralist account. After all, what we get to know when we know mathematics does not have to do with the nature of mathematical objects, but rather the structures (or patterns) they are part of.

---

[9] Note that there are two senses of incompleteness of a position in a pattern. First, there is an *ontological* sense, according to which there is no fact of the matter as to whether that position has certain properties or not. But there is also an *epistemic* sense, according to which we do not have complete information to decide whether a given position has or not certain properties. Clearly, the ontological sense entails the epistemic sense, but not vice versa; the ontological sense is thus more fundamental.

3.   Nominalism in Mathematics

After describing the Platonist proposals above, I will now consider some
of the nominalist alternatives. In particular, I will examine: mathematical
fictionalism (Field 1980, 1989), modal structuralism (Putnam 1967, Hellman
1989), and deflationary nominalism (Azzouni 2004). What all of these proposals have in common is the fact that they do not take mathematical objects
(or structures) to exist. In fact, they deny their existence. As a result, the main
difficulty faced by Platonism (the explanation of how we have mathematical
knowledge) vanishes. But other problems emerge as well. Along the way,
these nominalist proposals offer distinctive understandings of mathematics.

3.1   Mathematical fictionalism

In a series of works, Hartry Field provided an ingenious strategy for the
nominalization of science (Field 1980, 1989). As opposed to Platonist views,
in order to explain the usefulness of mathematics in science, Field does
not postulate the truth of mathematical theories. On his view, it is possible
to explain successful applications of mathematics with no commitment
to mathematical objects. Therefore, the indispensability argument, which
Field takes to be the only non-question-begging argument for Platonism,
is blocked. The nominalist component of Field's account derives from the
fact that no mathematical objects are assumed to exist; hence, mathematical
theories are false.[10] By devising a strategy that shows how to dispense with
mathematical objects in the formulation of scientific theories, Field rejects
the indispensability argument, and provides strong grounds for the articulation of a nominalist stance.

   Field's strategy depends on two interrelated moves. The first is to change
the aim of mathematics, which is not taken to be truth, but something
different. On his view, the proper norm of mathematics is *conservativeness*
(Field 1980, 16–19, 1989, 59). According to Field, a mathematical theory
is conservative if it is consistent with every internally consistent theory about
the physical world. Such theories about the physical world do not involve
any reference to, or quantification over, mathematical objects, such as sets,

---

[10]   As Field notes, strictly speaking, any existential mathematical statement is false, and any
universal mathematical statement is (vacuously) true.

functions, and numbers; they are called *nominalistic* theories (Field 1989, 58). And it is precisely because mathematics is conservative that, despite being false, it can be useful. Mathematics is useful because it shortens our derivations that do not involve reference to mathematical entities. After all, if a mathematical theory M is conservative, then a nominalistic assertion A about the physical world (i.e., an assertion that does not refer to mathematical objects) is implied by a body N of such assertions and M only if it is implied by N alone. That is, provided we have a sufficiently rich body of nominalistic assertions, by using mathematics, we do not obtain any new nominalistic consequences. Mathematics is only a useful instrument to help us in the derivations.

The outcome of this is that conservativeness can be employed only to do the required job if we have nominalistic premises to start with (Field 1989, 129).[11] The second move of Field's strategy is then to provide such nominalistic premises in one important and typical case, Newtonian gravitational theory. Field then elaborates on a work that has a respectable tradition, Hilbert's axiomatization of geometry (Hilbert 1971). What Hilbert provided was a synthetic formulation of geometry, which dispenses with metric concepts, and therefore does not include any quantification over real numbers. His axiomatization was based on concepts such as *point, betweenness,* and *congruence*. Intuitively speaking, we say that a point y is *between* the points x and z if y is a point in the line-segment whose endpoints are x and z. Also intuitively, we say that the line-segment xy is *congruent* to the line-segment zw if the distance from the point x to the point y is the same as that from the point z to w. After studying the formal properties of the resulting system, Hilbert proved a representation theorem. He showed, in a stronger mathematical theory, that given a model of the axiom system for space he had put forward, there is a function d from pairs of points onto non-negative real numbers such that the following "homomorphism conditions" are met:

(i) xy is congruent to zw iff $d(x, y) = d(z, w)$, for all points $x, y, z,$ and w;

(ii) y is between x and z iff $d(x, y) + d(y, z) = d(x, z)$, for all points $x, y,$ and z.

---

[11]   It is a confusion to argue against Field's view by claiming that if we add some bits of mathematics to a body of *mathematical* claims (not nominalistic ones), we can obtain new consequences that could not be achieved otherwise (Field 1989, 128). The restriction to *nominalistic* assertions is crucial.

As a result, if the function d is taken to represent distance, we obtain the expected results about congruence and betweenness. Thus, although we cannot talk about numbers in Hilbert's geometry (there are no such entities to quantify over), there is a metatheoretic result that associates assertions about distances with what can be said in the theory. Field calls such numerical claims *abstract counterparts* of purely geometric assertions, and they can be used to draw conclusions about purely geometrical claims in a smoother way. Indeed, because of the representation theorem, conclusions about space, that can be stated *without* real numbers, can be drawn far more easily than we could achieve by a direct proof from Hilbert's axioms. This illustrates Field's point that the usefulness of mathematics derives from shortening derivations.[12]

Roughly speaking, what Field established was how to extend Hilbert's results about space to space-time. Similarly to Hilbert's approach, instead of formulating Newtonian laws in terms of numerical functions, Field showed how they can be recast in terms of comparative predicates. For example, instead of adopting a function such as "the gravitational potential of x," which is taken to have a numerical value, he employed a comparative predicate such as "the difference in gravitational potential between x and y is less than that between z and w." Relying on a body of representation theorems (which plays the same role as Hilbert's representation theorem in geometry), Field established that several numerical functions can be "obtained" from comparative predicates. But in order to use those theorems, he first showed how to formulate Newtonian numerical laws (such as Poisson's equation for the gravitational field) only in terms of comparative predicates. The result is an *extended representation theorem*.[13]

---

[12]   For a discussion, see Field 1980, 24–9.

[13]   The theorem can be expressed as follows (Field 1989, 130–1): Let N be a theory formulated only in terms of comparative predicates (with no recourse to numerical function), and S be an (arbitrary) model of N, whose domain is constituted by space-time regions. For any model S of N, there are: a 1–1 spatio-temporal coordinate function f (unique up to a generalized Galilean transformation) mapping the space-time of S onto quadruples of real numbers; a mass density function g (unique up to a positive multiplicative transformation) mapping the space-time of S onto an interval of non-negative reals; and a gravitational potential function h (unique up to a positive linear transformation) mapping the space-time onto an interval of reals. Moreover, all these functions "preserve structure," in the sense that the comparative relations defined in terms of them coincide with the comparative relations used in N. Furthermore, if f, g, and h are taken as the denotation of the appropriate function, the laws of Newtonian gravitational theory in their functional form hold.

Using the representation theorem, Field can then quantify over space-time regions rather than real numbers in his formulation of Newtonian theory. Given his quantification over space-time regions, Field assumes a substantivalist view of space-time, according to which there are space-time regions that are not fully occupied (Field 1980, 34–6, 1989, 171–80). Given this result, the nominalist is allowed to draw nominalistic conclusions from premises involving N plus a mathematical theory T. After all, due to the conservativeness of mathematics, such conclusions can be obtained independently of T. Hence, what Field provided is a nominalization strategy, and since it reduces ontology, it seems promising for those who want to adopt a nominalist stance *vis-à-vis* mathematics.

Field can then adopt a fictionalist attitude about mathematical theories. Although mathematical objects (such as numbers) do not exist, Field can have a verbal agreement with the Platonist by introducing a fiction operator in the syntax of mathematical statements (see Field 1989). For example, despite the fact that the statement "There are infinitely many prime numbers" is false, since numbers do not exist, the statement "*According to arithmetic*, there are infinitely many prime numbers" is true. The fiction operator functions here in an analogous way to the corresponding fiction operator in literature. For instance, the statement "Sherlock Holmes lived in Baker Street" is false (or, at least, it lacks truth-value, depending on the theory of fictional names one adopts), since Sherlock Holmes does not exist. However, the corresponding statement with the fiction operator, "*According to the Holmes stories*, Sherlock Holmes lived in Baker Street," is true. Mathematical fictionalism explores this important connection between mathematics and fiction.

3.2   Modal structuralism

In recent years, Geoffrey Hellman has developed a program of interpretation of mathematics that incorporates two features: an *emphasis on structures* as the main subject-matter of mathematics, and a complete elimination of reference to mathematical objects by interpreting *mathematics in terms of modal logic* (as first suggested in Putnam 1967). Because of these features, his approach is called a *modal-structural* interpretation (Hellman 1989, vii–viii and 6–9).

But the approach is also supposed to meet two important requirements (Hellman 1989, 2–6). The first is that mathematical statements should have a truth-value, and thus "instrumentalist" readings of them are rejected from the outset (*semantic* component). The second is that "a reasonable account should be forthcoming of how mathematics does in fact apply to the material world" (Hellman 1989, 6).

In order to address these requirements, Hellman developed a general framework. The main idea is that although mathematics is concerned with the study of structures, this can be accomplished by focusing only on *possible* structures and not actual ones. Thus, the modal interpretation is not committed to mathematical structures. There is no reference to these structures as objects nor to any objects that happen to "constitute" these structures. And this is how the *ontological* commitment to such structures is avoided: the only claim is that the structures under consideration are *possible*.

In order to articulate this point, two steps are taken. The first is to present an appropriate translation schema in terms of which each ordinary mathematical statement S is taken as elliptical for a hypothetical statement, namely: that S *would hold* in a structure of the appropriate kind.[14] For example, if we are considering statements about natural numbers, the structures we are concerned with are sequences of natural numbers satisfying the usual principles for such numbers. The principles in question define the behavior of addition, subtraction, and other operations over numbers. In this case, each particular statement S is (roughly) translated as:

> □ $\forall X$ (X is a sequence of natural numbers satisfying the usual principles for them $\rightarrow$ S holds in X).

In other words, if there were sequences of natural numbers satisfying the usual principles for these numbers, then the statement S would hold in them. This is the *hypothetical component* of the modal-structural interpretation (for a detailed analysis and a precise formulation, see Hellman 1989, 16–24).

The *categorical component* constitutes the second step (Hellman 1989, 24–33). The idea is to assume that the structures of the appropriate kind are logically possible. In this case, we have:

> ◇ $\exists X$ (X is a sequence of natural numbers satisfying the usual principles for these numbers).

[14]   The modal-structural interpretation is formulated in a second-order modal language. However, in order not to be committed to a set-theoretical characterization of the modal operators, Hellman takes these operators as primitive (1989, 17 and 20–3).

In other words, it is possible that there are sequences of natural numbers satisfying the usual principles for such numbers. In this way, each mathematical statement *S* is translated into two modal statements. The first is that if there were structures of the appropriate kind, *S* would be true in such structures. The second is that it is possible that there are structures of that kind. Following this approach, truth-preserving translations of mathematical statements can be presented without ontological costs, given that only the *possibility* of the structures in question is assumed.

Hellman then shows how the practice of theorem proving can be regained in a modal-structural framework (roughly speaking, by applying the translation schema to each line of the original proof of the theorem under consideration). Moreover, by using this translation schema, he shows how arithmetic, the theory of real numbers, and set theory can be recovered (see Hellman 1989, 16–33, 44–7, and 53–93, respectively). In this way, Hellman is able to accommodate "virtually all the mathematics commonly encountered in current physical theories" (1989, 45–6).

We can now consider the issue of the application of mathematics. The main idea is to adopt the hypothetical component as the basis of the application, but now the structures to be entertained are those commonly used in particular branches of science. Two points need to be made.

The first is about the general form of applied mathematical statements (see Hellman 1989, 118–24). These statements involve three crucial components: the structures that are used in applied mathematics, the non-mathematical objects to which the mathematical structures are applied, and a statement of application that specifies the particular relations between the mathematical structures and the non-mathematical objects. The relevant mathematical structures can be formulated in a set theory. A set theory is a powerful mathematical theory in which one can formulate virtually all mathematical theories, including those used in applications. Let us call the set theory used in applied contexts *Z* (for Zermelo[15]). The non-mathematical objects of interest in the context of application can be expressed in *Z* as *Urelemente*, that is, as objects that are not sets. We will take "*U*" to be the statement that certain non-mathematical objects of interest are included as *Urelemente* in the structures of *Z*. Finally, "*A*" is the statement of application, describing the particular relations between the relevant mathematical structures of *Z* and the non-mathematical objects described in *U*. We can

---

[15]   In fact, Hellman uses what is called second-order Zermelo set theory. This is a very interesting set theory, since it can be axiomatized with finitely many axioms. Thus, we can identify *Z* with the conjunction of its axioms. I will denote this conjunction by $\wedge Z$.

now present the general form of an applied mathematical statement (Hellman 1989, 119):

$$\Box \; \forall X \, ((\wedge Z \; \& \; U)^X \rightarrow A).$$

This is, of course, the hypothetical component appropriately interpreted to express which relations would hold between certain mathematical structures (formulated as structures of $\wedge Z$) and the entities studied in the world (the *Urelemente*).

The second consideration examines in more detail the relationships between the physical (or the material) objects studied and the mathematical framework. Hellman calls them "synthetic determination" relations (1989, 124–35). More specifically, we have to determine which (synthetic) relations among non-mathematical objects can be taken, in the antecedent of an applied mathematical statement, as the basis for specifying "the actual material situation" (1989, 129). Hellman's proposal is to consider the models of a comprehensive theory T′. This theory embraces and links the vocabulary of the *applied* mathematical theory (T) and the synthetic vocabulary (S) in question, which intuitively fixes the actual material situation. It is assumed that T determines, up to isomorphism, a particular kind of mathematical structure (containing, for example, Z), and that T′ is an extension of T. In that case, a proposed "synthetic basis" will be adequate if the following condition holds:

> Let α be the class of (mathematically) standard models of T′, and let V denote the full vocabulary of T′: then S determines V in α iff for any two models m and m' in α, and any bijection ϕ between their domains, if ϕ is an S isomorphism, it is also a V isomorphism. (Hellman 1989, 132)

The introduction of isomorphism[16] in this context comes, of course, from the need to accommodate the preservation of structure between the (applied) mathematical part of the domain under study and the non-mathematical part. This holds in the crucial case in which the preservation of the synthetic properties and relations (S-isomorphism) by ϕ leads to the preservation of the applied mathematical relations (V-isomorphism) of the overall theory T′. It should be noted that the "synthetic" structure is not meant to capture the *full* structure of the mathematical theory in

---

[16]   Recall that an isomorphism is a transformation that preserves the structure under consideration (by preserving the various relations in the structures involved).

question, but only its *applied* part. (Recall that Hellman started with an *applied* mathematical theory T.)[17]

This is an important point. Mathematics is applied by establishing appropriate isomorphisms between (parts of) mathematical structures and those structures that represent the material situation. And this procedure is *justified*, since such isomorphisms establish the "sameness" of structure between the mathematical and the non-mathematical domains. But on the mathematical side, we are only considering *possible* mathematical structures (those expressed in the antecedent of an applied mathematical statement). Thus, as opposed to Quine's view, no commitment to actual mathematical structures is found even in the context of successful applications of mathematics.

3.3 Deflationary nominalism

Whereas both mathematical fictionalism and modal structuralism involved the reformulation of mathematical theories to avoid commitment to the existence of mathematical objects, no such reformulation is found in what can be called *deflationary nominalism*. The central move of this proposal is to emphasize that quantification over a certain object (such as a mathematical entity) is not enough to guarantee the commitment to the existence of such an object.

According to Jody Azzouni, two kinds of commitment should be distinguished: *quantifier* commitment and *ontological* commitment (see Azzouni 2004, 127; see also 49–122). We incur a quantifier commitment whenever our theories imply existentially quantified claims. But existential quantification, Azzouni insists, is not sufficient, in general, for ontological commitment. After all, we often quantify over objects we have no reason to believe exist, such as fictional entities. To incur an ontological commitment – that is, to be committed to the *existence* of a given object – a criterion *for* what exists needs to be met. There are, of course, various possible criteria for what exists (such as causal efficacy, observability, and so on). But the criterion

---

[17]   This can be illustrated with a simple example. Suppose that finitely many physical objects display a linear order. We can describe this by defining a function from those objects to an initial segment of the natural numbers. What is required by Hellman's synthetic determination condition is that this function and its description of the objects are captured only by the physical ordering among the objects. It is *not* claimed that the *full* natural number structure is thus captured.

Azzouni favors, as the one which all of us have collectively adopted, is *ontological independence* (2004, 99). What exist are the things that are onto-logically independent of our linguistic practices and psychological processes. The idea here is that if we have just made something up through our linguistic practices or psychological processes, there is no need for us to be committed to the existence of the objects in question. And typically, we would resist any such commitment.

Quine, as we saw, *identifies* quantifier and ontological commitments, at least in the crucial case of the objects that are indispensable to our best theories of the world. Such objects are those that cannot be eliminated through paraphrase and over which we have to quantify when we regiment the theories in question (in first-order logic). For Quine, these are exactly the objects to which we are ontologically committed. Azzouni insists that we should resist this identification. Even if the objects in our best theories are indispensable, even if we have to quantify over them in order to formulate our best theories of the world, this is not sufficient for us to be ontologically committed to them. After all, the objects we quantify over might be ontologically *dependent* on us – on our linguistic practices or psychological processes – and thus we might have just made them up. But, in this case, clearly there is no reason to be committed to their existence. However, for those objects that *are* ontologically independent of us, we *are* committed to their existence.

As it turns out, on Azzouni's view, mathematical objects are ontologic-ally dependent on our linguistic practices and psychological processes. And so it is not surprising that, even though they might be indispensable to our best theories of the world, still we are not ontologically committed to them. Hence, Azzouni is a nominalist.

But in what sense do mathematical objects depend on our linguistic practices and psychological processes? In the sense that, in mathematical practice, the sheer postulation of certain principles is enough: "A math-ematical subject with its accompanying posits can be created *ex nihilo* by simply writing down a set of axioms" (Azzouni 2004, 127). The only addi-tional constraint that sheer postulation has to meet, in practice, is that mathematicians should find the resulting mathematics "interesting." That is, briefly put, the consequences that follow from mathematical principles should not be obvious, nor should they be intractable. Thus, given that sheer postulation is (basically) enough in mathematics, mathematical objects have no epistemic "burdens." Azzouni calls such objects, or "posits," *ultra-thin* (2004, 127).

The same move that Azzouni makes to distinguish ontological commitment from quantifier commitment is also used to distinguish ontological commitment to *F*s from asserting the *truth* of "There are *F*s." Although mathematical theories used in science are (taken to be) true, this is not sufficient to commit us to the existence of the objects these theories are supposed to be about. After all, on Azzouni's picture, it might be *true* that there are *F*s, but to be ontologically committed to *F*s, a criterion for what exists needs to be met. As Azzouni points out:

> I take true mathematical statements as literally *true*; I forgo attempts to show that such literally true mathematical statements are *not* indispensable to empirical science, and yet, nonetheless, I can describe mathematical terms as referring to nothing at all. Without Quine's criterion to corrupt them, existential statements are innocent of ontology. (Azzouni 2004, 4–5)

In Azzouni's picture, as opposed to Quine's, ontological commitment is not signaled in any special way in natural (or even formal) language. We just do not read off the ontological commitment of scientific doctrines (even suitably regimented). After all, as noted, neither quantification over a given object (in a first-order language) nor formulation of true claims about such an object entails the existence of the latter. Only ontological independence does that.

Azzouni's proposal nicely expresses a view that should be taken seriously. And as opposed to other versions of nominalism, it has the significant benefit of aiming to take mathematical discourse literally. This is a benefit that, so far, only Platonist views could deliver.

4.   Conclusion

In this survey of central issues and proposals in the philosophy of mathematics, we saw with regard to the issue of the ontology of mathematics how Platonists and nominalists have offered importantly different conceptions. Platonists had the significant advantage of being able to take mathematical discourse literally; that is, they did not have to rewrite mathematical (and scientific) theories to avoid commitment to mathematical objects. With the possible exception of deflationary nominalism, all of the other nominalist views had to create a parallel discourse to accommodate mathematics. That is, they had to provide a reformulation of the theories in question in a nominalistically acceptable language (either by using a fiction operator or a suitable modal language).

Platonists, in turn, faced a significant challenge to make sense of the epistemological issue of how we can have knowledge of abstract mathematical objects and structures to which we have no causal access. As we saw, various epistemological strategies were devised. From reconstructive accounts in which mathematical objects emerge from logical concepts (in Fregean Platonism) through accounts based on the intuition of basic mathematical facts (in Gödelian Platonism) to the use of concrete templates as a vehicle to our knowledge of abstract patterns (in structuralist Platonism), Platonists have spent significant resources trying to make sense of mathematical knowledge. This was not a problem, however, that the nominalist faced. Bluntly put, if mathematical objects do not exist, we do not have to provide an account of how we come to know them.

However, understanding the application of mathematics then becomes a significant issue for the nominalist. If mathematical objects do not exist, how can we understand the success of the application of mathematics to the physical world? Nominalists addressed this issue directly, devising strategies to explain this success despite the non-existence of mathematical objects. They emphasized that mathematical theories need not be true to be good, as long as they are conservative (in mathematical fictionalism), or highlight the role played by possible structures in the application of mathematics (in modal structuralism). But for the deflationary nominalist, the problem of the application of mathematics is just an artifact – a philosophical creation of something that is not a philosophical issue (Azzouni 2000). Applied mathematics, just as its pure counterpart, involves finding out what follows from what. The trouble is that it is often not transparent what should be the consequences in each case, particularly those involving premises that describe aspects of the physical world. In the end, it does not look like we have here a special philosophical issue after all.

References

Azzouni, J. 2000. "Applying Mathematics: An Attempt to Design a Philosophical Problem." *The Monist* 83: 209–27.

Azzouni, J. 2004. *Deflating Existential Consequence: A Case for Nominalism.* New York: Oxford University Press.

Benacerraf, P. and H. Putnam (eds.). 1983. *Philosophy of Mathematics: Selected Readings*, 2nd edn. Cambridge: Cambridge University Press.

Boolos, G. 1998. *Logic, Logic, and Logic*. Cambridge, MA: Harvard University Press.

Colyvan, M. 2001. *The Indispensability of Mathematics*. New York: Oxford University Press.

da Costa, N.C.A., D. Krause, and O. Bueno. 2007. "Paraconsistent Logics and Paraconsistency." In D. Jacquette (ed.), *Philosophy of Logic*. Amsterdam: North-Holland, 791–911.

Field, H. 1980. *Science without Numbers: A Defense of Nominalism*. Princeton, NJ: Princeton University Press.

Field, H. 1989. *Realism, Mathematics and Modality*. Oxford: Basil Blackwell.

Frege, G. 1974. *Foundations of Arithmetic*, trans. J.L. Austin. Oxford: Basil Blackwell.

Gödel, K. 1964. "What is Cantor's Continuum Problem?" In P. Benacerraf and H. Putnam (eds.), *Philosophy of Mathematics: Selected Readings*, 2nd edn. Cambridge: Cambridge University Press,1983, 470–85.

Hale, B. and C. Wright. 2001. *The Reason's Proper Study*. Oxford: Oxford University Press.

Hellman, G. 1989. *Mathematics without Numbers: Towards a Modal-Structural Interpretation*. Oxford: Clarendon Press.

Hilbert, D. 1971. *Foundations of Geometry*. Translation of the 10th German edn., published in 1968. La Salle: Open Court.

Jacquette, D. (ed.). 2007. *Philosophy of Logic*. Amsterdam: North-Holland.

Maddy, P. 1990. *Realism in Mathematics*. Oxford: Clarendon Press.

Parsons, C. 2008. *Mathematical Thought and its Objects*. Cambridge: Cambridge University Press.

Priest, G. 2006. *In Contradiction*. 2nd edn. Oxford: Oxford University Press.

Putnam, H. 1967. "Mathematics without Foundations." *Journal of Philosophy* 64: 5–22. Reprinted in *Mathematics, Matter and Method*. Philosophical Papers, vol. 1. 2nd edn. Cambridge: Cambridge University Press, 43–59.

Putnam, H. 1971. *Philosophy of Logic*. New York: Harper and Row. Reprinted in *Mathematics, Matter and Method*. Philosophical Papers, vol. 1. 2nd edn. Cambridge: Cambridge University Press, 323–57.

Putnam, H. 1979. *Mathematics, Matter and Method*. Philosophical Papers, vol. 1. 2nd edn. Cambridge: Cambridge University Press.

Quine, W.V.O. 1960. *Word and Object*. Cambridge, MA: MIT Press.

Resnik, M. 1997. *Mathematics as a Science of Patterns*. Oxford: Clarendon Press.

Shapiro, S. 1991. *Foundations without Foundationalism: A Case for Second-order Logic*. Oxford: Clarendon Press.

Shapiro, S. 1997. *Philosophy of Mathematics: Structure and Ontology*. New York: Oxford University Press.

5   Philosophy of Probability

Aidan Lyon

In the philosophy of probability there are two central questions we are concerned with. The first is: what is the correct formal theory of probability? Orthodoxy has it that Kolmogorov's axioms are the correct axioms of probability. However, we shall see that there are good reasons to consider alternative axiom systems. The second central question is: what do probability statements mean? Are probabilities "out there" in the world, as frequencies, propensities, or some other objective feature of reality, or are probabilities "in the head," as subjective degrees of belief? We will survey some of the answers that philosophers, mathematicians and physicists have given to these questions.

1.   Introduction

The famous mathematician Henri Poincaré once wrote of the probability calculus: "if this calculus be condemned, then the whole of the sciences must also be condemned" (Poincaré 1902, 186).

Indeed, every branch of science makes extensive use of probability in some form or other. Quantum mechanics is well known for making heavy use of probability. The second law of thermodynamics is a statistical law and, formulated one way, states that the entropy of a closed system is most *likely* to increase. In statistical mechanics, a probability distribution known as the micro-canonical distribution is used to make predictions concerning the macro-properties of gases. In evolutionary theory, the concept of fitness is often defined in terms of a probability function (one such definition says that fitness is expected number of offspring). Probability also plays central roles in natural selection, drift, and macro-evolutionary models. The theory of three-dimensional random walks plays an important role in the biomechanical theory of a diverse range of rubbery materials: from the resilin in tendons that help flap insect wings, to arteries near the heart that regulate blood flow. In ecology and conservation biology, we see concepts like the expected time to extinction of a population. In economics, stochastic differential equations are used extensively to model all sorts of quantities: from inflation to investment flows, interest rates, and unemployment figures. And all of the sciences make extensive use of probability in the form of statistical inference: from hypothesis testing, to model selection, to parameter estimation, to confirmation, to confidence intervals. In science, probability is truly everywhere.

But the sciences do not have exclusive rights to probability theory. Probability also plays an important role in our everyday reasoning. It figures prominently in our formal theories of decision making and game playing. In fact, probability is so pervasive in our lives that we may even be tempted to say that "the most important questions of life, are indeed for the most part only problems of probability," as Pierre-Simon Laplace once did (Laplace 1814, 1).

In philosophy of probability, there are two main questions that we are concerned with. The first question is: what is the correct mathematical theory of probability? Orthodoxy has it that this question was laid to rest by Andrei Kolmogorov in 1933. But as we shall see in §2, this is far from true; there are many competing formal theories of probability, and it is not clear that we can single one of these out as *the* correct formal theory of probability.

These formal theories of probability tell us how probabilities behave, how to calculate probabilities from other probabilities, but they do not tell us what probabilities *are*. This leads us to the second central question in philosophy of probability: just what are probabilities? Or put another way: what do probability statements *mean*. Do probability claims merely reflect facts about human ignorance? Or do they represent facts about the world? If so, *which* facts? In §3, we will see some of the various ways in which philosophers have tried to answer this question. Such answers are typically called *interpretations of probability*, or philosophical theories of probability.

These two central questions are by no means independent of each other. What the correct formal theory of probability is clearly constrains the space of philosophical interpretations. But it is not a one-way street. As we shall see, the philosophical theory of probability has a significant impact on the formal theory of probability too.

2.   The Mathematical Theory of Probability

In probability theory, we see two types of probabilities: absolute probabilities and conditional probabilities. Absolute probabilities (also known as unconditional probabilities) are probabilities of the form "$P(A)$," while conditional probabilities are probabilities of the form "$P(A, B)$" – read as "the probability of A, given B."[1] These two types of probability can be defined in terms of each other. So when formalizing the notion of probability we have a choice: do we define conditional probability in terms of unconditional probability, or vice versa? The next section, §2.1, will focus on some of the various formal theories of probability that take absolute probability as primitive and define conditional probability in terms of the former. Then in §2.2, we will look at some of the various formal theories that take conditional probability as the primitive type of probability.

2.1   *Absolute probability as primitive*

Kolmogorov's theory of probability (Kolmogorov 1933) is the best known formal theory of probability and it is what you will learn if you take a course on probability. First, we start with a set of elementary events, which we will refer to by Ω. For example, if we are considering the roll of a die where the possible outcomes are the die landing with "one" face up, or with "two" face up, etc., then Ω would be the set $\{1, 2, 3, 4, 5, 6\}$. From this set of elementary events, we can construct other, less fine-grained events. For example, there is the event that an odd number comes up. We represent this event with the set $\{1, 3, 5\}$. Or, there is the event that some number greater than two comes up. We represent this event with set $\{3, 4, 5, 6\}$. In general, any event constructed from the elementary events will be a subset of Ω. The least fine-grained event is the event that *something* happens – this event is represented by Ω itself. There is also the event that cannot happen, which is represented by the empty set, $\varnothing$.[2]

---

[1]   Another common notation for conditional probability is $P(A \mid B)$, though some authors take this notation to have a very particular definition that does not correspond to our concept of conditional probability (see, e.g., Hájek 2003). Some authors also reverse A and B, so that $P(B, A)$ stands for the conditional probability of A given B, though this notation is not often used.

[2]   It is a theorem of standard set theory that the empty set is a subset of every subset.

In probability theory, we often want to work with the set of all events that can be constructed from Ω. In our die example, this is because we may want to speak of the probability of any particular number coming up, or of an even or odd number coming up, of a multiple of three, a prime number, etc. It is typical to refer to this set by F. In our example, if F contains every event that can be constructed from Ω, then it would be rather large. A partial listing of its elements would be: F = {∅, Ω, {1}, {2} . . . {6}, {1, 2, 3}, {4, 5, 6}, {1, 2}, {1, 2, 3, 4, 5} . . .}. In fact, there are a total of $2^6 = 64$ elements in F.

In general, if an event, A, is in F, then so is its complement, which we write as $\Omega \setminus A$. For example, if {3, 5, 6}, is in F, then its complement $\Omega \setminus$ {3, 5, 6} = {1, 2, 4} is in F. Also, if any two events are in F, then so is their union. For example, if {1, 2} and {4, 6} are in F, then their union {1, 2} ∪ {4, 6} = {1, 2, 4, 6} is in F. If a set, S, has the first property (i.e., if A is in S, then $\Omega \setminus A$ is in S), then we say S is *closed under Ω-complementation*. If S has the second property (i.e., if A and B are in S, then $A \cup B$ is in S), then we say S is *closed under union*. And if S has both of these properties, i.e., if it is closed under both Ω-complementation and union, then we say that S is an *algebra* on Ω. If a set, S, is an algebra, then it follows that it is also closed under intersection, i.e., that if A and B are in S, then $A \cap B$ is also in S.

In our die example, it can be seen that F (the set that contains all the subsets of Ω) is an algebra on Ω. However, there are also algebras on Ω that do not contain every event that can be constructed from Ω. For example, consider the following set: F = {∅, {1, 3, 5}, {2, 4, 6}, Ω}. The elements of this F would correspond to the events: (i) nothing happens; (ii) an odd number comes up; (iii) an even number comes up; and (iv) some number comes up. This is an important example of an algebra because it illustrates how algebras work. For example, not every "event" – intuitively understood – gets a probability. For instance, the event that the number two comes up gets no probability because {2} is not in F. Also note that even though this F does not contain every subset of Ω, it is still closed under union and Ω-complementation. For example, the union of {1, 3, 5} and {2, 4, 6} is Ω, which is in F, and the Ω-complement of, say, {1, 3, 5} is {2, 4, 6}, which is also clearly in F.

Once we have specified an algebra, F, we can then define a probability function that attaches probabilities to every element of F. Let P be a function from F to the real numbers, $\mathbb{R}$, that obeys the following axioms:

(KP1) $P(A) \geq 0$
(KP2) $P(\Omega) = 1$
(KP3) $P(A \cup B) = P(A) + P(B)$, if $A \cap B = \varnothing$

for every A and B in F. We call any function, P, that satisfies the above constraints a *probability function*.[3]

So far we have assumed that F only contains a finite number of events. But sometimes we want to work with infinitely many events (e.g., consider choosing a random point on a line; there are infinitely many points that can be chosen). When F is countably infinite,[4] we replace KP3 with:

$$(KP4) \quad P\left(\bigcup_{i=1}^{i=\infty} A_i\right) = \sum_{i=1}^{i=\infty} P(A_i)$$

on the condition that the intersection of any two of the A_i is the empty set. A simple example will help illustrate how these last two axioms work. Suppose we randomly choose a positive integer in a way such that the number 1 has a probability of 1/2 being chosen, the number 2 has a probability of 1/4 being chosen, and so on. In general, the number n has a probability of $1/2^n$ of being chosen. Let us write the event that the number n gets chosen as A_n (this will be the set $\{n\}$). So, for example, the event that some integer below 4 is chosen is: $A_1 \cup A_2 \cup A_3$ which is the set $\{1, 2, 3\}$. The probability of this event is then:

$$P(A_1 \cup A_2 \cup A_3) = P(A_1) + P(A_2) + P(A_3)$$
$$= 1/2 + 1/4 + 1/8$$

And the probability that *some* number is chosen is:

$$P\left(\bigcup_{i=1}^{i=\infty} A_i\right) = \sum_{i=1}^{i=\infty} P(A_i)$$

---

[3]   Also, for any Ω, F, and P that satisfy the above constraints, we call the triple (Ω, F, P) a *probability space*.

[4]   A set is countable if there is a one-to-one correspondence between it and some subset of the natural numbers, and uncountable if there is not. Some examples: any finite set is countable, the set of even integers is countable (and thus countably infinite), the set of rational numbers is countable (and countably infinite), the set of real numbers is *not* countable, and the set of real numbers between 0 and 1 is also not countable.

which can be expanded as:

$$P(A_1 \cup A_2 \cup A_3 \cup \ldots) = P(A_1) + P(A_2) + P(A_3) + \ldots$$
$$= 1/2 + 1/4 + 1/8 + \ldots$$
$$= 1$$

This fourth axiom – known as countable additivity – is by far the most controversial. Bruno de Finetti (1974) famously used the following example as an objection to KP4. Suppose you have entered a fair lottery that has a countably infinite number of tickets. Since the lottery is fair, each ticket has an equal probability of being the winning ticket. But there are only two ways in which the tickets have equal probabilities of winning, and on both ways we run into trouble. On the first way, each ticket has some positive probability of winning – call this positive probability ε. But then, by KP4, the probability that *some* ticket wins is ε added to itself infinitely many times, which equals infinity, and so violates KP2. The only other way that the tickets can have equal probability of winning is if they all have zero probability. But then KP4 entails that the probability that some ticket wins is 0 added to itself infinitely many times, which is equal to zero and so again KP2 is violated. It is a matter of either too much or too little!

Axioms KP1–4 define absolute probability functions on *sets*. However, many philosophers and logicians prefer to define probability functions on *statements*, or even other abstract objects instead. One reason for this is because Kolmogorov's axioms are incompatible with many philosophical interpretations of probability. For example, Karl Popper points out that the formal theory of probability should be sensitive to the needs of the philosophical theory of probability:

> In Kolmogorov's approach it is assumed that the objects a and b in $p(a, b)$ are sets (or aggregates). But this assumption is not shared by all interpretations: some interpret a and b as states of affairs, or as properties, or as events, or as statements, or as sentences. In view of this fact, I felt that in a formal development, no assumption concerning the nature of the "objects" or "elements" a and b should be made [. . .]. (Popper 1959b, 40)

A typical alternative to Kolmogorov's set-theoretic approach to probability is an axiom system where the bearers of probability are sentences (in §2.2, we shall see an axiom system inspired by Popper's work that makes fewer assumptions concerning what the bearers of probability are). For cases where there are only finitely many sentences, it is fairly easy to "translate" axioms KP1–3 into axioms that define probability functions on sentences. Instead of an algebra, F, we have a language, $\mathscr{L}$. $\mathscr{L}$ is a set of atomic sentences and their Boolean combinations. So, if A and B are in $\mathscr{L}$, then so is: their conjunction, $A \wedge B$, read as "A and B"; their disjunction $A \vee B$, read as "A or B"; their negations, e.g., $\neg A$, read as "not A"; their equivalence, $A \equiv B$, read as "A is equivalent to B"; and their material implications, e.g., $A \supset B$, read as "if A, then B." We also define a consequence relation $\vdash$ over the language $\mathscr{L}$. So, for example, $A \vdash B$ is read as "A entails B," or "B is a consequence of A." Because tautologies are always true, if A is a tautology, we write $\vdash A$, and since logical falsehoods are always false, if A is a logical falsehood, we write $\vdash \neg A$.[5] We then let P be a function from $\mathscr{L}$ to $\mathbb{R}$ that satisfies:

(P1) $P(A) \geq 0$
(P2) $P(A) = 1$, if $\vdash A$
(P3) $P(A \vee B) = P(A) + P(B)$, if $\vdash \neg(A \wedge B)$

for every A and B in $\mathscr{L}$. However, if we are interested in languages with infinitely many sentences, we cannot simply add on a fourth axiom for countable additivity as we did with axiom KP4. This is because we do not have infinite disjunction in our logic. A new logic has to be used if we wish to have a countable additivity axiom for probabilities attached to sentences.[6] Axioms P1–3 are what most philosophers use, so I will refer to them as the standard theory of probability (however, the reader should not take this to mean they are the "correct" axioms of probability).

We also have the notion of *conditional* probability to formalize. On the accounts just mentioned it is typical to define conditional probability in terms of unconditional probability. For example, in the standard theory of probability, conditional probability is defined in the following way:

---

[5]   This overview of classical logic is necessarily very brief. The reader who is not familiar with these ideas is encouraged to read the chapter on logic in the present volume.
[6]   See, e.g., Roeper and Leblanc 1999, 26, for details on how this can be done.

(CP) $P(A, B) \stackrel{\text{def}}{=} \dfrac{P(A \wedge B)}{P(B)}$, where $P(B) \neq 0$

Many have pointed out that this definition leaves an important class of conditional probabilities undefined when they should be defined. This class is comprised of those conditional probabilities of the form, $P(A, B)$ where $P(B) = 0$. Consider the following example, due to Émile Borel.[7] Suppose a point on the earth is chosen randomly – assume the earth is a perfect sphere. What is the probability that the point chosen is in the western hemisphere, given that it lies on the equator? The answer intuitively ought to be 1/2. However, CP does not deliver this result, because the denominator – the probability that the point lies on the equator – is zero.[8]

There are many responses one can give to such a problem. For instance, some insist that any event that has a probability of zero cannot happen. So the probability that the point is on the equator must be greater than zero, and so CP is not undefined. The problem though is that it can be proven that for any probability space with uncountably many events, uncountably many of these events *must* be assigned zero probability, as otherwise we would have a violation of the probability axioms.[9] This proof relies on particular properties of the real number system, $\mathbb{R}$. So some philosophers have said so much the worse for the real number system, opting to use a probability theory where the values of probabilities are not real numbers, but rather something more mathematically rich, like the hyperreals, $\mathbb{HR}$ (see, e.g., Lewis 1980, Skyrms 1980).[10]

Another response that philosophers have made to Borel's problem is to opt for a formal theory that takes conditional probability as the fundamental notion of probability (we will see some of these theories in §2.2). The idea is that by defining absolute probability in terms of conditional probability while taking the latter as the primitive probability notion to be

---

[7]   See Kolmogorov's discussion of this example in Kolmogorov 1933, 50–1.

[8]   It is zero because in this example probability corresponds to area. Since the equator has an area of zero, the probability that the chosen point lies on the equator is zero, even though this event is *possible*.

[9]   See, e.g., Hájek 2003 for the proof.

[10]   The hyper-reals are all the real numbers (1, 3/4, $\sqrt{2}$, $-\pi$, etc.), plus some more. For example, in the hyper-reals there is a number greater than zero but that is smaller than every positive real number (such a number is known as an *infinitesimal*).

axiomatized, conditional probabilities of the form $P(A, B)$ where $P(B) = 0$ can be defined.[11]

There are, also, other reasons to take conditional probability as the fundamental notion of probability. One such reason is that sometimes the unconditional probabilities $P(A \wedge B)$ and $P(B)$ are undefined while the conditional probability $P(A, B)$ *is* defined, so it is impossible to define the latter in terms of the former. The following is an example due to Alan Hájek (2003). Consider the conditional probability that heads comes up, given that I toss a coin fairly. Surely, this should be 1/2, but CP defines this conditional probability as:

$$P(\text{heads, I toss the coin fairly}) = \frac{P(\text{heads} \wedge \text{I toss the coin fairly})}{P(\text{I toss the coin fairly})}$$

But you have no information about how likely it is that I will toss the coin fairly. For all you know, I never toss coins fairly, or perhaps I always toss them fairly. Without this information, the terms on the right-hand side of the above equality may be undefined, yet the conditional probability on the left *is* defined.

There are other problems with taking absolute probability as the fundamental notion of probability (see Hájek 2003 for a discussion). These problems have led many authors to take conditional probability as primitive. However, this requires a new approach to the formal theory of probability. And so we now turn to theories of probability that take conditional probability as the primitive notion of probability.

2.2   Conditional probability as primitive

The following axiom system – based on the work of Alfred Rényi (1955) – is a formal theory of probability where conditional probability is the fundamental concept. Let Ω be a non-empty set, A be an algebra on Ω, and B be a non-empty subset of A. We then define a function, P, from $A \times B$ to $\mathbb{R}$ such that:[12]

[11]   Kolmogorov was also well aware of Borel's problem and in response gave a more sophisticated treatment of conditional probability using the Radon–Nikodym theorem. The details of this approach are beyond the scope of this survey. Suffice to say, though, this approach is not completely satisfactory either – see Hájek 2003 and Seidenfeld, Schervish, and Kadane 2001 for reasons why.

[12] $X \times Y$ is the *Cartesian product* of the sets X and Y. So, for example, if $X = \{x_1, x_2, x_3\}$ and $Y = \{y_1, y_2\}$, then $X \times Y = \{(x_1, y_1), (x_1, y_2), (x_2, y_1), (x_2, y_2), (x_3, y_1), (x_3, y_2)\}$.

(RCP1) $P(A, B) \geq 0$,

(RCP2) $P(B, B) = 1$,

(RCP3) $P(A_1 \cup A_2, B) = P(A_1, B) + P(A_2, B)$, if $A_1 \cap A_2 = \varnothing$

(RCP4) $P(A_1 \cap A_2, B) = P(A_1, A_2 \cap B) \cdot P(A_2, B)$

where the *A*s are in A and the *B*s are in B. Any function that satisfies these axioms is called a *Rényi conditional probability* function.[13] RCP3 is the conditional analogue of the KP3 finite additivity axiom and it also has a countable version:

(RCP5) $$P\left(\bigcup_{i=1}^{i=\infty} A_i, B \right) = \sum_{i=1}^{i=\infty} P(A_i, B)$$

on the condition that the A_i are mutually exclusive. RCP4 is the conditional analogue of CP, and absolute probability can then be defined in the following way:

(AP) $P(A) \stackrel{\text{def}}{=} P(A, \Omega)$

Popper, in many places, gives alternative axiomatizations of probability where conditional probability is the primitive notion (see, e.g., Popper 1938, 1955, 1959a, 1959b). The following set of axioms is a user-friendly version of Popper's axioms (these are adapted from Roeper and Leblanc 1999, 12):

(PCP1) $P(A, B) \geq 0$

(PCP2) $P(A, A) = 1$

(PCP3) $P(A, B) + P(\neg A, B) = 1$ if B is P-normal

(PCP4) $P(A \wedge B, C) = P(A, B \wedge C) \cdot P(B, C)$

(PCP5) $P(A \wedge B, C) \leq P(B \wedge A, C)$

(PCP6) $P(A, B \wedge C) \leq P(A, C \wedge B)$

(PCP7)   There is a D in O such that D is P-normal

for every A, B, and C in O. O is the set of "objects" of probability – they could be sentences, events, states of affairs, propositions, etc. An "object," B, is P-abnormal if and only if $P(A, B) = 1$ for every A in O, and it is P-normal if and only if it is not P-abnormal. P-abnormality plays the same

---

[13]   And any quadruple, (Ω, A, B, P), that satisfies the above conditions is called a *Rényi conditional probability space*.

role as logical falsehood. Any function, P, that satisfies the above axioms is known as a *Popper conditional probability function*, or often just as a *Popper function*, for short. This axiom system differs from Rényi's in that: (i) it is *symmetric* (i.e., if $P(A, B)$ exists, then $P(B, A)$ exists), and (ii) it is *autonomous*. An axiom system is autonomous if, in that system, probability conclusions can be derived only from probability premises. For example, the axiom system P1–3 is not autonomous, because, for instance, we can derive that $P(A) = 1$, from the premise that A is a logical truth.

2.3 Other formal theories of probability

We have just seen what may be the four most prominent formal theories of probability. But there are many other theories also on the market – too many to go into their full details here, so I will merely give a brief overview of the range of possibilities.

Typically it is assumed that the logic of the language that probabilities are defined over is classical logic. However, there are probability theories that are based on other logics. Brian Weatherson, for instance, introduces an intuitionistic theory of probability (Weatherson 2003). He argues that this probability theory, used as a theory of rational credences, is the best way to meet certain objections to Bayesianism (see §3.5.2). The defining feature of this formal account of probability is that it allows an agent to have credences in A and $\neg A$ that do not sum to 1, but are still additive. This can be done because it is not a theorem in this formal theory of probability that $P(A \vee \neg A) = 1$.

Another example of a "non-classical" probability theory is quantum probability. Quantum probability is based on a non-distributive logic, so it is not a theorem that $P((A \wedge B) \vee C) = P((A \vee C) \wedge (B \vee C))$. Hilary Putnam uses this fact to argue that such a logic and probability makes quantum mechanics less mysterious than it is when classical logic and probability theory are used (Putnam 1968). One of his examples is how the incorrect classical probability result for the famous two-slit experiment does not go through in quantum probability (see Putnam 1968 for more details and examples). See Dickson (2001) – who argues that quantum logic (and probability) is still a live option for making sense of quantum mechanics – for more details and references.

As we will see in §3.5, the probability calculus is often taken to be a set of rationality constraints on the credences (degrees of belief) of an individual. A consequence of this – that many philosophers find unappealing

– is that an individual, to be rational, should be logically omniscient. Ian Hacking introduces a set of probability axioms that relax the demand that an agent be logically omniscient (Hacking 1967).

Other formal theories of probability vary from probability values being *negative* numbers (see, e.g., Feynman 1987), to *imaginary* numbers (see, e.g., Cox 1955), to *unbounded* real numbers (see, e.g., Rényi 1970), to real numbered *intervals* (see, e.g., Levi 1980). Dempster-Shafer Theory is also often said to be a competing formal theory of probability (see, e.g., Shafer 1976). For more discussion of other formal theories of probability, see Fine (1973).

That concludes our survey of the various formal theories of probability. So far, though, we have only half of the picture. We do not yet have any account of what probabilities *are*, only how they *behave*. This is important because there are many things in the world that behave like probabilities, but are not probabilities. Take, for example, areas of various regions of a tabletop where one unit of area is the entire area of the tabletop. The areas of such regions satisfy, for example, Kolmogorov's axioms, but are clearly not probabilities.

3.   The Philosophical Theory of Probability

Interpretations of probability are typically categorized into two kinds: subjective interpretations and objective interpretations. Roughly, the difference is that subjective interpretations identify probabilities with the credences, or "degrees of belief," of a particular individual, while objective interpretations identify probability with something that is independent of any individual – the most common somethings being relative frequencies and propensities. The following is a brief survey of some of the interpretations of probability that philosophers have proposed. It is impossible to give a full and just discussion of each interpretation in the space available, so only a small selection of issues surrounding each will be discussed.

3.1   The classical interpretation

The central idea behind the classical interpretation of probability – historically the first of all the interpretations – is that the probability of an event is the ratio between the number of equally possible outcomes in which the event occurs and the total number of equally possible outcomes. This conception of probability is particularly well suited for probability statements concerning games of chance. Take, for example, a fair roll of a fair die. We quite naturally say that the probability of an even number coming up is 3/6 (which, of course, is equal to 1/2). The "3" is for the three ways in which an even number comes up (2, 4, and 6) and the "6" is for all of the possible numbers that could come up (1, 2, 3, 4, 5, and 6).

The idea of relating probabilities to equally possible outcomes can be found in the works of many great authors – e.g., Cardano (1663), Laplace (1814) and Keynes (1921). However, among these authors there is a considerable degree of variation in how this idea is fleshed out. In particular, they vary on how we are to understand what it means for events to be "equally possible." In the hands of some, the equally possible outcomes are those outcomes that are *symmetric* in some physical way. For example, the possible outcomes of a fair roll of a fair die might be said to be all equally possible because of the physical symmetries of the die and in the way the die is rolled. If we understand "equally possible" this way, then the classical interpretation is an *objective* interpretation. However, the most canonical understanding of the term "equally possible" is in terms of our knowledge (or lack thereof). Laplace is a famous proponent of this understanding of "equally possible":

> The theory of chance consists in reducing all the events of the same kind to a certain number of cases equally possible, that is to say, to such as *we* may be equally *undecided* about in regard to their existence, and in determining the number of cases favorable to the event whose probability is sought. (Laplace 1814, 6; emphasis added)

Understood this way, the classical interpretation is a *subjective* interpretation of probability. From now on, I will assume that the classical interpretation is a subjective interpretation, as this is the most popular understanding of the interpretation – see Hacking (1971) for a historical study of the notion of equal possibilities and the ambiguities in the classical interpretation.

If we follow Laplace, then the classical interpretation puts constraints on how we ought to assign probabilities to events. More specifically, it says we ought to assign equal probability to events that we are "equally undecided about." This norm was formulated as a principle now known as the Principle of Indifference by John Maynard Keynes:

> If there is no known reason for predicating of our subject one rather than another of several alternatives, then relative to such knowledge the assertions of each of these alternatives have an equal probability. (Keynes 1921, 42)

It is well known that the Principle of Indifference is fraught with paradoxes – many of which originate with Joseph Bertrand (1888). Some of these paradoxes are rather mathematically complicated, but the following is a simple one due to Bas van Fraassen (1989). Consider a factory that produces cubic boxes with edge lengths anywhere between (but not including) 0 and 1 m, and consider two possible events: (a) the next box has an edge length between 0 and 0.5 m or (b) it has an edge length between 0.5 and 1 m. Given these considerations, there is no reason to think either (a) or (b) is more likely than the other, so by the Principle of Indifference we ought to assign them equal probability: 1/2 each. Now consider the following four events: (i) the next box has a face area between 0 and 0.25 m^2; (ii) it has a face area between 0.25 and 0.5 m^2; (iii) it has a face area between 0.5 and 0.75 m^2; or (iv) it has a face area between 0.75 and 1 m^2. It seems we have no reason to suppose any of these four events to be more probable than any other, so by the Principle of Indifference we ought to assign them all equal probability: 1/4 each. But this is in conflict with our earlier assignment, for (a) and (i) are different descriptions of the same event (a length of 0.5 m corresponds to an area of 0.25 m^2). So the probability assignment that the Principle of Indifference tells us to assign depends on how we describe the box factory: we get one assignment for the "edge length" description, and another for the "face area" description.

There have been several attempts to save the classical interpretation and the Principle of Indifference from paradoxes like the one above, but many authors consider the paradoxes to be decisive. See Keynes (1921) and van Fraassen (1989) for a detailed discussion of the various paradoxes, and see Jaynes (1973), Marinoff (1994), and Mikkelson (2004) for a defense of the principle. Also see Shackel (2007) for a contemporary overview of the debate. The existence of paradoxes like the one above was one source of motivation for many authors to abandon the classical interpretation and adopt the frequency interpretation of probability.

3.2   The frequency interpretation

3.2.1   Actual frequencies

Ask any random scientist or mathematician what the definition of probability is and they will probably respond to you with an incredulous stare or, after they have regained their composure, with some version of the frequency interpretation. The frequency interpretation says that the probability of an outcome is the number of experiments in which the outcome occurs divided by the number of experiments performed (where the notion of an "experiment" is understood very broadly). This interpretation has the advantage that it makes probability empirically respectable, for it is very easy to measure probabilities: we just go out into the world and measure frequencies. For example, to say that the probability of an even number coming up on a fair roll of a fair die is 1/2 just means that out of all the fair rolls of that die, 50 percent of them were rolls in which an even number came up. Or to say that there is a 1/100 chance that John Smith, a consumptive Englishman aged 50, will live to 61 is to say that out of all the people like John, 1 percent of them live to the age of 61.

But which people are like John? If we consider all those Englishmen aged 50, then we will include consumptive Englishmen aged 50 and all the healthy ones too. Intuitively, the fact that John is sickly should mean we only consider consumptive Englishmen aged 50, but where do we draw the line? Should we restrict the class of those people we consider to those who are also named John? Surely not, but is there a principled way to draw the line? If there is, it is hard to say exactly what that principle is. This is important because where we draw the line affects the value of the probability. This problem is known as the *reference class problem*. John Venn was one of the first to notice it:

> It is obvious that every individual thing or event has an indefinite number of properties or attributes observable in it, and might therefore be considered as belonging to an indefinite number of different classes of things [. . .]. (Venn 1876, 194)

This can have quite serious consequences when we use probability in our decision making (see, e.g., Colyvan, Regan, and Ferson 2001). Many have taken the reference class problem to be a difficulty for the frequency interpretation, though Mark Colyvan et al. (2001) and Hájek (2007c) point out that it is also a difficulty for many other interpretations of probability.

The frequency interpretation is like the classical interpretation in that it identifies the probability of an event with the ratio of favorable cases to cases. However, it is unlike the classical interpretation in that the cases have to be *actual* cases. Unfortunately, this means that the interpretation is shackled too tightly to how the world turns out to be. If it just happens that I never flip this coin, then the probability of "tails" is undefined. Or if it is flipped only once and it lands "tails," then the probability of "heads" is 1 (this is known as a single case probability). To get around these difficulties many move from defining probability in terms of actual frequencies to defining it in terms of *hypothetical* frequencies. There are many other problems with defining probability in terms of actual frequencies (see Hájek 1997 for 15 objections to the idea), but we now move on to hypothetical frequencies.

3.2.2 Hypothetical frequencies

The hypothetical frequency interpretation tries to put some of the modality back into probability. It says that the probability of an event is the number of trials in which the event occurs divided by the number of trials, *if the trials were to occur*. On this frequency interpretation, the trials do not have to actually happen for the probability to be defined. So for the coin that I never flipped, the hypothetical frequentist can say that the probability of "tails" is 1/2 because this is the frequency we would observe, *if* the coin were tossed.

Maybe. But we definitely would not observe this frequency if the coin were flipped an odd number of times, for then it would be impossible to observe an even number of "heads" and "tails" events. To get around this sort of problem, it is typically assumed that the number of trials is countably infinite, so the frequency is a *limiting* frequency. Defenders of this type of view include Richard von Mises (1957) and Hans Reichenbach (1949). Consider the following sequence of outcomes of a series of fair coin flips:

THTTHTHHT ...

where *T* is for "tails" and *H* is for "heads." We calculate the limiting frequency by calculating the frequencies of successively increasing finite subsequences. So for example, the first subsequence is just *T*, so the frequency of "tails" is 1. The next larger subsequence is *TH*, which gives a frequency of 1/2. Then the next subsequence is *THT*, so the frequency becomes 2/3. Continuing on in this fashion:

THTT	3/4
THTTH	3/5
THTTHT	4/6
THTTHTH	4/7
THTTHTHH	4/8
THTTHTHHT	5/9

These frequencies appear to be settling down to the value of 1/2. If this is the case, we say that the limiting frequency is 1/2. However, the value of the limiting frequency depends on how we order the trials. If we change the order of the trials, then we change the limiting frequency. To take a simple example, consider the following sequence of natural numbers: (1, 2, 3, 4, 5, 6, 7, 8, 9, 10, 11, . . .). The limiting frequency of even numbers is 1/2. Now consider a different sequence that also has all of the natural numbers as elements, but in a different order: (1, 3, 5, 2, 7, 9, 11, 4, . . .). Now the limiting frequency of even numbers is 1/4. This means that the value of a limiting frequency is sensitive to how we order the trials, and so if probabilities are limiting frequencies, then probabilities depend on the order of the trials too. This is problematic because it seems probabilities should be independent of how we order the trials to calculate limiting frequencies.[14]

Another worry with the hypothetical frequency interpretation is that it does not allow limiting frequencies to come apart from probabilities. Suppose a coin, whenever flipped, has a chance of 1/2 that "tails" comes up on any particular flip. Although highly improbable, it is entirely possible that "tails" never comes up. Yet the hypothetical frequency interpretation says that this statement of 50 percent chance of "tails" *means* that the limiting frequency of "tails" *will* be 1/2. So a chance of 1/2 just means that "tails" has to come up at least once (in fact, half of the time). Many philosophers find this unappealing, for it seems that it is part of the concept of probability that frequencies (both finite and limiting frequencies) can come apart from probabilities.

One of the motivations for the move from the actual frequency interpretation to the hypothetical frequency interpretation was the problem of single-case probabilities. This was the problem that the actual frequency interpretation cannot sensibly assign probabilities to one-time-only events. This problem was also a main motivation for another interpretation of probability, the propensity interpretation.

3.3   The propensity interpretation

The propensity interpretation of probability originates with Popper in Popper 1957, and was developed in more detail in Popper 1959b. His motivation

---

[14]   This problem is similar to the problem in decision theory where the expected utility of an action can depend on how we order the terms in the expected utility calculation. See Nover and Hájek 2004 for further discussion.

for introducing this new interpretation was the need, that he saw, for a theory of probability that was objective, but that could also make sense of single-case probabilities – particularly the single-case probabilities which he thought were indispensable to quantum mechanics. His idea was (roughly) that a probability is not a frequency, but rather it is the tendency, the disposition, or the *propensity* of an outcome to occur.

Popper, who was originally a hypothetical frequentist, developed the propensity theory of probability as a slight modification of the frequency theory. The modification was that instead of probabilities being properties of sequences (viz., frequencies), they are rather properties of the conditions that generate those sequences, when the conditions are repeated: "This modification of the frequency interpretation leads almost inevitably to the conjecture that probabilities are dispositional properties of these conditions – that is to say, propensities" (Popper 1959b, 37). And earlier: "Now these propensities turn out to be *propensities to realize singular events*" (Popper 1959b, 28; author's emphasis).

Perhaps the best known and most influential objection to Popper's original propensity interpretation is due to Paul Humphreys, and is known as Humphreys' paradox – though Humphreys himself did not intend the objection to be one against the propensity interpretation (Humphreys 1985). The objection, in a nutshell, is that propensities are not symmetric, but according to the standard formal theory of probability, probabilities are.[15] For example, it is often possible to work out the probability of a fire having been started by a cigarette given the smoking remains of a building, but it seems strange to say that the smoking remains have a *propensity*, or *disposition*, for a cigarette to have started the fire. The standard reaction to this fact has been "if probabilities are symmetric and propensities are not, then too bad for the propensity interpretation." Humphreys, however, intended his point to be an objection to the standard formal theory of probability (Humphreys 1985, 557) and to the whole enterprise of interpreting probability in a way that takes the formal theory of probability as sacrosanct:

It is time, I believe, to give up the criterion of admissibility [the criterion that a philosophical theory of probability should satisfy "the" probability calculus]. We have seen that it places an unreasonable demand upon one plausible construal of propensities. Add to this the facts that limiting

---

[15] Remember from §2.2, a formal theory of probability is symmetric if whenever $P(A, B)$ is defined, $P(B, A)$ is also defined.

relative frequencies violate the axiom of countable additivity and that their probability spaces are not sigma-fields unless further constraints are added; that rational degrees of belief, according to some accounts, are not and cannot sensibly be required to be countably additive; and that there is serious doubt as to whether the traditional theory of probability is the correct account for use in quantum theory. Then the project of constraining semantics by syntax begins to look quite implausible in this area. (Humphreys 1985, 569–70)

In response to Humphreys' paradox, some authors have offered new formal accounts of propensities. For example, James Fetzer and Donald Nute developed a probabilistic causal calculus as a formal theory of propensities (see Fetzer 1981). A premise of the argument that leads to the paradox is that probabilities are symmetric. But as we saw in §2.2, there are formal theories of probability that are *asymmetric* – Rényi's axioms for conditional probability, for instance. A proponent of Popper's propensity interpretation could thus avoid the paradox by adopting an asymmetric formal theory of probability. Unfortunately for Popper, though, his own formal theory of probability is symmetric.

There are now many so-called propensity interpretations of probability that differ from Popper's original account. Following Donald Gillies, we can divide these accounts into two kinds: long-run propensity interpretations and single-case propensity interpretations (Gillies 2000b). Long-run propensity interpretations treat propensities as tendencies for certain conditions to produce frequencies identical (at least approximately) to the probabilities in a sequence of repetitions of those conditions. Single-case propensity interpretations treat propensities as dispositions to produce a certain result on a specific occasion. The propensity interpretation initially developed by Popper (1957, 1959b) is both a long-run and single-case propensity interpretation. This is because Popper associates propensities with repeatable "generating conditions" to generate singular events. The propensity interpretations developed later by Popper (1990), and David Miller (1994, 1996), can be seen as only single-case propensity interpretations. These propensity interpretations attribute propensities not to repeatable conditions, but to entire states of the universe. One problem with this kind of propensity interpretation is that probability claims are no longer testable, a cost noted by Popper himself (1990, 17). This is because probabilities are now properties of entire states of the universe – events that are not repeatable – and Popper believed that to test a probability claim, the event needs to be repeatable so that a frequency can be measured.[16]

For a general survey and classification of the various propensity theories, see Gillies (2000b); and see Eagle (2004) for 21 objections to them.

3.4   *Logical probability*

In classical logic, if ⊢ B, then we say A entails B. In model-theoretic terms, this corresponds to every model in which A is true, B is true. The logical interpretation of probability is an attempt to generalize the notion of entailment to *partial* entailment. Keynes was one of the earliest to hit upon this idea:

> Inasmuch as it is always assumed that we can sometimes judge directly that a conclusion *follows from* a premiss, it is no great extension of this assumption to suppose that we can sometimes recognize that a conclusion *partially follows from*, or stands in a relation of probability to a premiss. (Keynes 1921, 52)

On this interpretation "$P(B, A) = x$" means A entails B to degree x. This idea has been pursued by many philosophers – e.g., William Johnson (1921), Keynes (1921), though Rudolf Carnap gives by the far the most developed account of logical probability (e.g., Carnap 1950).

By generalizing the notion of entailment to partial entailment, some of these philosophers hoped that the logic of *deduction* could be generalized to a logic of *induction*. If we let c be a two-place function that represents the confirmation relation, then the hope was that:

$$c(B, A) = P(B, A)$$

For example, the observation of ten black ravens deductively entails that there are ten black ravens in the world, while the observation of five black ravens only partially entails, or *confirms*, that there are ten black ravens, and the observation of two black ravens confirms this hypothesis to a lesser degree.

---

[16]   This is, perhaps, not the only way in which a probability claim can be tested. For example, it may be possible to test the claim "this coin has a chance 0.5 to land heads when flipped" by investigating whether or not the coin is physically symmetrical.

One seemingly natural way to formalize the notion of partial entailment is by generalizing the model theory of full entailment. Instead of B being true in *every* model in which A is true, we relax this to there being some percentage of the models in which A is true. So "$P(B, A) = x$," which is to say, "A partially entails B to degree x" is true, if the number of models where B and A are true, divided by the number of models where A is true, is equal to x.[17] If we think of models as like "possible worlds," or possible outcomes then this definition is the same as the classical definition of probability. We might suspect then that the logical interpretation shares some of the same difficulties (in particular, the language relativity of probability) that the classical interpretation has. Indeed, this is so (see, e.g., Gillies 2000a, 29–49).

Carnap maintains that $c(B, A) = P(B, A)$, but investigates other ways to define the probability function, P. In contrast to the approach above, Carnap's way of defining P is purely syntactic. He starts with a language with predicates and constants, and from this language defines what are called *state descriptions*. A state description can be thought of as a maximally specific description of the world. For example, in a language with predicates F and G, and constants a and b, one state description is $Fa \wedge Fb \wedge Gb \wedge \neg Ga$. Any state description is equivalent to a conjunction of predications where every predicate or its negation is applied to every constant in the language. Carnap then tried to define the probability function, P, in terms of some measure, m, over all of the state descriptions. In Carnap 1950, he thought that such a measure was unique. Later on, in Carnap 1963, he thought there were many such measures. Unfortunately, every way Carnap tried to define P in terms of a measure over state descriptions failed for one reason or another (see, e.g., Hájek 2007b).

Nearly every philosopher now agrees that the logical interpretation of probability is fundamentally flawed. However, if they are correct, this does not entail that a formal account of inductive inference is not possible. Recent attempts at developing an account of inductive logic reject the sole use of conditional probability and instead measure the degree to which evidence E confirms hypothesis H by how much E affects the probability of H (see, e.g., Fitelson 2006). For example, one way to formalize the degree to which E supports or confirms H is by how much E *raises* the probability of H:

---

[17]   Unconditional, or absolute, probability, $P(A)$, is understood as the probability of A given a tautology T, so $P(A) = P(A, T)$ in which case $P(A)$ is just the number of models where A is true divided by the total number of models (since a tautology is true in every model).

$$c(H, E) = P(H, E) - P(H)$$

This is one such measure among many.[18] The function c, or some other function like it, may formally capture the notion of evidential impact that we have, but these functions are defined in terms of *probabilities*. So an important and natural question to ask is: what are these probabilities? Perhaps the most popular response is that these probabilities are subjective probabilities, i.e., the credences of an individual. According to this type of theory of confirmation (known as Bayesian confirmation theory), the degree to which some evidence confirms a hypothesis is relative to the epistemic state of an individual. So E may confirm H for one individual, but disconfirm H for another. This moves us away from the strictly objective relationship between evidence and hypothesis that the logical interpretation postulated, to a more subjective one.

3.5   The subjective interpretation

While the frequency and propensity interpretations see the various formal accounts of probability as theories of how frequencies and propensities behave, the subjective interpretation sees them as theories of how people's beliefs *ought* to behave. We can find this idea first published by Frank Ramsey (1931) and de Finetti (1931a, 1931b). The normativity of the "ought" is meant to be one of ideal epistemic rationality. So subjectivists traditionally claim that for one to be ideally epistemically rational, one's beliefs must conform to the standard probability calculus. Despite the intuitive appeal of this claim (which by the way is typically called *probabilism*), many have felt the need to provide some type of argument for it. Indeed, there is now a formidable literature on such arguments. Perhaps the most famous argument for probabilism is the Dutch Book Argument.

3.5.1   The Dutch Book Argument
A *Dutch book* is any collection of bets that collectively guarantee a sure monetary loss. An example will help illustrate the idea. Suppose Bob assigns a credence of 0.6 to a statement, A, and a credence of 0.5 to that statement's negation, $\neg A$. Bob's credences thus do not satisfy the probability calculus since his credence in A and his credence in $\neg A$ sum to 1.1. Suppose further

---

[18]   See Eells and Fitelson 2002 for an overview of some of the other possible measures.

that Bob bets in accordance with his credences, that is, if he assigns a credence of x to A, then he will buy a bet that pays $y if A, for at most $xy. Now consider the following two bets:

- Bet 1: This bet costs $0.6 and pays $1 if A is true.
- Bet 2: This bet costs $0.5 and pays $1 if $\neg A$ is true.

Bob evaluates both of these bets as fair, since the expected return – by his lights – of each bet is the price of that bet.[19] But suppose Bob bought both of these bets. This would be apparently equivalent to him buying the following bet:

- Bet 3: This bet costs $0.6 + $0.5 = $1.1 and pays $1 if A is true, and $1 if $\neg A$ is true (i.e., the bet pays $1 no matter what).

If Bob were to accept Bet 3, then Bob would be guaranteed to lose $0.1, no matter what. The problem for Bob is that he evaluates Bet 1 and Bet 2 as both individually fair, but by purchasing both Bet 1 and Bet 2, Bob effectively buys Bet 3, which he does not evaluate as fair (since his expected return on the bet is less than the price of the bet).

There is a theorem called the Dutch Book Theorem which, when read informally, says that if an agent has credences like Bob's – i.e., credences that do not obey axioms P1–3 – then there is always a Dutch book that the agent would be willing to buy. So having credences that do not obey axioms P1–3 results in you being susceptible to a Dutch book. Conversely, there is a theorem called the Converse Dutch Book Theorem which, when also read informally, says that if an agent has credences that *do* obey P1–3, then there is no Dutch book that that agent would be willing to buy. Taken together these two theorems give us:

(DBT & CDBT)   An agent is susceptible to a Dutch book if and only if the agent has credences that violate axioms P1–3.

Then with the following rationality principle:

[19]   To work out the expected return of a bet we multiply the probability of each pay-off by the value of that pay-off and sum these numbers together. For example, in Bet 1 there is only one pay-off – $1, when A is true – so we multiply that by Bob's credence in A, 0.6, so the expected return is $0.6.

(RP)   If an agent is ideally epistemically rational, then that agent is not susceptible to a Dutch book.

we get the following result:

(C)   If an agent is ideally epistemically rational, then that agent's credences obey axioms P1–3.

This is known as the Dutch Book Argument. It is important that CDBT is included, because it blocks an obvious challenge to RP. Without CDBT one might claim that it is *impossible* to avoid being susceptible to a Dutch book, but it is still possible to be ideally epistemically rational. CDBT guarantees that it is possible to avoid a Dutch book, and combined with DBT it entails that the only way to do this is to have one's credences satisfy the axioms P1–3.

There are many criticisms of the Dutch Book Argument – too many to list all of them here, but I will mention a few.[20] One criticism is that it is not clear that the notion of rationality at issue is of the right kind. For example, David Christensen writes:

> Suppose, for example, that those who violated the probability calculus were regularly detected and tortured by the Bayesian Thought Police. In such circumstances, it might well be argued that violating the probability calculus was imprudent, or even "irrational" in a practical sense. But I do not think that this would do anything toward showing that probabilistic consistency was a component of rationality in the epistemic sense relevant here. (Christensen 1991, 238)

In response to this worry, some have offered what are called depragmatized Dutch book arguments, in support of probabilism (see, e.g., Christensen 1996). Others have stressed that the Dutch Book Argument should not be interpreted literally and rather that it merely dramatizes the inconsistency of a system of beliefs that do not obey the probability calculus (e.g., Skyrms 1984, 22, and Armendt 1993, 3).

Other criticisms focus on the assumptions of the Dutch Book and Converse Dutch Book Theorems. For instance, the proofs of these theorems assume that if an agent evaluates two bets as both fair when taken

---

[20]   For a detailed discussion of these and other objections see Hájek 2007a.

individually, then that agent will, and should, also consider them to be fair when taken collectively. This assumption is known as the package principle (see Schick 1986 and Maher 1993 for criticisms of this principle). The standard Dutch Book Argument is meant to establish that our credences ought to satisfy axioms P1–3, but what about a countable additivity axiom? Dutch Book Arguments that try to establish a countable additivity axiom as a rationality constraint rely on a countably infinite version of the package principle (see Arntzenius, Elga, and Hawthorne 2004 for objections to this principle).

These objections to the Dutch Book Argument – and others – have led some authors to search for other arguments for probabilism. For instance, Patrick Maher argues that if you cannot be *represented* as an expected utility maximizer, relative to a probability and utility function, then you are irrational (Maher 1993). Some have argued that one's credences ought to obey the probability calculus because for any non-probability function, there is a probability function that better matches the relative frequencies in the world, no matter how the world turns out. This is known as a calibration argument (see, e.g., van Fraassen 1984). James Joyce argues for probabilism by proving that for any non-probability function, there is a probability function that is "closer" to the truth, no matter how the world turns out (Joyce 1998). This is known as a gradational accuracy argument. For criticisms of all these arguments see Hájek (forthcoming).

Suppose for the moment that it has been established that one's credences ought to satisfy the probability axioms. Are these the *only* normative constraints on credences? One feature our beliefs have is that they *change* over time, especially when we learn new facts about the world. And it seems that there are rational and irrational ways of changing one's beliefs. In fact, perhaps most probabilists believe that there are rational and irrational ways to respond to evidence, beyond simply remaining in synch with the probability calculus. One particularly large subgroup of these probabilists are known as Bayesians.

3.5.2 *Bayesianism*

Orthodox Bayesianism is the view that an agent's credences: should at all times obey the probability axioms; should change only when the agent acquires new information; and, in such cases, the agent's credences should be updated by Bayesian Conditionalization. Suppose that an agent has a prior credence function Cr_{old}. Then, according to this theory of updating, the agent's posterior credence function, Cr_{new}, after acquiring evidence E, ought to be:

(BC) $Cr_{new}(H) = Cr_{old}(H, E)$

for every H in $\mathscr{L}$. BC is said to be a *diachronic* constraint on credences, whereas, for example, P1–3 are said to be *synchronic* constraints on credences. There are Dutch Book Arguments for why credences ought to be diachronically constrained by Bayesian Conditionalization (see, e.g., Lewis 1999). Arguments of this type suppose that BC is a *rationality* constraint, and that violations of it are a type of inconsistency. Christensen argues that since the beliefs are changing across time, violations of BC are not, strictly speaking, inconsistencies (Christensen 1991).

One important criticism of orthodox Bayesianism, due to Richard Jeffrey, is that it assumes that facts are always acquired (learned) with full certainty (Jeffrey 1983). Critics argue that it should at least be *possible* for evidence that you are not entirely certain of to impact your credences. For this reason, Jeffrey developed an alternative account for how credences should be updated in the light of new evidence, which generalized BC to account for cases when we acquire evidence without full certainty. Jeffrey called his theory *Probability Kinematics*, but it is now known as *Jeffrey Conditionalization*. According to Jeffrey Conditionalization, an agent's new credence in H after acquiring some information that has affected the agent's credence in E should be:

(JC) $Cr_{new}(H) = Cr_{old}(H, E)Cr_{new}(E) + Cr_{old}(H, \neg E)Cr_{new}(\neg E)$

for every H in $\mathscr{L}$.[21] Notice that the right-hand side of JC contains both the old and new credence function, whereas BC only had the old credence function. At first glance this may give the impression that JC is circular. It is not, though. Initially you have a probability in E and $\neg E$, $Cr_{old}(E)$ and $Cr_{old}(\neg E)$, respectively. Then you acquire some information that causes you to change your credences concerning E and $\neg E$ (and only these statements). These new credences are $Cr_{new}(E)$ and $Cr_{new}(\neg E)$, respectively. JC then tells you how the information you acquired should affect your other credences, given how that information affected your credences concerning E and $\neg E$.

So, Bayesianism is a theory of epistemic rationality that says our credences at any given time should obey the probability calculus and should be updated by conditionalization (either BC or JC). However, some insist that there is still more to a full theory of epistemic rationality.

---

[21]   Actually, this is a special case of Jeffrey Conditionalization. The general equation is: $Cr_{new}(H)$ $= \Sigma_i\, Cr_{old}(H, E_i)Cr_{new}(E_i)$, where the E_i are mutually exclusive and exhaustive in $\mathscr{L}$.

3.5.3   Objective and subjective Bayesianism

Within the group of those probabilists who call themselves Bayesians is another division between so-called objective Bayesians and subjective Bayesians. As we saw in the previous sections, Bayesians believe that credences should obey the probability calculus and should be updated according to conditionalization, when new information is obtained. So far, though, nothing has been said about which credence function one should have *before* any information is obtained – apart from the fact that it should obey the probability calculus.

Subjective Bayesians believe there ought to be no further constraint on initial credences. They say: given that it satisfies the probability calculus, no initial credence function is any more rational than any other. But if subjective Bayesians believe any coherent initial credence function is a rational one, then, according to them, a credence function that assigns only 1s and 0s to all statements – including statements that express contingent propositions – is also a rational credence function. Many philosophers (including those that call themselves subjective Bayesians) balk at this idea and so insist that any initial credence function must be *regular*. A regular credence function is any probability function that assigns 1s and 0s only to logical truths and falsehoods; all contingent sentences must be assigned strictly intermediate probability values.[22] The idea roughly is that an initial credence function should not assume the truth of any contingency, since nothing contingent about the world is known by the agent.

However, we may worry that this is still not enough, for a credence function that assigns a credence of, say, 0.9999999 to some contingent sentence (e.g., that the earth is flat) is still counted as a rational initial credence function. There are two responses that Bayesians make here. The first is to point to so-called Bayesian convergence results. The idea, roughly, is that as more and more evidence comes in, such peculiarities in the initial credence function are in a sense "washed out" through the process of repeated applications of conditionalization. More formally, for any initial credence function, there is an amount of possible evidence that can be conditionalized on to ensure the resulting credence function is arbitrarily close to the truth.

---

[22]   There are some technical difficulties with this condition of regularity because it is impossible (in some formal accounts of probability) for there to be a regular probability function over uncountably many sentences (or sets, or propositions, or whatever the bearers of probability are). See, e.g., Hájek 2003 for discussion.

See Earman (1992, 141–9) for a more rigorous and critical discussion of the various Bayesian convergence results.

The second response to the original worry that some Bayesians make is that there are in fact further constraints on rational initial credence functions. Bayesians who make this response are known as objective Bayesians. One worry with the prior that assigned 1s and 0s to contingent statements was that such a prior does not truly reflect our epistemic state – we do not know anything about any contingent proposition before we have learned anything, yet our credence function says we do. A similar worry may be had about the prior that assigns 0.9999999 to a contingent statement. This type of prior reports an overwhelming confidence in contingent statements before anything about the world is known. Surely such blind confidence cannot be rational. Reasoning along these lines, E.T. Jaynes, perhaps the most famous proponent of objective Bayesianism, claims that our initial credence function should be an accurate description of how much information we have:

> [A]n ancient principle of wisdom – that one ought to acknowledge frankly the full extent of his ignorance – tells us that the distribution that maximizes *H* subject to constraints which represent whatever information we have, provides the most honest description of what we know. The probability is, by this process, "spread out" as widely as possible without contradicting the available information. (Jaynes 1967, 97)

The quantity *H* is from information theory and is known as the Shannon entropy.[23] Roughly speaking, *H* measures the information content of a distribution. According to this view, in the case where we have no information at all, the distribution that provides the most honest description of our epistemic state is the uniform distribution. We see then that the principle that Jaynes advocates – which is known as the Principle of Maximum Entropy – is a generalization of the Principle of Indifference. This version of objective Bayesianism thus faces problems similar to those that plague the logical and classical interpretations. Most versions of objective Bayesianism ultimately rely on some version of the Principle of Indifference and so suffer a similar fate. As a result, subjective Bayesianism with the condition that a prior should be regular is perhaps the most popular type of Bayesianism amongst philosophers.

---

[23] To learn more about information theory and Shannon entropy, see Shannon and Weaver 1962.

3.5.4   *Other norms*

At this stage, we have the following orthodox norms on partial beliefs:

1. One's credence function must always satisfy the standard probability calculus.
2. One's credence function must only change in accordance with conditionalization.
3. One's initial credence function must be regular.

However, there are still more norms that are often said to apply to beliefs. One important such norm is David Lewis' Principal Principle (LPP) (1980). Roughly, the idea behind this principle is that one's credences should be in line with any of the objective probabilities in the world, if they are known. More formally, if $Ch_t(A)$ is the chance of A at time t (e.g., on a propensity interpretation of probability this would be the propensity at time t of A to obtain), then:

$$(\text{LLP}) \quad Cr(A, Ch_t(A) = x \wedge E) = x$$

where E is any proposition, so long as it is not relevant to A.[24] LPP, as originally formulated by Lewis, is a synchronic norm on an agent's *initial* credence function, though LPP is commonly used as synchronic constraint on an agent's credence function at any point in time.

Another synchronic norm is van Fraassen's Reflection Principle (VFRP) (1995). Roughly, the idea behind this principle is that if, upon reflection, you realize that you will come to have a certain belief, then you ought to have that belief *now*. More formally, the Reflection Principle is:

$$(\text{VFRP}) \quad Cr_{t_1}(A, Cr_{t_2}(A) = x) = x$$

where $t_2 > t_1$ in time.

Another more controversial norm is Adam Elga's principle of indifference for indexical statements, used to defend a particular solution to the Sleeping Beauty Problem. The problem is that Sleeping Beauty is told by scientists on Sunday that they are going to put her to sleep and flip a fair

---

[24]   Actually, strictly speaking E should be what Lewis calls *admissible*. See Lewis 1980 for details on this issue.

coin. If the coin lands "tails," they will wake her on Monday, wipe her memory, put her back to sleep, and wake her again on Tuesday. If the coin lands "heads," they will simply wake her on Monday. When Sleeping Beauty finds herself having just woken up, what should her credence be that the coin landed "heads"? According to Lewis, it should be 1/2 since this is the chance of the event and LPP says Sleeping Beauty's credence should be equal to the chance. According to Elga, there are three possibilities: (i) she is being woken for the first time, on Monday; (ii) she is being woken for the second time, on Tuesday; or (iii) she is being woken for the first time, on Monday. All of these situations are indistinguishable from Sleeping Beauty's point of view, and Elga argues that an agent should assign equal credence to indistinguishable situations – this is his indifference principle. So, according to Elga, Sleeping Beauty should assign equal probability to each possibility, and so her credence that the coin landed "heads" ought to be 1/3. See Elga (2000) for more on the Sleeping Beauty Problem, and Weatherson (2005) for criticism of Elga's version of the Principle of Indifference.

4. Conclusion

In a short amount of space we have covered a lot of territory in the philosophy of probability. In §2, we considered various formal theories of probability. We saw that not only are there rival theories to Kolmogorov's axioms, but these rivals arguably have desirable features that Kolmogorov's axioms lack. In §3, we saw some of the various interpretations of probability and some of the issues connected with each interpretation. The discussion of each interpretation was necessarily brief, but each of these interpretations suffers from one problem or another. In fact, the failures of each interpretation have motivated some to take probability as a primitive, undefined concept (e.g., Sober forthcoming). We see, then, that despite the ubiquity of probability in our lives, the mathematical and philosophical foundations of this fruitful theory remain in contentious dispute.[25]

[25] Thanks to Alan Hájek for many discussions and feedback on earlier drafts, and to Jens Christian Bjerring, Mark Colyvan, Fritz Allhoff, John Matthewson, Joe Salerno, Mike Smithson, and Weng Hong Tang for helpful comments on an earlier draft.

References

Armendt, B. 1993. "Dutch Books, Additivity and Utility Theory." *Philosophical Topics* 21.1: 1–20.

Arntzenius, F., A. Elga, and J. Hawthorne. 2004. "Bayesianism, Infinite Decisions, and Binding." *Mind* 113.450: 1–34.

Bertrand, J. 1888. *Calcul des Probabilités.* Paris: Gauthier-Villars.

Cardano, G. 1663. *The Book on Games of Chance.* English translation in O. Ore, *Cardano: The Gambling Scholar.* New York: Dover, 1965 [1953].

Carnap, R. 1950. *Logical Foundations of Probability.* Chicago: University of Chicago Press.

Carnap, R. 1963. *Replies and Systematic Expositions.* La Salle, IL: Open Court.

Christensen, D. 1991. "Clever Bookies and Coherent Beliefs." *Philosophical Review* 100.2: 229–47.

Christensen, D. 1996. "Dutch-Book Arguments Depragmatized: Epistemic Consistency for Partial Believers." *Journal of Philosophy* 93.9: 450–79.

Colyvan, M., H.M. Regan, and S. Ferson. 2001. "Is it a Crime to Belong to a Reference Class?" *Journal of Political Philosophy* 9.2: 168–81.

Cox, D.R. 1955. "A Use of Complex Probabilities in the Theory of Stochastic Processes." *Proceedings of the Cambridge Philosophical Society* 51: 313–19.

de Finetti, B. 1931a. "On the Subjective Meaning of Probability." In P. Monari and D. Cocchi (eds.), *Induction and Probability.* Bologna: Clueb, 1993.

de Finetti, B. 1931b. "Probabilism." Published in English in *Erkenntnis* 31 (1989): 169–223.

de Finetti, B. 1974. *Theory of Probability,* vol. 1. Wiley Classics Library. Chichester: John Wiley and Sons, 1990.

Dickson, M. 2001. "Quantum Logic is Alive $\wedge$ (It Is True $\vee$ It Is False)." *Philosophy of Science* 68.3: S274–87.

Eagle, A. 2004. "Twenty-One Arguments Against Propensity Analyses of Probability." *Erkenntnis* 60: 371–416.

Earman, J. 1992. *Bayes or Bust?* Cambridge, MA: MIT Press.

Eells, E. and B. Fitelson. 2002. "Symmetries and Asymmetries in Evidential Support." *Philosophical Studies* 107.2: 129–42.

Elga, A. 2000. "Self-Locating Belief and the Sleeping Beauty Problem." *Analysis* 60.2: 143–7.

Fetzer, H. 1981. *Scientific Knowledge: Causation, Explanation, and Corroboration. Boston Studies in the Philosophy of Science.* Dordrecht: Reidel.

Feynman, R.P. 1987. "Negative Probability." In B.J. Hiley and F. David Peat (eds.), *Quantum Implications: Essays in Honour of David Bohm,* New York: Routledge.

Fine, T. 1973. *Theories of Probability.* New York: Academic Press.

Fitelson, B. 2006. "Logical Foundations of Evidential Support." *Philosophy of Science* 73: 500–12.

Gillies, D. 2000a. *Philosophical Theories of Probability*. New York: Routledge.

Gillies, D. 2000b. "Varieties of Propensity." *British Journal for the Philosophy of Science* 51: 807–35.

Hacking, I. 1967. "Slightly More Realistic Personal Probability." *Philosophy of Science* 34.4: 311–25.

Hacking, I. 1971. "Equipossibility Theories of Probability." *British Journal for the Philosophy of Science* 22.4: 339–55.

Hájek, A. 1997. "'Mises Redux' – Redux: Fifteen Arguments against Finite Frequentism." *Erkenntnis* 45: 209–27.

Hájek, A. 2003. "What Conditional Probability Could Not Be." *Synthese* 137.3: 273–323.

Hájek, A. 2007a. "Dutch Book Arguments." In P. Anand, P. Pattanik, and C. Puppe (eds.), *The Oxford Handbook of Corporate Social Responsibility*. Oxford: Oxford University Press.

Hájek, A. 2007b. "Interpretations of Probability." Stanford Encyclopedia of Philosophy. Available at http://plato.stanford.edu/archives/fall2007/entries/probability-interpret/ (accessed March 18, 2008).

Hájek, A. 2007c. "The Reference Class is Your Problem Too." *Synthese* 156: 563–85.

Hájek, A. Forthcoming. "Arguments For – or Against – Probabilism." In F. Huber and C. Schmidt-Petri (eds.), *Degrees of Belief*. Oxford: Oxford University Press.

Humphreys, P. 1985. "Why Propensities Cannot Be Probabilities." *Philosophical Review* 94: 557–70.

Jaynes, E.T. 1967. "Foundations of Probability Theory and Statistical Mechanics." In M. Bunge (ed.), *Popular Electronics*. Berlin: Springer-Verlag.

Jaynes, E.T. 1973. "The Well Posed Problem." *Foundations of Physics* 4.3: 477–92.

Jeffrey, R.C. 1983. *The Logic of Decision*. Chicago: University of Chicago Press.

Johnson, W.E. 1921. *Logic*. Cambridge: Cambridge University Press.

Joyce, J. 1998. "A Non-Pragmatic Vindication of Probabilism." *Philosophy of Science* 65: 575–603.

Keynes, J.M. 1921. *A Treatise on Probability*. London: Macmillan.

Kolmogorov, A.N. 1933. *Foundations of Probability*. New York: Chelsea Publishing Company, 1950.

Laplace, P. 1814. *A Philosophical Essay on Probabilities*. New York: Dover, 1951.

Levi, I. 1980. *The Enterprise of Knowledge*. Cambridge, MA: MIT Press.

Lewis, D. 1980. "A Subjectivist's Guide to Objective Chance." In R.C. Jeffrey (ed.), *Studies in Inductive Logic and Probability*, vol. 2. Berkeley: University of California Press.

Lewis, D. 1999. "Why Conditionalize?" In D. Lewis (ed.), *Papers in Metaphysics and Epistemology*, vol. 2. Cambridge: Cambridge University Press.

Maher, P. 1993. *Betting on Theories*. Cambridge: Cambridge University Press.

Marinoff, L. 1994. "A Resolution of Bertrand's Paradox." *Philosophy of Science* 61.1: 1–24.

Mikkelson, J.M. 2004. "Dissolving the Wine/Water Paradox." *British Journal for the Philosophy of Science* 55: 137–45.

Miller, D.W. 1994. *Critical Rationalism. A Restatement and Defence.* La Salle, IL: Open Court.

Miller, D.W. 1996. "Propensities and Indeterminism." In A. O'Hear (ed.), *Karl Popper: Philosophy and Problems.* Cambridge: Cambridge University Press.

Nover, H. and A. Hájek. 2004. "Vexing Expectations." *Mind* 113: 237–49.

Poincaré, H. 1902. *Science and Hypothesis.* New York: Dover, 1952.

Popper, K.R. 1938. "A Set of Independent Axioms for Probability." *Mind* 47.186: 275–7.

Popper, K.R. 1955. "Two Autonomous Axiom Systems for the Calculus of Probabilities." *British Journal for the Philosophy of Science* 6.21: 51–7.

Popper, K.R. 1957. "The Propensity Interpretation of the Calculus of Probability and the Quantum Theory." In S. Körner (ed.), *The Colston Papers,* no. 9. London: Butterworth Scientific Publications, 65–70.

Popper, K.R. 1959a. *The Logic of Scientific Discovery.* New York: Basic Books.

Popper, K.R. 1959b. "The Propensity Interpretation of Probability." *British Journal for the Philosophy of Science* 10.37: 25–42.

Popper, K.R. 1990. *A World of Propensities.* Bristol: Thoemmes.

Putnam, H. 1968. "Is Logic Empirical?" In R. Cohen and M. Wartofsky (eds.), *Boston Studies in the Philosophy of Science,* vol. 5. Dordrecht: Reidel, 216–41.

Ramsey, F. 1931. "Truth and Probability." In R.B. Braithwaite (ed.), *Foundations of Mathematics and Other Essays.* London: Routledge & Kegan Paul, 156–98.

Reichenbach, H. 1949. *The Theory of Probability.* Berkeley: University of California Press.

Rényi, A. 1955. "On a New Axiomatic Theory of Probability." *Acta Mathematica Academiae Scientiarum Hungaricae* 6: 286–335.

Rényi, A. 1970. *Foundations of Probability.* San Francisco, CA: Holden-Day, Inc.

Roeper, P. and H. Leblanc. 1999. *Probability Theory and Probability Logic.* Toronto: University of Toronto Press.

Schick, F. 1986. "Dutch Bookies and Money Pumps." *Journal of Philosophy* 83.2: 112–19.

Seidenfeld, T., M.J. Schervish, and J.B. Kadane. 2001. "Improper Regular Conditional Distributions." *Annals of Probability* 29.4: 1612–24.

Shackel, N. 2007. "Bertrand's Paradox and the Principle of Indifference." *Philosophy of Science* 74: 150–75.

Shafer, G. 1976. *A Mathematical Theory of Evidence.* Princeton: Princeton University Press.

Shannon, C.E. and W. Weaver. 1962. *The Mathematical Theory of Communication.* Champaign, IL: University of Illinois Press.

Skyrms, B. 1980. *Causal Necessity*. New Haven, CT: Yale University Press.

Skyrms, B. 1984. *Pragmatics and Empiricism*. New Haven, CT: Yale University Press.

Sober, E. 2008. "Evolutionary Theory and the Reality of Macro Probabilities." In E. Eells and J. Fetzer (eds.), *Probability in Science*. La Salle, IL: Open Court.

van Fraassen, B. 1984. "Belief and Will." *Journal of Philosophy* 81.5: 235–56.

van Fraassen, B. 1989. *Laws and Symmetry*. Oxford: Oxford University Press.

van Fraassen, B. 1995. "Belief and the Problem of Ulysses and the Sirens." *Philosophical Studies* 77: 7–37.

Venn, J. 1876. *The Logic of Chance*. 2nd edn. London: Macmillan and Co.

von Mises, R. 1957. *Probability, Statistics and Truth*. New York: Macmillan.

Weatherson, B. 2003. "From Classical to Intuitionistic Probability." *Notre Dame Journal of Formal Logic* 44.2: 111–23.

Weatherson, B. 2005. "Should We Respond to Evil with Indifference?" *Philosophy and Phenomenological Research* 70: 613–35.

Unit 3

Philosophy of the Natural Sciences

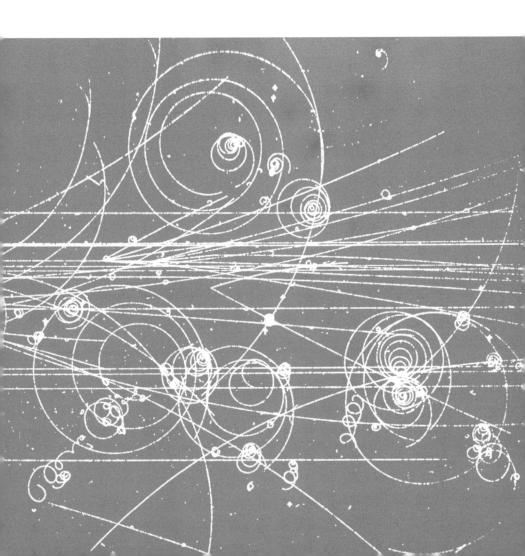

6   Philosophy of Physics

Richard DeWitt

 We have always tended to look to our basic sciences for insight into more broadly philosophical questions, for example, questions such as "what sort of universe do we inhabit?" The previous 100 years saw the development of two new approaches in physics, these being relativity theory and quantum theory. As it turns out, both of these new (and now well established) theories have substantial implications for some of our most basic philosophical questions as to the nature of the universe. The main goal of the current essay is to provide an overview of these two new branches of physics, with an eye toward seeing the implications each has for our most longstanding views on the universe.

1.   Introduction

Since the beginnings of western science, with the investigations of the ancient Greeks roughly 2,500 years ago, what we tend to think of as primarily scientific issues and primarily philosophical issues have had strong influences on one another. To take just one example, certain broadly philosophical views on the sort of universe we inhabit have tended to be informed by the fundamental sciences of the time. So in the western world for much of the period from the ancient Greeks to the 1600s, the universe was broadly conceived of as a teleological universe, that is, a universe with natural goals and functions. And the universe was viewed as roughly analogous to an organism. In the same way, say, that the parts of an organism act as they do to fulfill goals (to pump blood, to digest food, and so on), so too natural objects behaved as they did because of natural, goal-directed tendencies. And this teleological view of the universe reflected the teleology of the sciences of the time.

With the scientific changes in the 1600s, and the increasingly mechanistic approach (that is, a non goal-directed approach) of fundamental sciences such as Newtonian physics, the general conception of the universe changed to a more mechanistic universe. No longer was the universe thought to be analogous to an organism; rather, the prevailing view came to be that the universe was viewed as like a machine, sort of a clock-like universe. And again, this mechanistic view of the universe reflected the mechanistic tendencies of the fundamental sciences of the time.

In general, our broadly philosophical views on what sort of universe we inhabit has tended to be informed, as it should be, by our current best science. And notably, two of the key developments in twentieth-century science turn out to have surprising implications for the philosophical question as to the sort of universe we inhabit.

Early in the twentieth century, two important new theories arose in physics. The first of these was Einstein's theory of relativity, and the second was quantum theory. As suggested, both of these new theories have non-trivial consequences for certain deeply held beliefs about the sort of universe we inhabit, and one of the main goals of this essay will be to develop some of the philosophical implications of these theories. In what follows, we will begin by looking at relativity theory, with an eye toward understanding some of the key implications of this theory, and then turn to quantum theory and some of its implications.

2.   The Theory of Relativity

In 1905, Albert Einstein (1879–1955) published a paper containing the heart of what would come to be known as the Special Theory of Relativity.[1] The special theory, as the name suggests, applies so long as certain special circumstances apply (more on this below). Somewhat more than 10 years later, in 1916, Einstein published the General Theory of Relativity, which, again as the name suggests, is a general theory not restricted to the special circumstances needed to use the special theory.

Both the special and general theories have intriguing consequences for some long-held beliefs – for example, beliefs about the nature of space and time. Many of these consequences can be illustrated with just the special theory. This is convenient, as the special theory, although it has

---

[1]   Einstein 1905 is the original publication in German. Einstein 1952 is a standard English translation of this work, and several other of Einstein's key papers dealing with relativity.

surprising and counter-intuitive implications, is itself not particularly difficult to understand. In contrast, the general theory is substantially more difficult.

We will first take a look at special relativity and its implications. Then, in a somewhat shorter section, we will look briefly at the general theory, as well as some of the philosophical implications of the general theory.

2.1 The basic principles of the special theory of relativity

At its core, what came to be called the special theory of relativity is based on two fundamental principles. One of these principles is what Einstein termed the "principle of relativity," and the other is what is often referred to as the principle of the constancy of the velocity of light. Let us begin by getting clear on these basic principles. In the 1905 paper mentioned above, Einstein sums up the principle of relativity as the principle that "the laws of electrodynamics and optics will be valid for all frames of reference for which the equations of mechanics hold good" (Einstein 1952, 37). When Einstein speaks of "frames of reference for which the equations of mechanics hold good," he has in mind what are now generally referred to as "inertial reference frames," or just "inertial frames." An example might help illustrate.

Picture yourself on a long, straight stretch of interstate highway, with plenty of other cars. Some cars are going the same direction as you, some in the opposite direction, some faster, some slower. Usually, on such a highway, cars accelerate, decelerate, change lanes, and so on. But for this illustration, imagine no one is accelerating, decelerating, or changing lanes. In short, each car is moving in a straight line, and each car is moving along the highway with a uniform velocity, that is, neither accelerating nor decelerating.

In this example, each car will be an inertial reference frame (or at least approximately so – strictly speaking, we would have to assume perfectly uniform motion). Roughly, the key idea is that we are dealing with straight-line motion at uniform speed. Notice that the different cars will be different reference frames, that is, different points of view. For example, if you are moving along the highway at 60 miles per hour, and a car passes you at 70 miles per hour, your car and the passing car will be different points of view (for example, from your perspective objects along the highway, such as telephone poles, are passing by your window at one rate, while those objects will be passing by the window of the other car at a different rate). But you and the other car will both be inertial frames, and notably, from each perspective, the other is moving in a straight line with uniform speed.

In the 1905 paper, Einstein was primarily concerned with issues involving a branch of physics known as electrodynamics – which is primarily concerned with electrical and magnetic phenomena – and thus he speaks of the laws of electrodynamics. But the principle of relativity can be (and usually is) generalized to include all laws of physics, that is, the basic idea is that the laws of physics are the same in all inertial reference frames.

This basic idea behind the principle of relativity – that the laws of physics are the same in all inertial reference frames – can be rephrased in a variety of ways. One way to rephrase the principle of relativity, and one which I think makes it easier to flesh out the key implications of the principle, is as follows:

> *The Principle of Relativity:* If A and B are two inertial reference frames (i.e., A and B are moving relative to one another in a straight line at uniform speed) and if identical experiments are carried out in A and B, then the results of the experiments will be identical.

The other basic principle of the special theory of relativity is what is often termed the "principle of the constancy of the velocity of light," which hereafter for convenience I will abbreviate to PCVL. As Einstein phrases it in the 1905 paper, "light is always propagated in empty space with a definite velocity c which is independent of the state of motion of the emitting body" (Einstein 1952, 38). This amounts to saying that if one measures the speed of light in a vacuum, the value will always be the same, regardless of, say, whether the source of the light or the measuring device is in motion.

At bottom, these two principles are the only basic principles of special relativity. From them one can deduce some surprising consequences for our usual notions involving space and time. These consequences are the main focus of the next section.

2.2   *Implications of the principle of relativity and the PCVL*

In the 1905 paper, Einstein treats the principle of relativity and the PCVL as postulates, from which he deduces "a simple and consistent theory of . . . moving bodies" (Einstein 1952, 38). The reference to a "consistent theory" is worth noting, as the general belief at the time was that the principle of relativity and the PCVL were inconsistent with one another. However, as Einstein demonstrates, these two principles only seem to be inconsistent. And indeed, when one first encounters the implications of the principle of relativity and the PCVL, there tends to be a sense that these implications in some way must be inconsistent. But again, the appearance of inconsistency turns out to be illusory.

The three most fundamental implications of the principle of relativity and the PCVL involve our usual notions of space, time, and whether events are or are not simultaneous.[2] Let me begin by stating briefly these three key implications, and then illustrate them with some examples.

Time dilation: Suppose A and B are inertial references frames, that is, A and B are in motion relative to one another in a straight line at uniform speed. Then from the perspective of reference frame A, time in B moves more slowly by a factor of

$$\sqrt{1 - \left(\frac{v}{c}\right)^2}$$

and from the perspective of reference frame B, time in A moves more slowly by the same factor. (This factor, incidentally, is known as the *Lorentz-Fitzgerald equation.* The variable *v* represents the velocity involved, and *c*, as always, represents the speed of light, which for the examples to follow will be approximated to 300,000 kilometers/second.)

Length contraction: Again suppose A and B are inertial references frames. Then from the perspective of A, space in B contracts, in the direction of motion, by a factor of

$$\sqrt{1 - \left(\frac{v}{c}\right)^2}$$

---

[2] Mermin (1968) provides a thorough, yet accessible, demonstration of how these implications can be derived from the basic principles of special relativity.

and from the perspective of reference frame B, space in A contracts, in the direction of motion, by the same factor.[3]

Relativity of simultaneity: Again suppose A and B are inertial reference frames. Then events that are simultaneous from the perspective of one reference frame will not be simultaneous from the perspective of the other reference frame.[4]

Some examples may help illustrate these implications. Let us begin with an example involving time dilation. Suppose A and B are inertial reference frames, in motion relative to each other at 180,000 k/s (i.e., .6 * c). For ease of reference, suppose Angela is in reference frame A, and Betty is in reference frame B. Suppose each has two highly accurate clocks, separated by some distance. We will refer to Angela's clocks as A_1 and A_2, and Betty's clocks as B_1 and B_2. Finally, suppose that when Angela measures the distance between her two clocks A_1 and A_2, she finds them to be 10 kilometers apart, and that when

---

[3]   The equation given here assumes that we are dealing with distances in the direction of motion, for example, that we are dealing with the distance between two objects x and y, where x and y lie along the direction of motion. For length contractions in other directions, let $\theta = 0°$ represent the direction of motion and $\theta = 90°$ represent the direction perpendicular to the direction of motion. Then distances at an angle θ between 0° and 90° shrink by a factor of

$$\frac{\sqrt{1 - \left(\frac{v}{c}\right)^2}}{\sqrt{1 - \sin^2\theta\left(\frac{v}{c}\right)^2}}.$$

[4]   More specifically: suppose x and y are two events in reference frame B, that is, from the perspective of reference frame A, x and y are moving just as reference frame B is moving. Assume also that x and y lie along the direction of motion, and that with respect to the direction of motion, x is in the front (i.e., if we draw a line between x and y, the direction of motion is in the direction from y to x). Finally, suppose that from within reference frame B, the events x and y are simultaneous. Then from reference frame A, x and y will not be simultaneous. In particular, from the perspective of A, event y will occur before x by the amount

$$\frac{\left(\frac{lv}{\sqrt{1 - \left(\frac{v}{c}\right)^2}}\right)}{c^2}$$

(In this equation, *l* is the distance, from the perspective of reference frame A, between x and y.)

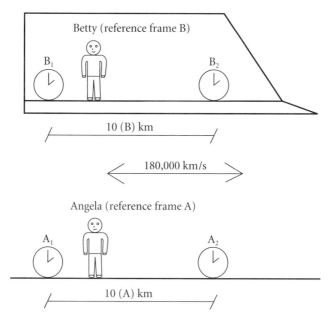

Figure 6.1   Illustration for special theory of relativity

Betty measures the distance between her two clocks B_1 and B_2, she finds them to be 10 kilometers apart. The scenario is summarized in Figure 6.1 above.

From Angela's reference frame, for every 10 minutes that elapse for her (that is, for every 10 minutes that pass on clocks A_1 and A_2), only

$$10 * \sqrt{1 - \left(\frac{v}{c}\right)^2}$$

= 8 minutes elapse in reference frame B (that is, only 8 minutes pass on clocks B_1 and B_2). That is, from Angela's perspective, Betty's clocks are moving more slowly, Betty is aging more slowly, and, in general, time is moving more slowly in reference frame B.[5]

---

[5]   It is worth noting that time dilation becomes greater as the velocity increases. That is, as the velocity decreases, the time dilation decreases. So for the sorts of velocities we experience, say involving airplanes, the time dilation is very, very small (though still measurable with modern technology). As the velocity approaches a limit of the speed of light, the time dilation factor approaches a limit of 0, and thus time moves slower and slower for higher and higher velocities.

Recall that the time dilation is symmetrical. That is, from Betty's perspective in reference frame B, her time is moving perfectly normally, and it is time in reference frame A that is moving more slowly. So, for example, from Betty's perspective, for every 10 minutes that pass for her, that is, for every 10 minutes that elapse on clocks B_1 and B_2, only

$$10 * \sqrt{1 - \left(\frac{v}{c}\right)^2}$$

= 8 minutes elapse in reference frame A (that is, only 8 minutes pass on clocks A_1 and A_2). So from Betty's perspective, it is Angela's clocks that are moving more slowly, Angela who is aging more slowly, and, in general, time in reference frame A is moving more slowly.

In short, from the perspective of Angela's reference frame, time in A is moving normally and time in B is moving slowly. But from the perspective of Betty's reference frame, time in B is moving normally and time in A is moving more slowly. In short, from the point of view of each of the reference frames, time is moving normally in that reference frame and more slowly in the other.

This leads to what is often referred to as the "twin paradox." Suppose Angela and Betty are twins, that is, they were born at the same moment. Suppose, at some point, Betty embarks on a space journey at high speeds, again, let us say 180,000 k/s, and such that Betty is moving, relative to Angela, in a straight line at uniform speed. Then while Betty is in motion relative to Angela, from Angela's reference frame, time is moving more slowly for Betty, and hence for Angela, she (Angela) is the older twin. But again recall that, so long as we are dealing with inertial reference frames, the effect on time is symmetrical, so that from Betty's reference frame, time is moving more slowly for Angela, so she (Betty) is the older twin. In short, from Angela's perspective she, Angela, is the older twin, and from Betty's perspective she, Betty, is the older twin. In other words, from their respective references frames, each of them will be older than the other.

Who is right in this case, that is, who is really the older twin? Importantly, neither is any more right or wrong than the other. To see this, recall one of the basic postulates of special relativity, the principle of relativity. Recall that Angela and Betty are in inertial reference frames, and hence according to the principle of relativity, if Angela and Betty conduct identical experiments, the results of those experiments will be identical. What this means is that there is no empirical evidence that Betty can produce to show that her point of view is the "right" point of view. Again, this is simply because any evidence Betty might produce can be exactly replicated by Angela. And likewise, there is no evidence Angela can produce to show that hers is the "right" point of view.

As the above example illustrates, one implication of special relativity is that time moves at different rates in different reference frames, and thus it is possible to have twins who are no longer the same age. And moreover, as the so-called twin paradox illustrates, it is possible to have twins where, from each twin's perspective, he or she is the older twin. And as a more general implication (this is an implication of the principle of relativity itself), there is no "right" point of view, that is, there is no privileged reference frame – there are simply different reference frames, but no one is any more "right" than any other.

Thus far, we have seen that special relativity has some surprising consequences for our usual conception of time. As noted above, relativity also has implications for our usual conceptions involving space and distances. Consider again the scenario illustrated above in Figure 6.1. Recall that when Betty measures the distance between her two clocks B_1 and B_2, she finds them to be 10 kilometers apart. But if Angela measures the distance between Betty's clocks B_1 and B_2, she will find the distance to be only

$$10 * \sqrt{1 - \left(\frac{v}{c}\right)^2}$$

= 8 meters.[6] That is, from the perspective of Angela's reference frame, distances in reference frame B have contracted.

As with the effects on time, the effects on distances are again symmetrical. Recall again that when Angela measures the distance between her two clocks A_1 and A_2, she finds them to be 10 kilometers apart. But if Betty measures the distance between A_1 and A_2, she will find those clocks to be only

$$10 * \sqrt{1 - \left(\frac{v}{c}\right)^2}$$

[6] As noted earlier, this amount of contraction assumes that we are dealing with distances that lie along the direction of motion. As described in an earlier footnote, the contraction for distances other than in the direction of motion involves a somewhat more complex equation. Hereafter we will assume, for such examples, that we are dealing with distances and events lying along the direction of motion.

= 8 meters apart. That is, from Betty's reference frame, her distances are perfectly normal, and it is the distances in Angela's reference frame that are contracting.

And here again, neither Angela nor Betty is any more "right" than the other. It is just that the distances between points, or the amount of space an object takes up, are different when measured from different reference frames. And again, no reference frame is any more right than any other.

Most people I know, when first introduced to these implications of special relativity, have the sense that this cannot be a consistent picture of space and time (I recall having the same sense, when I first came across relativity, that there must be some sort of contradiction lurking in special relativity). For example, consider again the twin paradox. How could Angela and Betty *both* be the older twin? How could Angela be older than Betty, and Betty be older than Angela? Does that not have to be contradictory?

One key to seeing how this is not genuinely contradictory is to keep in mind that there is no privileged reference frame – for example, no reference frame from which to say who is "really" older and who is "really" younger. From relativity theory, we now understand that we must always speak of motion from a point of view, that is, motion relative to a reference frame. And likewise, we now understand that we must speak of time and space relative to a reference frame. We cannot say that Angela is older than Betty, but rather, that from Angela's reference frame, Angela is older than Betty. And from Betty's reference frame, Betty is older than Angela.

There is no reference frame within which Angela and Betty are both older than the other, and that fact should alleviate some of the sense of contradiction. Another key factor in why there is no contradiction is the relativity of simultaneity. Recall that one of the implications of special relativity is that Angela and Betty will disagree on whether two events are or are not simultaneous. As it turns out, this disagreement about when events occur is central in seeing why there is no contradiction. The details of this are lengthy, such that we do not have space to go into all these details. But we can see an outline of how the relativity of simultaneity factors in.[7]

Suppose Angela and Betty both monitor some timekeeping device, say the clock in London known as Big Ben (perhaps they have arranged a video feed of Big Ben to be sent to their respective locations). And suppose that when the hands of Big Ben read midnight on New Year's Eve

---

[7] See DeWitt 2004, 213–19, for a more detailed explanation of how the relativity of simultaneity figures into this scenario.

of the current year, Angela records that, from her reference frame, she is exactly 25 years old. Angela likewise checks on Betty's age, and finds that it is less than 25 years. (For the sake of a concrete number, suppose Angela finds Betty to be exactly 24 years old.) Thus, Angela notes that from her reference frame, she is the older twin.

Another precisely equivalent way to describe Angela's situation is as follows: from Angela's reference frame, the event of Big Ben reading midnight on New Year's Eve, and the event of Angela turning 25 years old, are simultaneous events. And the event (again from Angela's reference frame) of Big Ben reading midnight, and the event of Betty turning 24 years old, are simultaneous events. And this is, from Angela's reference frame, why Angela is (again from that reference frame) the older twin.

But recall that Betty will disagree with Angela about whether events are simultaneous. From Betty's reference frame, she might agree with Angela that the event of Big Ben reading midnight on New Year's Eve is simultaneous with her (Betty) turning 24 years old. But from Betty's reference frame, the event of Big Ben reading midnight, and the event of Angela turning 25, are *not* simultaneous. In fact, from Betty's reference point, the event of Big Ben reading midnight on New Year's Eve will be simultaneous with Angela being less than 24 years old (I'll omit the exact figure for Angela's age). So in other words, the events that are simultaneous from Betty's reference frame are her (Betty) being 24 when Big Ben strikes midnight, and Angela being less than 24 when Big Ben strikes midnight. So from her reference frame, when Big Ben strikes midnight on New Year's Eve, she (Betty) is the older twin.

Again, for simplicity I am omitting the mathematical details of this example. But suffice it to say that, when one applies all the implications of relativity to this example, including time dilation, length contraction, and the relativity of simultaneity, what emerges is an entirely consistent account of space, time, and simultaneity. It is, though, an account that runs counter to long-held intuitions about the nature of space, time, and simultaneity.

2.3   The basic principles of the general theory of relativity

In 1916, Einstein published the general theory of relativity.[8] As we saw above, the special theory of relativity applies so long as special circumstances are

---

[8]   The original source is Einstein 1916. Einstein 1920 is a less technical account by Einstein of the special and general theories of relativity.

met, in particular, so long as the situation involves inertial reference frames. In contrast, the general theory of relativity, as the name suggests, is a general theory, in that it applies to all circumstances.

As with special relativity, general relativity is based on two fundamental principles. Let us begin by describing these two principles.

The first principle is what Einstein often referred to as the "general principle of relativity," but which is now more often termed the *principle of general covariance*. The basic idea is reasonably straightforward. Recall that in the special theory of relativity, the principle of relativity was the principle that the laws of physics are the same in all inertial reference frames. That is, so long as certain special circumstances hold, for example, so long as we are dealing with reference frames that are moving relative to one another in a straight line with uniform speed, then the laws of physics will be the same in both reference frames.

The principle of general covariance removes the special circumstances of applying only to inertial reference frames. As Einstein phrased the principle in the 1916 paper, the "laws of physics must be of such a nature that they apply to systems of reference in any kind of motion" (Einstein 1952, 113).[9] In other words, the laws of nature are the same in all reference frames.

The second basic principle on which the general theory of relativity is based is usually termed the *principle of equivalence*. The principle of equivalence says, roughly, that effects due to gravity and effects due to acceleration are indistinguishable. To see the basic idea here, consider an example (one that Einstein himself liked to use to illustrate this principle).

Suppose we again use Angela and Betty as examples, and suppose that Angela and Betty are both in enclosed, windowless rooms, say about the size of an elevator. Suppose Angela, in her room, is on the surface of the earth. Betty's room, in contrast, is in deep space, far enough from any planets or stars so as to experience no effects from such bodies. Moreover, we will imagine that Betty, in her room, is being accelerated "up" (that is, in the direction of a line running from the floor to the ceiling), at 9.8 meters per second squared (i.e., the acceleration due to gravity on the earth's surface).

Now consider what sorts of observational effects Angela and Betty will observe. Angela notices that she feels a "pull" in the direction of her floor, and Betty notices that she feels the exact same sort of pull toward the floor of her room. Angela jumps, and notices that she quickly falls back toward

---

[9]   This is an English translation of the original 1916 paper.

her floor. Betty notices the exact same effect when she jumps. Angela holds an object out at arm's length, drops it, and notices that it moves toward the floor of her room, accelerating toward the floor at a rate of 9.8 meters per second squared. Betty likewise drops a similar object, and notices that it accelerates toward the floor of her room at 9.8 meters per second squared.

Traditionally, we would describe the effects Angela observes as resulting from Angela being in the gravitational field of the earth; and that gravitational field explains the pull she feels, why she falls back down when she jumps, why dropped objects accelerate downwards as they do, and so on. Betty, in contrast, is not in the presence of any gravitational fields, and we would traditionally describe the effects she experiences as being the result of her being accelerated upwards.

But it seems a striking coincidence that the effects of a gravitational field, on the one hand, and the effects of acceleration on the other, would be exactly the same (and indeed, on Newtonian physics, these sorts of identical effects had to be treated as a coincidence). Einstein's principle of equivalence in essence rejects the view that these identical effects are due to different causes, and instead treats situations we would traditionally describe as involving a gravitational field, and situations we would traditionally describe as involving acceleration, as indistinguishable.

At bottom, these two principles – the principle of general covariance and the principle of equivalence – are the foundational principles of general relativity. After articulating these principles in the 1916 paper, Einstein's main task was to provide the key equations that would satisfy the requirements of these principles. These equations (usually referred to as the Einstein field equations) are quite complex, but the basic idea is that solutions to these equations indicate how space, time, and matter influence one another, and these equations are the mathematical core of general relativity. In the next section, we will take an overview of some of the implications of general relativity.

2.4 Implications of the general theory of relativity

In the earlier sections on special relativity, we saw that special relativity has interesting consequences for our traditional views of space and time. The implications of general relativity for space and time are similar to those of special relativity (which is not surprising, given that the conditions under which special relativity applies are a subset of the more general conditions – that is, any conditions – under which general relativity applies).

Consider again Angela and Betty, in reference frames A and B respectively, where A and B are moving relative to one another. Then the same effects on space and time will follow from general relativity as from special relativity. That is, from Angela's perspective, time is moving more slowly, and distances have contracted, in Betty's reference frame; whereas from Betty's reference frame, time is moving more slowly for Angela, and Angela's distances have contracted.

In addition to the effect of motion on space and time, in general relativity gravitational effects also influence space and time. Note that "gravitational effects" will include both effects due to being in the presence of a large body, as well as effects due to acceleration (recall the principle of equivalence, according to which these two situations – being in the presence of a large body versus being in an accelerated reference frame – are indistinguishable).

So, for example, suppose we revisit the twin paradox discussed above. Recall that in the earlier scenario, Angela and Betty (in reference frames A and B respectively) are moving relative to one another in a straight line with uniform speed. In such a situation involving inertial reference frames, we saw that the effects of space and time are symmetrical, so that, for example, from their respective reference frames, Angela and Betty each consider themselves the older twin.

Now suppose we remove the restriction that A and B be inertial frames, and we envision the situation as involving the two rejoining each other and comparing ages. To make the situation a bit more concrete (and the twin paradox is often described using the scenario below), suppose we consider a trip to the nearest star system, which is the Alpha Centauri system, roughly 4.5 light years from the earth. Suppose at the start of the trip Angela and Betty are both 20 years old, and that Betty travels to, and returns from, the Alpha Centauri system, traveling at an average speed of nine-tenths the speed of light.

Notice that the earth and spacecraft will be two different reference frames, moving relative to one another, and also experiencing different gravitational effects relative to one another. For example, Betty, on the spacecraft, will experience substantial G forces not experienced by Angela. ("G forces" refer to the sorts of effects we feel, for example, when accelerating or decelerating rapidly in a car, that is, we feel the sense of being pushed back into our seats, or of being thrown forward. Since Betty's velocity is so much greater than that of a car, the G forces she experiences will be much more substantial than any we could experience in a car.) Betty will experience such G forces as she accelerates away from the earth; then as she nears the star system, she will again experience substantial G forces as she decelerates. She will again experience such forces as she accelerates back toward the earth, and as she decelerates as she nears the earth.

Notice that in this scenario, Angela and Betty will both agree that Betty has experienced gravitational forces (which we would usually describe as resulting from her acceleration and deceleration) that Angela has not experienced. And on general relativity, such forces (as well as motion) affect the passage of time. In particular, in this scenario, Angela and Betty will both agree that the forces experienced by Betty will have resulted in less time passing for Betty, so in this case, both will agree that Betty is the younger twin.

I will omit the calculations, but the bottom line is that, from Angela's reference frame (on earth), Betty's trip to Alpha Centauri and back will have taken approximately 10 years. So upon Betty's return, Angela is 30 years old. In contrast, time (and distances) will not be the same from Betty's reference frame, and only 5 years will have elapsed in that reference frame during the trip to Alpha Centauri and back. So upon her return, whereas Angela is 30 years old, Betty is only 25 years old.

Again, in special relativity, which is restricted to inertial reference frames, the effects on space and time are symmetrical. But in general relativity, in cases where the reference frames are not inertial frames, one still finds that time, space, and simultaneity are affected, but the effects are no longer symmetrical. And in the scenario above, because both Angela and Betty agree that Betty has experienced forces that Angela does not experience, they will both agree that Betty is the younger twin.

In short, we find in general relativity the same sorts of surprising effects on space, time, and simultaneity as we found in special relativity, with the difference that the effects are not necessarily symmetrical. That is, how much time passes, how much space an object occupies and what the distance is between points, and whether events are or are not simultaneous, varies from one reference frame to another.

Another curious consequence of general relativity has to do with the curvature of spacetime. If you have not encountered the notion before, "spacetime" can sound like a complex subject, but the basic idea is quite straightforward. If we want to specify a location in space, a common way to do so is by specifying the location using three coordinates, one each for the x, y, and z axes (this assumes that we have specified an origin point, that is, the location of the (0,0,0) coordinate, but here we will take that for granted). Suppose in addition to the usual three spatial coordinates, we also want to specify a time, so that, for example, we can specify that an event took place at such and such a location at such and such a time. An easy way to do this is to take the usual three spatial coordinates, and add an additional coordinate representing time. We would then be dealing with a 4-tuple (x, y, z, t), where x, y, and z represent the usual spatial coordinates, and t represents time. At bottom, that is all there is to the notion of spacetime; that is, it is simply a way to specify both spatial and temporal coordinates.

So, for example, instead of thinking of an object as moving through space, we can think of it as moving through a system of coordinates which track both locations in space as well as locations in time, that is, spacetime. And one of the more interesting implications of general relativity is that a large body, such as the sun, will cause a curvature in the region of spacetime around it.

This notion of the curvature of spacetime has an interesting consequence for our traditional view of gravity. Since Newton (1643–1727), we have tended to view gravity as a mutually attractive force between bodies. And it is this mutually attractive force, say between Mars and the sun, that we have taken as key to the understanding of the motion of Mars about the sun.

But in general relativity, there is no notion of a mutually attractive gravitational force. Instead, a planet such as Mars moves in a straight line (that is, the shortest path between two points). But this is a shortest path through a spacetime that is curved by the presence of the sun. And in particular, the curvature of spacetime is such that Mars, moving on a straight line through curved spacetime, moves on what appears to be an ellipse about the sun. In short, one of the broadest consequences of general relativity is that it provides an alternative account to the usual Newtonian concept of gravity. And (more on this below), the account provided by general relativity is generally taken to be a better account than the usual Newtonian account. In short, although most of us are raised with the Newtonian account of gravity as a mutually attractive force between bodies, our current best theory replaces this Newtonian account with an account not involving attractive forces, but rather, that involves objects (for example, planets) moving in straight lines through a curved spacetime.

This is not by any means an exhaustive catalog of the implications of general relativity. But the above does provide a sampling of some of the implications relativity has for some of our most common views, for example on space, time, simultaneity, and gravity. Of course, it is one thing to have a theory that has unusual implications, and another thing to have empirical support for that theory. The bottom line is that the empirical support for relativity is quite strong, and in the next section, we take a brief look at some of this empirical support.

2.5   Empirical tests of relativity

In this brief section, we take a quick look at some of the empirical work that supports the general theory of relativity. The bottom line is that the implications of general relativity have been well established empirically, and the main goal of this brief section is to highlight some of these empirical results.

One type of substantial support for general relativity comes from a source you may be familiar with, and perhaps use. Consider a GPS (global position system) unit, which is the heart of the sorts of navigational systems that are becoming increasingly popular in cars. The success of such GPS units provides substantial empirical confirmation for general relativity. At heart, these GPS navigational units depend upon the tenets of general relativity. The basic idea is that a GPS unit triangulates your position based on signals from numerous orbiting GPS satellites. The success of the GPS system hinges crucially on extremely accurate determinations of times and distances, and in a GPS system, the determination of times and distances is based on the equations of general relativity. In this sense, then, the success of GPS systems, such as those in navigational systems in cars, provides daily confirming evidence for general relativity.

There are also a number of more traditional tests of relativity, and here we will mention just a few.

Given advances over the past 100 years in timekeeping devices, one of the easiest implications of relativity to test is the predicted relativistic effects on time. One can, for example, take two identical, highly accurate atomic clocks, and put one on a plane while keeping one on the ground. From the average speed of the plane together with the duration of the flight, it is reasonably easy to predict, based on general relativity, what the differences should be between the amount of time that passes on the clock on the ground and the clock on the plane. The first such tests of general relativity were done over 50 years ago, and in the past half-century, such tests have continued to confirm the predictions of general relativity.

The earliest, and perhaps most famous, test of general relativity involves the bending of starlight. As mentioned above, one of the implications of general relativity is that spacetime will be curved by the presence of a massive body such as the sun. As a result, starlight passing near the sun should appear, from our perspective on earth, to curve. Starlight near the sun usually cannot be seen (because it is washed out by the brightness of the sun), but solar eclipses provide opportunities to observe starlight near the sun. And not long after the general theory of relativity was published, a convenient solar eclipse provided a nice opportunity to put the theory to the test. The details of the test are somewhat complex,[10] but the consensus of the scientific community was that the observations of starlight, in particular the bending of starlight passing near the sun, were in accordance with the predictions of general relativity, thereby providing early confirming evidence for the theory.

There are countless other examples of confirming evidence, but I will close this section with one more. At the end of the 1916 paper, Einstein shows that general relativity does a nice job explaining what had become a bit of an astronomical puzzle. Since the 1600s, it has been recognized that planets orbit in ellipses. In the 1800s, astronomers had begun to notice that the perihelion of Mercury – that is, the point in Mercury's orbit closest to the sun – moved slightly during each orbit. The overall effect is that the perihelion of Mercury (and other planets as well, as it turns out) slowly moves about the sun. The elliptical orbit of planets is nicely explained by Newtonian physics, but the movement of the perihelion of planets is not. But in the 1916 paper, Einstein shows that the movement of the perihelion of a planet is to be expected on general relativity, and he moreover provides the calculations, based on general relativity, showing how much the perihelion of Mercury would be expected to move each year. The prediction was again in close keeping with the observed movement of the perihelion of Mercury, and this too provided good confirming evidence for general relativity.

As mentioned, there are countless other empirical examples that provide support for general relativity. In short, there is not much question that, as counter-intuitive as many of these seem to be, the general relativistic consequences for space, time, simultaneity, the curvature of spacetime and the like, have been well confirmed.

---

[10] See Laymon 1984 for a discussion of some of the complexities involved.

3. Quantum Theory

Another twentieth-century development with surprising implications is quantum theory. Roughly, this is a branch of physics that is primarily used for situations involving atomic-or-smaller levels. For example, if your main line of work is particle physics, say investigating the behavior of sub-atomic particles such as electrons, then quantum theory will be a major tool in your toolbox.

Quantum theory covers a lot of territory, certainly too much for us to cover in a single essay. What we can do, however, is get a sense of the sorts of questions quantum theory raises concerning our views of the universe. We will first take a look at some basic experimental results that raise deeply puzzling and difficult questions about the nature of reality. These results, and some of the philosophical questions raised by them, are the subject of §3.1. The results discussed in §3.1 are largely what led to the need for a new theory, where that need was filled by the development of quantum theory. In §3.2, then, we take a brief look at the historical development of quantum theory.

As noted at the outset of this essay, we have always looked to our basic sciences for insight into what the world is like, that is, for insight into the philosophical question of what sort of universe we inhabit. Whereas the experiments discussed in §3.1 raise some difficult questions about the nature of reality, in §3.3 we look at some more recent results that shed light on this reality question. In particular, the topics discussed in that section, notably Bell's theorem and the Aspect experiments, show that a broad class of answers to this reality question are no longer an option. In particular, in §3.3 we will see how a quite standard view of reality, one dating back to at least the ancient Greeks, is not compatible with recent experimental results from quantum theory.

3.1   Some puzzling experimental results

In the 1800s, and continuing into the early 1900s, physicists encountered a number of experimental phenomena that did not fit cleanly into the existing theoretical framework. It is largely such phenomena that led to the development of quantum theory. In this section, we will look at a handful of such experiments, and try to get a sense of some of the puzzling aspects of these experiments.[11]

We will begin with one of these classic experimental results, as well as one of the earliest, namely, what has come to be called the "two-slit" experiment.[12]

Suppose we ask the question of whether entities such as electrons are particles or waves. At first glance, since waves and particles have different sorts of experimental effects, it seems that it should be relatively easy to answer the question. For example, suppose we set up an experimental arrangement involving a source of electrons (an "electron gun"), and shoot electrons toward a barrier with two slits. On the side of the barrier away from the electron gun, we will set up a screen capable of recording any electrons that hit it. Depending on whether electrons are particles or waves, we should get two quite separate experimental results.

First, if electrons are discrete particles, then we would expect the barrier to block all the electrons except those that pass through the slits, and as a result, we would expect the recording screen to record a "particle effect" pattern of electrons, with electron hits corresponding to the regions of the screen in line with the slits. Figure 6.2 illustrates this.

In contrast, if electrons are waves, then the two slits should have the effect of splitting each wave into two waves. These two waves would then continue on toward the recording screen, overlapping with one another as they approach the screen. When waves overlap in this way, they produce an interference pattern, which in this case would result in alternating bands of dark and light on the screen. The dark bands will correspond to areas where the waves interacted in a "constructive" manner, much like the way some waves at the beach will add together into a larger wave, and the light bands will correspond to the areas where the waves interacted with one another in a "destructive" way, again much like the way some waves at the beach will interact with one another and effectively cancel each other out. Such "wave effect" patterns of interference are well known and well

---

[11]   I should note that I am presenting a sort of "idealized" look at some of these experimental results, which is geared primarily toward illuminating some of their puzzling aspects. An alternative approach, more closely tied to the actual historical experiments that led to quantum theory, would be more historically informative but would not serve to illuminate these puzzling aspects as well. Baggott 2004 and Cushing 1998 are good sources for a more historical account.

[12]   Historically, this experiment dates back to Thomas Young in the first decade of the 1800s. Young originally performed the experiment using a beam of light, though it was later found that the same sort of results could be found with sub-atomic particles such as electrons. See Baggott 2004 for a good overview of these early developments.

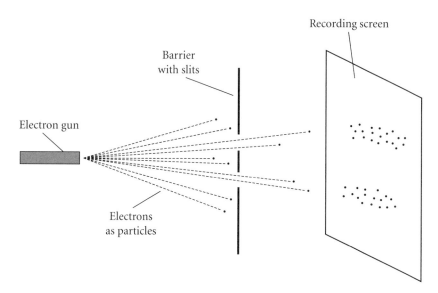

Figure 6.2   Electrons as particles

studied within physics, and if electrons are waves, we would expect this sort of wave effect. Again, Figure 6.3 might help illustrate this.

When this basic, two-slit experiment is carried out, the recording screen registers what above I was calling a "wave effect," that is, an interference pattern characteristic of waves. This result, then, seems to suggest that electrons are waves. By itself, this result is not surprising. Wave phenomena are common, and there are branches of physics well equipped to handle wave phenomena.

What is very puzzling, though, is what happens when we consider a slightly modified version of the two-slit experiment. For example, suppose we take the two-slit experiment described above, and behind each slit we place a passive electron detector. (This would be a device that will detect the passage of an electron through its respective slit, but in a way that should not interfere with the electron. This would be analogous to the way you might passively detect people that pass by your window, that is, you can record their presence without interfering with them.) The setup would be as illustrated in Figure 6.4.

Given that the electron detectors seem to be passive detectors, we would expect the same result as in the basic two-slit experiment, that is, we would expect a wave effect. But when we run this experiment, we get a clear

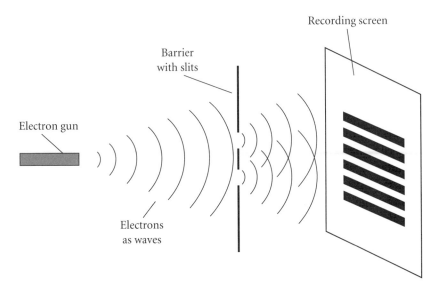

Figure 6.3   Electrons as waves

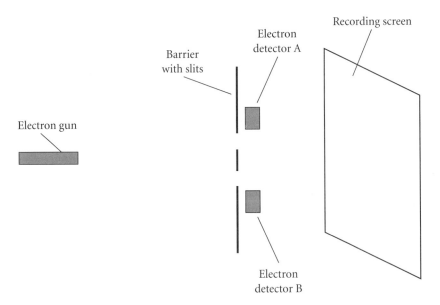

Figure 6.4   Two-slit experiment with detectors

particle effect. That is, we get a piling-up pattern on the recording screen, exactly as if the electrons were particles, with each particle having passed through one or the other slit.

Moreover, we can switch at will between the wave effect and particle effect simply by turning the electron detectors on and off. That is, so long as the electron detectors are turned off, we see a wave effect on the screen. When we flip the switch and turn the detectors on, the wave effect is immediately replaced with a particle effect. Flip the switch off, and we are back to the wave effect. And so on. In addition, if we carefully check the results of the electron detectors, we find that they never detect an electron passing through the slits simultaneously, but rather, it is as if each electron passes through one or the other of the slits, but never both simultaneously (note that, if electrons are waves, we would expect the wave to pass through both slits simultaneously).

What I have described so far are simply experimental results, that is, if we set up experiments as described above, we will get the results described. What if we go a bit beyond the experiments, and consider a philosophical question concerning reality? In particular, what could really be going on in such experiments? What sort of reality could produce these sorts of results? On the one hand, the wave effect we find in the basic two-slit experiment seems like it could only be produced if the electron is really a wave, and moreover, a wave that passes through both slits simultaneously. In contrast, the particle effect we find when the detectors are turned on, together with the fact that the detectors never record an electron passing through both slits simultaneously, suggests that electrons are not waves at all, but particles.

To push this reality problem a bit further, recall again that the wave effect we find in the basic two-slit arrangement could seemingly only be produced if electrons pass through both slits simultaneously. And the particle effect we see when the detectors are turned on, and the behavior of the detectors, could seemingly only be produced if electrons are passing through one slit or the other but never both slits simultaneously. But how could whether the detectors are turned on or off affect whether an electron passes through both slits simultaneously, or through only one slit? Speaking anthropomorphically for a moment, how could an electron "know" whether the detectors are on or off?

There are no agreed-upon answers to these reality questions.[13] The experimental outcomes of these and similar experiments are unequivocal,

---

[13] See Baggott 2004, Herbert 1985, or DeWitt 2004 for an overview of some of the options that have been proposed over the years in response to these sorts of reality questions.

but the more philosophical issue of what is "really" going on, of what sort of reality could produce these experimental facts, remains deeply puzzling. These and other puzzling results helped lead to the development of quantum theory. Eventually quantum theory would be developed in a form that was able to "handle" these sorts of experimental results, in the sense that the theory would be able to make very accurate predictions about what result to expect in such experiments. But quantum theory itself, although probably the most successful theory we have ever had in terms of making accurate predictions, does not address these sorts of reality questions.

This is a good point to bring up a topic introduced in Chapter 2 of this book, namely, the issue of instrumentalist and realist attitudes towards theories.[14] Again, an instrumentalist is one who looks to a theory primarily to make accurate predictions, without concern for whether the theory reflects the way things "really" are. One who takes a realist approach, on the other hand, wants a theory not only to make accurate predictions, but also to provide a picture or model of reality.

It is probably safe to say that the majority of physicists working with quantum theory tend to take an instrumentalist attitude toward the theory, that is, they look to the theory to provide accurate predictions (which it does), without worrying about the sorts of reality questions discussed above. And it is certainly true that the majority of texts for college-level physics classes on quantum theory take an instrumentalist approach; I have yet to find an exception. This is a perfectly reasonable and understandable attitude for a working physicist, or a text on quantum theory, to take. For example, a college-level course in quantum theory has to focus on getting the student to acquire the necessary quantum tools, and there is barely time for that in a single semester, much less time to worry about these more philosophical questions.

But as noted in the introduction to this essay, we have always looked to our basic sciences to shed light on our more philosophical views on what reality is like. What is different about recent results involving quantum theory, as illustrated by the two-slit experiments above, is that these results do not seem to allow for any sort of a "normal" picture of reality. Moreover, some more recent results have shed additional light on the reality question, largely by ruling out a large class of possible models of reality. Before turning to those more recent results, which will be the subject of §3.3 below, we will take a quick overview of the historical development of quantum theory.

---

[14]   See Chapter 2, pp. 21–4.

3.2   A brief overview of the development of quantum theory

Quantum theory was developed in the first several decades of the twentieth century, and as suggested earlier, largely grew out of attempts to handle unexpected experimental results. In the early years of the twentieth century, developments in quantum theory tended to be piecemeal attempts to solve, or at least make progress on, individual problems, rather than being approaches that arose out of any unified theoretical framework. This early "piecemeal" approach, from roughly 1900 to 1930, is often referred to as the "old" quantum theory. This old quantum theory is in contrast to the "new," or mature, quantum theory that arose in the late 1920s and early 1930s, which was an account that provided a more unified approach capable of handling a wide range of quantum phenomena.

As an illustration of this early piecemeal approach, consider some early attempts to handle one of the major outstanding problems at the turn of the century, which was a problem involving black body radiation. By way of background, a "black body" is an idealized object that absorbs all radiation directed toward it (such a body would absorb all light, and so appear black). A heated black body will also emit radiation, much as an approximate black body, say a coil burner on an electrical stove, will emit radiation when heated (e.g., a heated coil burner will emit radiation in the form of light, glowing red-hot as it heats up).

Although a black body is an idealized object, it is possible to construct devices that will emit the same pattern of radiation as a heated black body. In this way physicists could produce the pattern of radiation emitted by such a device, and compare it with the pattern predicted by the existing physics. The problem was that, for certain areas of the spectrum, the predicted pattern, and the actual pattern, were way, way off.[15] In short, there was something badly amiss with the existing understanding of radiation.

In 1900, the physicist Max Planck (1858–1947) proposed a modification to the usual account of radiation. Predictions concerning black body radiation, based on Planck's modification, now matched up exactly with the experimental data. However, Planck's modification arose from trying to make the predictions fit the data, rather than arising out of any broader

---

[15]   In particular, the discrepancy between the actual pattern of radiation and the predicted pattern came in the ultraviolet region of the spectrum, and so the problem was often referred to as the "ultraviolet catastrophe."

theoretical framework. Basically, Planck was doing the best he could, trying to get the theory to match the data, by whatever means possible.

Although this is only one example, it is typical of those early years. Physicists tried various approaches to various problems, with nothing unifying the approaches. One approach might help make progress on the problem of black body radiation, another seemingly unrelated approach might help with problems of understanding the structure of the atom, yet another for issues involving radioactive decay, yet another for phenomena involving newly discovered x-rays, and so on. In short, the period was marked by physicists trying whatever seemed to work, for whatever problem they were focused on, with no broader theoretical structure unifying the approaches.

As mentioned, this piecemeal approach would continue until about the late 1920s. During this time, Werner Heisenberg (1909–76) and Erwin Schrodinger (1887–1961) independently arrived at approaches that would eventually provide the mathematical foundations of the mature quantum theory.[16] With the new mathematics, Schrodinger was able to show that some of the early piecemeal accounts, proposed without any theoretical foundation and proposed mainly because they made the theory fit the data, were consequences of the new mathematics.

To understand this better, consider again the discussion above of Einstein and special relativity. As described earlier, the unusual implications for space, time, and simultaneity can be derived mathematically from the basic principle of relativity. In this way, those basic principles provide the theoretical foundation for the relativistic account of space, time, and simultaneity. In a roughly similar way, now Schrodinger is able to show that some of the earlier piecemeal approaches to problems in quantum theory can be seen as mathematical consequences of his mathematics. In this way, for the first time there is a mathematical foundation for the earlier piecemeal approaches.

By the mid-1930s, the mathematical foundations of quantum theory had been developed in essentially the same way as they exist today. For example, the mathematics one finds at the core of quantum theory in modern classes on the subject is, for all practical purposes, the same mathematics as that developed in the late 1920s and early 1930s.

---

[16] Heisenberg's and Schrodinger's accounts appeared at first to be different, but turned out to be essentially mathematically equivalent. Schrodinger's approach was of a sort more familiar to most physicists, and partly because of this his approach became the more widely used.

The details of the mathematics are beyond the scope of this essay. But in outline, and as indicated, quantum theory is, as are most basic theories in physics, a mathematically based theory. Moreover, quantum theory is used in essentially the same way other theories in physics are used. In particular, the mathematics of quantum theory allows one to make predictions about the outcomes of measurements one might perform on a particular system (for example, whether one would expect a wave or particle effect in a two-slit experiment, or the position or momentum of an electron, or the polarization attributes of light that has passed through a certain type of tilter, and so on). Also as with typical theories in physics, quantum theory allows one to make predictions about how a system will evolve over time. So in this sense – primarily that quantum theory is a mathematically based theory used to make predictions about the outcome of experiments and about how a system will evolve over time – quantum theory is not appreciably different from other theories in physics.

Yet from the start, the mathematics of quantum theory developed in the 1920s and 1930s did not seem to fit well with some of our most widely held views on what, deep down, the universe is like. Einstein, for example, objected all his life to quantum theory.[17] Einstein did not object to quantum theory because he had any reservations about the predictive success of the mathematics of quantum theory (in terms of making accurate predictions, quantum theory is probably the most successful theory we have ever had). Rather, Einstein objected to quantum theory because he did not think the picture of reality suggested by the theory could possibly be accurate.[18]

At bottom, Einstein's concerns about quantum theory mainly concerned a substantial tension between quantum theory and our usual intuitions about the sort of universe we inhabit. Although these concerns are not the only ones raised by quantum theory,[19] in what remains of this essay we will focus on these concerns. As it turns out, in the early 1960s the physicist John Bell was able to clarify where precisely this tension between quantum theory, on the one hand, and our usual intuitions about reality, on the other,

[17]   Einstein, Podolsky, and Rosen 1935 is a classic paper on these matters.
[18]   See Chapter 2 of this book for more on the differences between realist and instrumentalist approaches to theories.
[19]   One very puzzling issue about quantum theory is what is termed the "measurement problem." Because of space limitations, and the fact that it is difficult if not impossible to accurately explain the measurement problem in a brief space, we are not touching upon it at all in this essay. Baggott 2004 and Herbert 1985 are good sources on this subject. A more detailed discussion, more closely tied to the mathematics of quantum theory, can be found in Hughes 1989.

resided. This aspect of Bell's work is now commonly referred to as *Bell's theorem* or *Bell's inequality*. The main focus of the next section will be to provide an overview of Bell's theorem, together with some relevant experimental results that have substantial implications for our usual views on the nature of the universe.

3.3   Bell's theorem and the Aspect experiments

John Bell (1928–1990) was a physicist with a long-held interest in the broader implications of quantum theory. One particular interest of Bell's involved the fact that, from the early days of quantum theory, the theory seemed to be at odds with the broadly held view that we live in a "local" universe. Bell's most widely known result, Bell's theorem, focuses on the locality issue, and our first preliminary task will be to clarify what is meant by describing the universe as local.

From the beginnings of western science with the ancient Greeks, we have been convinced that one event or object can only affect another event or object if there is some sort of connection or communication between them. And at bottom, this is what is meant by saying the universe is "local," that is, one thing can only influence something in its local region of the universe, or alternatively, one thing cannot influence something with which it has no contact or communication. To use a phrase that dates back to the ancient Greeks, we have always been convinced that we live in a universe in which there is no "action at a distance." This view, that the universe is a local universe, is often referred to as the "locality assumption."

An example might help to clarify. Suppose you dial the number of a friend on your cell phone, and within a few seconds, your friend's cell phone rings. In this scenario, the chain of connections is rather complex, but the chain of events exists and is reasonably well understood. In broad outlines, your pressing the buttons on your phone causes certain electrical changes within your phone. Some of these changes result in an electromagnetic signal, complex but well understood, being sent by your phone. That signal travels at the speed of light from your phone to a transmission tower, and that transmission tower in turn passes on other electromagnetic signals, until eventually the signals reach your friend's phone, where they in turn cause changes within your friend's phone resulting in your friend's phone ringing. Again, the chain of events is complex, but in each case, one event influences only events in which there is some sort of connection between them.

This example illustrates the way we typically think the universe works. As another example, one which we will return to later, suppose we are in a room, about the size of a large laboratory. In front of us is a device with a movable dial, along with something that looks like a miniature stoplight, with red, yellow, and green lights. We begin to tinker with the dial, moving it to various positions, and as we do so we notice that our moving the dial affects whether the red, yellow, or green light on our stoplight is lit.

Now we notice that on the far side of the room is a similar looking stoplight, again with red, yellow, and green lights. And we notice that as we tinker with the dial on our stoplight, thereby affecting what light is lit on our stoplight, the lit light on the other stoplight is likewise influenced. In particular, we notice there is a very strong correlation between what light is lit on our stoplight, and what light is lit on the other stoplight. We observe this correlation for some time, until there is no doubt about it: when we move the dial on our device, it influences not only what light is lit on our stoplight but also what light is lit on the stoplight on the far side of the room.

In such a scenario, we have strong intuitions that there must be some sort of connection or communication between our device and the stoplight on the far side of the room. We might suspect there are wires under the floor connecting our device with the other stoplight, or we might suspect there is a wireless signal (maybe of the sort involved in the cell phone example above) that allows communication from our device to the far stoplight. But at any rate, we tend to be convinced that there must be some sort of communication or connection between our device and the far stoplight. And the reason for our conviction is just the locality assumption: we are convinced that what happens at one location (our device) cannot influence what happens at a distant location (the stoplight on the far side of the room) unless there is some sort of communication or connection between the two.

Again, we have been convinced, since at least the ancient Greeks, that this is the sort of universe we inhabit. That is, we have been convinced that we inhabit a universe in which the locality assumption holds.

And it is here that Bell's theorem comes in. Bell was able to show that the standard version of quantum theory (for example, the version taught in almost every university and the standard version used by physicists thousands of times a day) is incompatible with the locality assumption.[20]

---

[20] Bell 1964 is the original source. Other discussions by Bell on these matters can be found in Bell 1988. See DeWitt 2004, 283–98, for a non-technical exposition of Bell's theorem.

In particular, Bell's theorem shows that there are possible experimental scenarios in which predictions based on the locality assumption, and predictions based on quantum theory, contradict. In other words, Bell showed that quantum theory and the locality assumption cannot both be correct.

At the time Bell published the theorem in the 1960s, it was not possible, for technical reasons, actually to perform an experiment of the sort just outlined. That is, we can think of Bell's theorem as pointing to a design of an experiment that would show, between the locality assumption and quantum theory, that one of them was mistaken. But the experiment itself could not be carried out at the time.

However, over the next several decades, physicists found increasingly sophisticated ways to carry out the sort of experiment just outlined, that is, a Bell-type experiment that would indicate which of the two, the locality assumption or quantum theory, was mistaken. One of the most important sets of such experiments was carried out by Alain Aspect in the early 1980s, and when I refer to the "Aspect experiments," I will have in mind this set of experiments.

In outline, Aspect set up a series of experiments that, conceptually at least, are similar to the sort of setup described above involving the stoplights. Bell, of course, did not use miniature stoplights. Rather, his experimental setup involved devices that recorded the spin states of electrons passing through devices that would register those spin states. The indications of the various spin states can be thought of as roughly analogous to the lighting of the red, yellow, or green light in the stoplight example. This device also had the equivalent of a dial, and in a way roughly analogous to the way that moving the dial influenced which of the stoplight's lights lit, changing the dial in the Aspect experiment influenced which spin state was recorded on one device.

Likewise, in Aspect's experiment, there was a similar device on the far side of the room, which recorded the spin states of electrons passing through it. A key part of this setup was that Aspect was able to assure that there was no communication or connection between the two devices.[21] So if the locality assumption is correct, moving the dial on the one device could not influence the readings on the other device.

Yet in the Aspect experiments, changing the dial on the one device did influence the readings on the other device, even though there was no

---

[21]   The details of Aspect's experiments can be found in Baggott 1992 and Baggott 2004. (This latter work is a substantially expanded and modified version of the former.)

possibility of any communication or connection between the two devices. So in the Aspect experiments, contrary to the locality assumption, events at the one location did influence events at the other location.

Since the original Aspect experiments in the early 1980s, the results have been replicated by a number of laboratories, using a number of different experimental arrangements. The results of these experiments are robust, in the sense that as in the original Aspect experiments, the locality assumption is consistently seen to be mistaken.

The bottom line is that, as a matter of empirical fact, there is no longer much question that we live in a universe in which the locality assumption does not hold. Again, this result is directly at odds with a belief we have held deeply since at least the ancient Greeks. It does not seem possible for reality to be this way, yet it is – we live in a universe in which events at one location can influence events at another location, even though there is no sort of connection or communication between the two locations.

No one has any idea of how such influence is possible, only that it does in fact occur. The discovery that we live in such a universe is one of the most puzzling results of modern science.

4. Conclusion

As noted at the beginning of this essay, we have always looked to our sciences to inform our more broadly philosophical views as to the nature of the universe. And the branch of science we now refer to as "physics" has, for most of our history, been considered in some sense to be the most fundamental of the sciences, and thus I think we have tended to look predominantly to physics for insights into these more broadly philosophical questions as to the nature of reality.

Part of the reason for this view of physics as fundamental stems, I think, from the fact that we have always paid a good deal of attention to the question of what the fundamental constituents of the universe are. This question – what are the fundamental constituents of the universe – was one of the first questions addressed by the earliest of the ancient Greek philosopher/scientists. And since those first philosopher/scientists of the fifth and sixth century BCE, the question has remained a central one. Physics is the field of science that most directly addresses this question, and this is part of the reason why we tend to view physics as fundamental.

In addition, we have, since about the time of Newton in the late 1600s, come to view the physical sciences in a unified way, with physics investigating phenomena at the most basic level (for example, quantum theory investigating phenomena primarily at the sub-atomic level). We tend to view chemistry as investigating phenomena at a somewhat higher level, at the level of entities, for example, atoms and elements, composed out of the more basic entities investigated by branches of physics such as quantum physics. And we likewise tend to view biology as investigating phenomena at a yet higher level, yet still such that we view biological entities as fundamentally composed of the same entities investigated at lower levels.

In general, we tend to view the physical sciences as unified in the sense of investigating the same world, albeit at different levels. And since physics is typically viewed as the branch of science investigating the most basic level, it is perhaps not surprising that we have tended to look especially closely at the implications physics has for our philosophical questions as to what sort of universe we inhabit.

And within physics, the development of relativity theory and quantum theory has turned out to have a substantial impact on our views on the nature of the universe. However, I would not want to leave the impression that these are the only areas of physics with philosophical implications,[22] nor would I want to leave the impression that the philosophical issues discussed above are the only philosophical issues involving relativity and quantum theory.[23] But the implications of physics for our views on the nature of the universe were a central focus of this essay. And that is much of the reason why this essay has focused on these two branches of physics.

As we saw above, relativity and quantum theory both have substantial implications for some of our more broadly philosophical questions. Relativity theory, as we saw, has surprising implications for many of our traditional views, for example, our traditional views on the nature of space and time. We have long assumed that space and time are independent of one's point of view. For example, we tended to think that 10 minutes was 10 minutes, for everyone everywhere. Likewise, a distance of 10 meters was 10 meters regardless of one's perspective.

---

[22]   There are countless other aspects of physics with substantial philosophical implications. For just two examples, see Lange 2002 or Jones 1991.

[23]   For example, in addition to the measurement problem mentioned in footnote 19, there are substantial questions as to whether relativity theory and quantum theory are compatible with one another. See Maudlin 1994 for a good investigation into this question.

But we have found that time passes at different rates for different reference frames, and that distances likewise will differ depending on one's frame of reference. We also saw that relativity has other surprising and counter-intuitive implications, for example, the curvature of spacetime, the replacement of the traditional Newtonian view of gravity as a mutually attractive force, the relativity of simultaneity, and the like.

But beyond these particular implications of relativity, I think there is a broader implication. Relativity vividly illustrates how wrong we can be about beliefs that seem so obvious. Before being introduced to relativity theory, everyone I know (myself included) took it as an obvious fact that time and space were independent of one's perspective. It seemed just obvious and unquestionable that time moved along at the same rate for everyone, so that, for example, it was absurd to think that two twins could wind up being of substantially different ages. In short, relativity should, I think, force us to be more cautious about the degree of confidence we have in beliefs that seem obviously correct.

As for quantum theory, the philosophical implications of it are, I think, even more dramatic. Since we began systematically inquiring into the universe over 2,500 years ago, we have always been interested in the general question of the sort of universe we inhabit. And as noted above, we have tended, quite reasonably, to look to our basic sciences to inform our views about the universe.

Through all the scientific changes from ancient Greek science through most of the twentieth century, one belief that has held constant is our conviction that we live in a local universe, that is, a universe in which the locality assumption holds. But the implications of Bell's theorem and the Aspect-style experiments show that we have been wrong about this. We do not live in a universe in which the locality assumption holds. As noted above, no one knows how the universe can be this way; we only know that it is this way.

In short, relativity theory and quantum theory have forced us to rethink some of our most basic and long-held beliefs. The philosophical implications discussed above are only a small part of the more broadly philosophical issues that arise not just in relativity theory and quantum theory, but in other areas of modern science as well. Shakespeare seems to have got it right when he wrote, 400 years ago, that "there are more things in heaven and earth, Horatio, than are dreamt of in your philosophy."[24]

---

[24] From *Hamlet*, Act 1, Scene 5, lines 187–8. See, for example, Shakespeare 2003, 67.

References

Aspect, A., J. Dalibard, and G. Roger. 1982. "Experimental Test of Bell's Inequalities Using Time-Varying Analyzers." *Physical Review Letters* 49: 1804.

Baggott, J. 1992. *The Meaning of Quantum Theory*. Oxford: Oxford University Press.

Baggott, J. 2004. *Beyond Measure*. Oxford: Oxford University Press.

Bell, J. 1964. "On the Einstein-Podolsky-Rosen Paradox." *Physics* 1: 195.

Bell, J. 1988. *Speakable and Unspeakable in Quantum Mechanics: Collected Papers on Quantum Philosophy*. Cambridge: Cambridge University Press.

Cushing, J. 1998. *Philosophical Concepts in Physics: The Historical Relation between Philosophy and Scientific Theories*. Cambridge: Cambridge University Press.

DeWitt, R. 2004. *Worldviews: An Introduction to the History and Philosophy of Science*. Malden, MA: Blackwell Publishing.

Einstein, A. 1905. "On the Electrodynamics of Moving Bodies." *Annalen der Physik* 17.

Einstein, A. 1916. "The Foundations of the General Theory of Relativity." *Annalen der Physik* 49.

Einstein, A. 1920. *Relativity: The Special and General Theory*. New York: Henry Holt.

Einstein, A. 1952. *The Principle of Relativity*. F. Davis (ed.). New York: Dover Publications.

Einstein, A., B. Podolsky, and N. Rosen. 1935. "Can Quantum-Mechanical Description of Physical Reality be Considered Complete?" *Physical Review* 48: 777.

Herbert, N. 1985. *Quantum Reality: Beyond the New Physics*. New York: Doubleday.

Hughes, R. 1989. *The Structure and Interpretation of Quantum Mechanics*. Cambridge, MA: Harvard University Press.

Jones, R. 1991. "Realism About What?" *Philosophy of Science* 58: 185–202.

Kragh, H. 1999. *Quantum Generations: A History of Physics in the Twentieth Century*. Princeton, NJ: Princeton University Press.

Lange, M. 2002. *An Introduction to the Philosophy of Physics: Locality, Fields, Energy, and Mass*. Oxford: Blackwell Publishers.

Laymon, R. 1984. "The Path from Data to Theory." In J. Leplin (ed.), *Scientific Realism*. Berkeley, CA: University of California Press, 108–23.

Leplin, J. (ed.). 1984. *Scientific Realism*. Berkeley, CA: University of California Press.

Maudlin, T. 1994. *Quantum Non-Locality and Relativity*. Oxford: Blackwell Publishers.

Mermin, D. 1968. *Space and Time in Special Relativity*. New York: McGraw-Hill.

Shakespeare, W. 2003. *Hamlet*. New Folger Library Shakespeare. New York: Washington Square Press.

7   Philosophy of Chemistry

Joachim Schummer

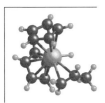

This essay provides an introduction to the philosophy of chemistry with a focus on ontological, epistemological, methodological, and ethical issues. Rather than surveying the vast and diverse literature, I address four questions: What is chemistry about? Is chemistry reducible to physics? Are there fundamental limits to chemical knowledge? And is chemical research ethically neutral? The answers to these questions follow two threads: radical change and dealing with real-world complexity. First, I argue that chemistry is essentially about radical change that cannot adequately be captured by physics; and because radical change enables unlimited synthesis, chemical knowledge is fundamentally incomplete and chemical research ethically relevant in a particular sense. Second, chemistry deals with real-world complexity by adjusting the material world in the laboratory to its classificatory concepts and by following methodological pluralism, both of which pose fundamental limits to understanding the world outside the laboratory, including predictions of how its synthetic products behave in that world. Unlike the ideal of universalism, the methodology of chemistry provides a kind of pragmatic patchwork knowledge, which is paradigmatic of most experimental laboratory sciences and which requires reconsidering standard philosophy of science approaches as well as the ethical dimensions of science and technology.

1.   Introduction

It would seem that philosophy of chemistry emerged only recently. In the early 1990s philosophers and chemists began to meet in many different countries to discuss philosophical issues of chemistry – at first in isolated national groups but soon cultivating international exchange through regular meetings and the publication of two journals (*Hyle* and *Foundations of*

Chemistry) devoted to the philosophy of chemistry. While the social formation is indeed a recent phenomenon, which is still in progress, the philosophical topics have a much longer history, which in some cases predates
chemistry. One could even argue that ancient Greek natural philosophy
started with profoundly chemical questions about the elemental constitu-
tion of the world and about how to provide explanations for the sheer
unlimited material variety and its wondrous changes, in which, for instance,
water becomes solid or gaseous; wood turns into fire, smoke, and ashes;
stones change into metals; food transforms into the human body; or
certain materials convert a sick body into a healthy body.

In fact, there is an almost continuous philosophical tradition focused
on such questions. Because Aristotle's natural philosophy, which was
centered on his theory of elements, was influential far into the eighteenth
century, it provided the basis for much of chemical philosophy. The
meticulous arts of performing desirable material changes in the laboratory,
particularly alchemy and metallurgy, were deeply involved in pondering
metaphysical and methodological issues, out of which not only modern
chemistry but also the experimental method emerged, which influential
figures like Francis Bacon popularized.

Although the seventeenth century brought about a fundamental split into
the mathematical and experimental sciences and many famous philosophers
were inclined towards the mathematical tradition, philosophical discussions
of chemical issues did not stop then. For instance, Immanuel Kant, at least
in his posthumous works, wrote extensively on chemistry, as did G.W.F.
Hegel, F.W.J. Schelling, and particularly Friedrich Engels, whose dialectical
materialism later inspired twentieth-century generations of philosophers
in communist countries to reflect on chemistry. Outstanding nineteenth-
and twentieth-century chemists, from Justus Liebig to Pierre Duhem,
Wilhelm Ostwald, and Michael Polanyi, were heavily engaged in philosophical
issues, although their influence gradually faded as philosophy of science
established itself as an independent branch of philosophy in the twentieth
century. Particularly in German- and English-speaking countries, professional philosophers of science became almost exclusively focused on the
mathematical tradition, with favorite topics in statistics, mathematical logic,
relativity theory, and quantum mechanics. While their work has without
doubt been important to theoretical physics, they mistakenly consider this
peculiar research field to be exemplary or representative of all the sciences.
Apart from communist countries, the situation was different perhaps only
in France, where two chemically trained philosophers, Émile Meyerson and

Gaston Bachelard, were most influential in shaping French *épistémologie* and philosophy of science. In most countries, however, the gap left by philosophers of science was largely filled by chemists and historians of science, like Thomas Kuhn who developed his theory of paradigm changes on the model of the chemical revolution. The narrow focus of professional philosophers of science was only slowly opened, particularly through the philosophy of biology movement since the 1970s. Other philosophies of the special sciences followed soon, one of which is philosophy of chemistry.

In this essay I will not try to review all the recent and past works in the philosophy of chemistry (for review articles, see Schummer 2003a, 2006), because the topics are far too diverse and many require detailed chemical background knowledge. Instead, I discuss four issues that together might serve as an introduction to the philosophy of chemistry and at the same time give an idea of its scope. The four issues, which are chosen so that they build on each other and inspire further thinking, and which are necessarily a personal selection, are: What is chemistry about? Is chemistry reducible to physics? Are there fundamental limits to chemical knowledge? And is chemical research ethically neutral?

2.   What is Chemistry about?

Philosophers, like children, tend to ask plain questions such as: what is chemistry about? What is its specific subject matter that distinguishes chemistry from other sciences? Dictionaries tell us that chemistry is about substances, chemical reactions, molecules, and atoms – but what is a substance, a chemical reaction, a molecule, and an atom, and how do these concepts relate to each other? Unlike substances in philosophy, a chemical substance is a piece of matter of any size, form, and state of aggregation with clearly defined and unique chemical properties that are qualitatively different from the chemical properties of other substances. A chemical property of a substance is its ability to change into other substances under certain conditions, and such changes from one substance to another are called chemical reactions. Because a substance is defined through its specific chemical reactions and a chemical reaction is defined through the specific substances involved, we end up in circular definitions: reactions define substances and substances define reactions. Can we escape the circle by giving priority to either substances or reactions?

The seemingly innocent question of what chemistry is about prompts us to decide between two opposing metaphysical traditions: substance philosophy and process philosophy. Substance philosophers claim priority for entities, things, or substances and consider changes, like motion in space, to be only secondary attributes of entities. In chemistry, however, change is essential rather than secondary; and it is radical because through chemical reactions all properties radically change. This suggests that process philosophy would be more suitable here, because it gives priority to processes and considers entities only as temporary states. Moreover, process philosophers can point to the fact that in the natural world there are no fixed and isolated chemical substances but only permanent chemical change of matter. However, in order to describe these changes precisely we need concepts that grasp the various states of change, for which the concept of chemical substances appears to be most suitable.

Chemists have solved the puzzle in a way that sheds light on the manifold uses of experimentation in science. Because, as process philosophy correctly says, there are no fixed and isolated chemical substances in the natural world, chemists make them in the laboratory and put them in bottles, so that they are pure, isolated, and remain stable for further investigation. The material world is thus adjusted to the conceptual needs. However, the experimental trick works only through a quasi-operational definition of chemical substances, according to which a chemical substance is the result of perfect purification, which includes thermodynamic operations such as distillation. It happened that only the results of such purification procedures meet the definition of chemical substances, that only they have clearly defined and unique chemical properties that qualitatively differ from those of other substances.[1] The trick thus yields substances that are characterized through their chemical changeabilities, which combines aspects of both substance and process philosophies. Once such chemical substances are produced, they can also be characterized and later recognized by other properties, such as optical and thermodynamic properties.

Chemists have used the same experimental strategy to develop an operational hierarchy of matter that formally resembles the metaphysical hierarchy known since Aristotle. Every technique that takes materials apart defines a part–whole relationship between the end products and the starting

---

[1] There are some exceptions, however, as always in the chemical world, particularly the so-called berthollides (for more details, see Schummer 1998). On the other hand, the quasi-operational approach allows solving the philosophical puzzle of natural kinds.

material. Thus, a material that can be taken apart by purification is, by definition, a mixture and the resultant materials are its component substances; a material that cannot be taken apart is, by definition, a chemical substance.

There are two other sets of separation techniques that each define a part–whole relationship between materials. A mixture that can be taken apart into different materials by mechanical means, such as sorting or cutting, is a heterogeneous mixture; otherwise it is a homogeneous mixture. A chemical substance that can be taken apart by chemical means, including electrochemical processes, is a compound; otherwise it is a chemical element. At the same time the chemical separation defines the elemental composition of a compound, which is an important chemical property.

Overall, this results in an operationally defined four-level hierarchy, from chemical elements, to compounds, to homogeneous and heterogeneous mixtures. The hierarchy allows characterizing both a material and its changes through its composition on the lower levels. For instance, a compound is characterized by its elemental composition and a homogeneous mixture by its composition of substances.

Because chemistry is about radical change, it needs to deal with fundamental problems, as the following example illustrates. Assume you want to characterize something through its specific changes: as long as you do not perform the change, you have no certain idea about that change, but once you have performed the change, the thing you want to characterize no longer exists. Again, the logical puzzle is solved experimentally in chemistry. Because any material, from homogeneous mixtures down in the hierarchy to elements, cannot by definition be changed through mechanical separations, one can mechanically take small pieces from such a material and perform chemical test changes on these samples. The operational hierarchy guarantees that the chemical characteristics of all samples are exactly the same as those of the entire piece of material.

Thus far we have dealt only with substances and reactions. What about atoms and molecules? Because these are widely conceived as the true microscopic components of all materials, many argue that chemistry is ultimately about atoms and molecules rather than about substances. Investigating substances and chemical reactions is only a means to develop a better understanding of atoms and molecules and their dynamical behavior and reconfigurations that we perceive as chemical change. One could also argue that all our knowledge about atoms and molecules is only a means to understand better and then explain and predict the chemical behavior of substances.

While all chemical knowledge actually starts with the artificial creation of pure chemical substances and then continues with investigating them in the laboratory, the two positions differ only in what kind of knowledge they consider means and ends of chemistry.[2] The first position (which one might call theoreticism) takes the knowledge of substances as means for the knowledge of atoms and molecules to be considered an end in itself. For the second position (experimentalism) the knowledge of atoms and molecules is only a theoretical means for the proper end of understanding the behavior of substances. And because substances are artificially produced in the laboratory to suit our conceptual needs, one can also assume a third position, which one might call realism in the original sense because, unlike idealism, it acknowledges a fundamental difference between our concepts and the world. This position takes our knowledge of substances, whether reinforced by theoretical knowledge or not, only as a means to develop a better understanding of our messy material world, which includes both our natural environment and the chemical processes that happen in all kinds of industries.

Of course the three positions express different views about the end of science in general, and they usually come from different areas of science – here, theoretical, experimental, and applied science. However, in chemistry the difference between theoreticism and experimentalism is more complicated than an introductory textbook of chemistry might suggest. That is because there is no one-to-one relationship between substances and molecules, such that each substance would consist of a single kind of molecule. Indeed, the concept of molecules works only for certain substances as a useful model approximation. If we assume that substances consist somehow of atoms, the molecular model singles out certain groups of atoms that on time average stick a bit closer together with each other than with other atoms. This model works quite well with many organic substances and gases, but fails, for instance, with simple substances like water, metals, or salts, for most purposes. In liquid water one can single out hundreds or thousands of different kinds of molecules, depending on one's accuracy and time average, such that pure water would be a complex molecular mixture. In metals and salts, all atoms stick together in the same way such that each piece would consist of a single molecule. Hence, rather than talking

---

[2]   In philosophy of science these two positions are sometimes called scientific realism and instrumentalism, which in my view is a misleading terminology, because both views are instrumentalist with regard to the other kind of knowledge.

of molecules, a more generic concept is that of interatomic structures of substances.

Interatomic structures of substances are dynamic entities, even if we disregard quantum mechanics for the sake of simplicity. To take water again as an example, the structure continuously changes on a time scale of less than a trillionth of a second. We might be able to identify some hundred kinds of preferred structures that recur on time average, but others appear if we only slightly change the temperature. Also, for those organic substances where the molecular model works quite well, interatomic distances and angles change with temperature. Theoreticism is thus confronted with severe conceptual problems because the classical chemical concepts no longer work. If, in theoretical terms, a chemical reaction is defined by a change of the interatomic structure, pure substances would be complex mixtures that undergo permanent chemical reactions; and changes of temperature that do not change the substance identity would induce radical chemical reactions on interatomic structure. The problem of theoreticism is that it lacks useful concepts of kinds, both for entities and processes. If such concepts are introduced by virtue of model approximations, theoreticism would have to concede that chemistry is ultimately about its own models about the world rather than about the material world itself, i.e., only about what theoreticians are doing. Compare that with experimentalism, which cannot only acknowledge such models as useful intellectual tools but can also claim that its own concepts perfectly fit at least a part of the material world, even if that part is artificially produced in the laboratory.

However, experimentalism also smacks of self-satisfaction because it creates and focuses on the laboratory systems that best fit its conceptual framework. If the goal of science is to understand the world that we all live in, then realism is the only viable position, such that theoretical and experimental laboratory investigations are only useful means to that end.[3] That is even more important, if chemistry, as many think, is about developing an understanding of our material world in order to improve it according to human needs.

[3] Note that theoreticism, experimentalism, and realism also differ with regard to our original question if entities or processes have ontological priority. Since ancient atomism, theoreticism has, at least before quantum mechanics, always favored substance philosophy and tried to reduce any change to motion in space. Experimentalism combines both substance philosophy and process philosophy and experimentally adjusts part of the material world to the conceptual needs of substance philosophy, whereas realism is forced to acknowledge the omnipresence of change.

3. Is Chemistry Reducible to Physics?

In recent philosophy of chemistry, the issue of whether chemistry is reducible to physics has been vividly debated. The debate was originally inspired by older bold claims like that of the mathematician Paul Dirac from 1929, according to whom the whole of chemistry would be reducible to quantum mechanics and thus would be part of physics. Insofar as such claims express disciplinary chauvinism as a means to acquire social prestige and intellectual hegemony, or just the frequent disciplinary narrow-mindedness that ignores everything outside one's discipline, they should not much concern philosophy. On the other hand, insofar as such claims belong to the general position of physicalism – according to which physics would be fundamental to any science, including biology, the social sciences, and psychology – they express a metaphysical worldview that, in its generality, is beyond the scope of philosophy of chemistry, although philosophers of chemistry can make specific and useful contributions to such debates. Furthermore, if the claim is about the explanatory and predictive scope of a specific theory, it is up to scientists rather than philosophers to assess the exact limits of the theory by checking the thesis against experimental findings and rejecting unfounded claims according to established scientific standards. The remaining job of philosophers – both of chemistry and physics, because the reductionist claim is about the relation between chemistry and physics – largely consists in clarifying the underlying concepts and in checking for hidden assumptions and blind spots.

Because there are many different versions of reductionism, conceptual distinctions are necessary. *Metaphysical* or *ontological reductionism* claims that the supposed objects of chemistry are actually nothing other than the objects of quantum mechanics and that quantum-mechanical laws govern their relations. In its strong, eliminative, version, metaphysical reductionism even states that there are no chemical objects proper. Microstructural essentialism reformulates eliminative metaphysical reductionism in semantic terms by employing a certain theory of meaning and reference to claim that the proper meaning of chemical substance terms, such as "water," is nothing other than the (quantum-mechanical) microstructure of the substance. However, as was shown above, it makes a difference if the objects of chemistry are substances or interatomic structures, such that giving up substances, as eliminative reductionism and its semantic twin claim, would be giving up chemistry as we know it.

Even if substances have an interatomic structure, the fact that a theory can be used to describe the structure and to develop useful explanations does not mean that it "owns" interatomic structures. There are other important theories to describe interatomic structures, such as classical chemical structure theory which is much more useful to explain chemical properties, as we will see below. Moreover, anti-reductionists argue that theoretical entities are determined by their corresponding theory, such that theoretical entities of different theories cannot simply be identified. For instance, from the different meanings of the term 'electron' in quantum electrodynamics and in chemical reaction mechanisms one might conclude that the term 'electron' has different references, which rules out ontological reductionism.

Epistemological or *theory reductionism* claims that all theories, laws, and fundamental concepts of chemistry can be derived from first principle quantum mechanics as the more basic and more comprehensive theory. That claim has prompted many technical studies on the difficulties of quantum mechanics to derive the classical concept of molecular structure and the chemical law that underlies the periodic system of elements. Moreover, because most of the successful applications of quantum mechanics to chemical problems include model assumptions and concepts taken from chemistry rather than only first principles, their success can hardly support epistemological reductionism. Apart from such technical matters, quantum mechanics cannot derive chemistry's classificatory concepts of substances and reactions, and it cannot explain, cannot even compete with, chemical structure theory, which has been developed since the mid-nineteenth century in organic chemistry to classify, explain, predict, and synthesize substances.

Methodological reductionism, while acknowledging the current failure of epistemological reductionism, recommends applying quantum-mechanical methods to all chemical problems, because that would be the most successful approach in the long run (approximate reductionism). However, the mere promise of future success is hardly convincing unless a comparative assessment of different methods is provided.

By modifying the popular notion that "the whole is nothing but the sum of its parts," two further versions of reductionism have been developed. *Emergentism* acknowledges that new properties of wholes (say, of water) emerge when the parts (say, oxygen and hydrogen) are combined, but concedes that the properties of the whole can be explained or derived from the relations between the parts (i.e., epistemological reductionism).

Supervenience, in a simple version, means that, although epistemological reductionism might be wrong, the properties of a whole asymmetrically depend on the properties of the parts, such that every change of the properties of the whole is based on changes of the properties of or the relations between the parts, but not the other way round. If applied to the reduction of chemistry to quantum mechanics (i.e., to chemical entities as wholes and quantum-mechanical entities as parts), emergentism and supervenience presuppose elements of epistemological or ontological reductionism, such that the criticism of these positions applies accordingly.

The discussion of reductionism distracts from the fact that chemistry and physics have historically closely developed with many fruitful interdisciplinary exchanges without giving up their specific disciplinary focus. For instance, chemistry greatly benefits from quantum mechanics, because that is the only theory we have to explain electromagnetic, mechanical, and thermodynamic properties of materials. However, when it comes to chemical properties, the properties that define chemical substances and which chemists are mostly interested in, quantum mechanics is extremely poor, such that chemists here rely almost exclusively on chemical structure theory. Rather than focusing on reductionism, with its underlying notion of a Theory of Everything, it seems more useful to discuss the strengths and weaknesses of different theories for different purposes. For instance, quantum mechanics helps analyze the optical properties that chemists routinely use in all kinds of spectroscopies to understand the kind of time-averaged interatomic structures that chemists are interested in. If these structures can successfully be translated into chemical structure theory, however, it is chemical structure theory rather than quantum mechanics that provides information about chemical properties.

Chemical structure theory, which has been continuously developed since the mid-nineteenth century, is more like a rich sign language than a depiction of individual physical structures. It is one of the hidden assumptions of reductionism that both kinds of structures are the same. However, chemical structure theory encodes types of chemical reactivities according to chemical similarities in characteristic groups of atoms and it has numerous general rules for how these groups can interact and be reconfigured to describe chemical reactions. The important difference to physical structures, which are described in terms of individual space coordinates, is that it describes both the structures and their reconfigurations in general concepts that are chemically meaningful. Despite its recourse to general concepts, the language is rich enough to distinguish clearly between hundreds of millions of substances and their chemical properties. Once the chemical structure of a substance is known, chemical structure theory allows both identifying the substance and predicting its chemical properties. Moreover, because chemical properties describe radical change of substances, these predictions enable one to make new, unknown substances in the laboratory, such that predictions guide the production of novelty. This is nowadays successfully performed several million times per year, which makes chemical structure theory one of the most powerful predictive tools of science.

One of the blind spots of reductionism, or physicalism for that matter, is that sciences other than physics deal with different issues and subject matters that require entirely different kinds of methodologies, concepts, and theories. In chemistry, which deals with substances and radical change, classification and synthesis are at least as important as analysis, or its physics counterpart of a quantitatively accurate and true description of the world as it is. Classification is not only a matter of building useful empirical or operational concepts. It also requires theoretical approaches that include or can deal with classificatory concepts and substantial change, otherwise the theories cannot address the issues that are to be explained or predicted. Chemical theories need to deal with hundreds of millions of different substances and hundreds of thousands of kinds of reactions. Theoretical physics, on the other hand, stands out among the sciences because, apart from particle physics, it intentionally lacks classificatory concepts.

Furthermore, because radical change is essential to chemistry, synthesis is an integral part of chemistry both on the experimental and theoretical level. That is not simply because synthesis can provide useful compounds, although this option has historically shaped much of chemistry. Chemical properties are revealed only through synthesis (i.e., by chemical reactions that change one substance into another under controlled laboratory conditions). Accordingly, a chemical theory that is expected to make predictions must be able to predict syntheses, and the only way to test the predictions is, of course, by way of synthesis. Again, synthesis is not part of the methodology of physics, at least as mainstream philosophers of physics conceive it, so that the model of physics would miss a central part of chemical concepts, theories, and methods. However, since many physicists along with chemists engage in materials science to produce new useful materials, the methodology of experimental physics might approach that of chemistry.

4.   Are There Fundamental Limits to Chemical Knowledge?

An important epistemological task of philosophy of science consists in understanding the limits of scientific knowledge on a general level. Again, it is up to scientists to check the limits of a specific theory or model in order to avoid unjustified scientific claims that lead people astray by unfounded promises. Unfortunately, such promises increasingly appear, with the public struggle for funding and public attention, in popularizations of science, and sometimes even in the disguise of philosophy. The epistemological task consists in scrutinizing a scientific approach, its concepts and methods, for implicit assumptions that limit the scope or validity of its epistemic results. Such an analysis may provide not only an epistemological assessment of the scientific approach but also answers to the more ambitious question of whether complete and perfect knowledge is ever possible or not. In the following, I discuss three issues that each shed light on the limits of chemical knowledge: the concept of pure substances, methodological pluralism, and the proliferation of chemical objects.

As has been discussed in the previous sections, chemistry is based on the concept of chemical substances – experimentally in characterizing, classifying, and producing materials and in describing chemical change, as well as theoretically in explaining, classifying, and predicting materials and chemical change through structure theory. However, chemical substances are idealizations in two regards that each pose limits to chemical knowledge. First, although chemical substances are experimentally produced through purification techniques and as such are real entities, perfect purity is a conceptual ideal that can never be fully reached in practice. Thus, any real substance as an object of experimental investigation contains impurities, whereas any conceptual description needs to assume perfect purity or a well defined mixture of pure substances. Because even very small amounts of impurities can drastically change chemical properties, through catalytic activity, there is always the risk that the gap between concepts and objects leads to misconceptions and wrong conclusions. On the other hand, because chemists know well about the problem, they can take particular care about possible impurities that they assume are relevant in each case.

Second, and more importantly, the pure substances that chemists produce and put in bottles for chemical investigations do not exist outside the laboratory. Instead, the materials outside the laboratory are messy and mostly under continuous transformation and flux. Any material sample of, say, a soil, a plant, or even sea water, can be analyzed into hundreds or thousands of substances of different amounts, depending on one's analytic accuracy. And before it became a sample, the piece of matter was in continuous flux and interaction with its environment and hardly a perfect homogeneous mixture. The problem is not to describe all that; rather the problem is that any accurate description of material phenomena outside the laboratory turns into an endless list of facts. Moreover, if a mixture contains more than five or ten substances, the theoretical reasoning of chemistry fails because of over-complexity. Hence, the conceptual framework of chemistry is not very suitable to describe the real material world, but still it is the best we have for that purpose. The way chemists deal with such real-world issues is, again, by making assumptions about what is relevant and what not by focusing on specific questions for which the relevance of factors can be estimated or controlled.

Once relevance aspects shape the kind of facts one considers and the kind of knowledge one pursues, the abstract ideal of complete and perfect knowledge is given up. The fragmentation into different knowledge domains according to different relevance aspects then seems unavoidable, and new domains grow as new questions become relevant. While that might to some degree be true of all the experimental sciences, in contrast to theoretical physics, it is characteristic of chemistry as the prototype of experimental laboratory sciences and as by far the biggest discipline.[4] In contrast to the ideal of a universal Theory of Everything, which has been important in theoretical physics, chemistry is guided by a pragmatist pluralism of methods. Not only does each sub-discipline of chemistry develop its own kinds of methods, concepts, and models tailored to specific substance classes and types of chemical change, but also within each particular research field there is, even for the same experimental system, a variety of different models at hand that serve different purposes. One might argue that this is because the right universal approach has not yet been found. However, methodological pluralism seems to be rather a characteristic of chemistry that allows flexibly dealing with complexity by splitting up approaches according to what matters in each case. Rather than being a surrogate of universal theories, methodological pluralism is an epistemological approach in its own right. It requires that the quality of a model is not judged by

---

[4]   Note that, in quantitative terms of publications, chemistry is almost as big as all the other sciences together (Schummer 2006).

standards of truth and universality but, instead, by its usefulness and the precision by which its scope of applications is limited. A model in chemistry is a theoretical tool to address specific questions, which is useless if you do not know for which kind of systems and research questions it can reasonably be used.

Methodological pluralism produces a kind of patchwork knowledge rather than universal knowledge. The advantage is that it allows incorporating new kinds of knowledge without fundamental crisis, by extending the patchwork. Moreover, it can deal with relevance aspects, which the claim to universal knowledge cannot. Because patchwork knowledge can always be extended, by including new kinds of knowledge and new relevance aspects, the scientific endeavor is open-ended in both dimensions. Therefore, the idea of complete and perfect knowledge, and all its derived epistemological concepts that might be useful to apply to the notion of universal knowledge, is meaningless in chemistry.

Further support for the last conclusion, that chemical knowledge can never be perfect and complete, comes from an analysis of the concept of chemical properties (i.e., from the specific subject matter of chemistry). All material properties are dispositions: they describe the behavior of materials under certain contextual conditions, such as mechanical forces, heat, pressure, electromagnetic fields, chemical substances, biological organisms, ecological systems, and so on. Because a property is defined by both the behavior and the contextual conditions, we can freely invent new properties by varying the contextual conditions to increase the scope of possible knowledge almost at will. Chemical properties stand out because the important contextual factor is of the same kind as the object of investigation, both being chemical substances, such that chemical properties are strictly speaking dispositional relations. A chemical property of a substance is defined by how it behaves together with one or more other substances, and the important behaviors are those of chemical transformation, although the lack of transformation (i.e., chemical inertness) is sometimes also important. If a new, hitherto unknown substance results from the transformation, it can be made subject to further investigations, by studying its reactivity with all known substances, which in turn may result in many hitherto unknown substances to be studied, and so on. The procedure results in exponential growth of substances, not just in theory but also historically over the past two centuries, and there is no fundamental limit to an endless proliferation in the future. Because each substance increases the scope of possible chemical knowledge, chemical knowledge can never be complete.

Even worse, one can argue that the synthesis of new substances increases the scope of possible knowledge (the number of undetermined properties) much faster than the scope of actual knowledge (the number of known properties). If we call the difference between possible knowledge and actual knowledge non-knowledge, chemistry produces through synthesis much more non-knowledge than knowledge, as the following simplified calculation illustrates. Assume we have a system of n different substances, then the number of all possible chemical properties corresponds to the number of all combinations from pairs to n-tuples (times the variations in concentration and other contextual conditions, which will be neglected here). While the synthesis of a new substance increases the scope of actual knowledge only by a single property (the reaction from which the substance resulted), it increases the scope of possible knowledge or undetermined chemical properties according to simple combinatorics by

$$\sum_{k=2}^{n+1} \frac{(n+1)!}{k!(n+1-k)!} - \sum_{k=2}^{n} \frac{n!}{k!(n-k)!}.$$

For instance, if the original system consists of 10 substances, which corresponds to 1,013 possible properties, the synthesis of a single new substance creates 1,023 new possible properties. Thus, while the actual knowledge increases only by one property, non-knowledge grows by 1,022 undetermined properties. If the system consists of 100 substances, a single new substance increases non-knowledge by 10^{30} undetermined properties, and so on. One might criticize the calculation as being too simplistic, but a more precise calculation, which additionally considers variations in concentration and other contextual conditions, would bring about an even faster growth of non-knowledge.

The epistemological problem or paradox is ultimately rooted in the peculiarities of the chemical subject matter (i.e., in radical change) and therefore unknown in other sciences. Rather than depicting the world as it is, chemistry develops an understanding of the world by changing the world. Because the changes are radical in that they create new entities, any such step of understanding increases the complexity of the world and thus makes understanding more difficult. We will see below that this paradox of understanding also poses specific ethical issues.

5.   Is Chemical Research Ethically Neutral?

Chemical knowledge has always been mysterious and suspicious in western societies because it is knowledge of radical change. Early Jewish and Christian mythology, particularly the apocryphal Book of Enoch, associates chemical knowledge with the secret knowledge of primordial Creation that the Fallen Angels had once betrayed to humans. Up to the eighteenth century, performing chemical changes was routinely accused of modifying divine Creation against God's will, and some people think so even today. On the other hand, the prospects of radical change have always fueled fantasies of changing the material world at will according to human needs or specific economic interests, from alchemy to the chemical industry and current visions of nanotechnology. Since thoughtless industrial chemical production has caused severe environmental problems, through pollution, accidents, and unsafe products, anything related to chemistry is publicly considered with suspicion. Many consider the archetypical mad scientist, the chemist Victor Frankenstein in Mary Shelley's novel, emblematic of the modern academic-industrial endeavor of chemistry.

It would be wrong to disregard the specific cultural embedding of chemistry from a philosophical point of view, because it has essentially shaped ethical views of chemistry. After all, ethics is a branch of philosophy, such that ethics of chemistry is a natural branch of philosophy of chemistry. From the fact that, for instance, mathematics is rather poor in ethical issues but rich in logical issues, it would be mistaken to conclude that the focus of all philosophy of science is on logic. Each discipline has its own variety of issues that call for philosophical treatment. Although this essay does not include an ethical analysis of chemistry for the sake of brevity (see Schummer 2001), it prepares such an analysis by some conceptual clarifications that are focused on the issue of whether chemical synthesis is ethically neutral or not (i.e., if it can be made subject to justified moral judgments).

At first it is useful to point out the distinction between the academic discipline of chemistry and the chemical industry, of which only the former concerns us here. The chemical industry, as any industry, is definitely not ethically neutral because it deliberately acts according to (non-epistemic) values, and its actions have direct positive and negative consequences for human beings. The important question is whether chemical research that synthesizes new chemical substances is ethically neutral. Strictly speaking, no scientific research is ethically neutral insofar as it produces knowledge about the world that could enable people to perform ethically relevant actions. Those can be either actions to prevent harm, such as when understanding the causes of stratospheric ozone depletion by chlorofluorocarbons enables one to take effective measures against the depletion; or actions to cause harm, such as when understanding the biochemical metabolism of human beings allows one to choose a more effective poison. On this general level, because scientific knowledge enables effective actions, scientists have a particular responsibility for the kind of knowledge they pursue. Apart from and above that, is there anything that makes the synthesis of new substances ethically relevant?

We are accustomed to making a distinction between science and technology, including technological research or engineering sciences. In this view, science describes the natural world and makes true discoveries of the world, whereas technology changes the world by producing artifacts and making useful inventions for change. In this view, technology is, unlike science, ethically relevant above the general level because, like industry, it deliberately acts according to values of usefulness and directs its actions accordingly. Because chemical synthesis meets that definition of technology, it would seem that chemical synthesis is essentially a technology rather than a science and therefore ethically relevant above the general level.

However, the distinction between science and technology includes two related problematic assumptions, which incidentally have their roots in the cultural background mentioned at the beginning of this section. First, it assumes that science cannot, by definition, be about understanding radical change, because that is the domain of technology. However, if the goal of science is describing and understanding nature, the assumption is equivalent to the thesis that there are no radical changes in nature so that there is no place for such a science. The underlying philosophical view has been known since antiquity as the opposite of process philosophy, and its Christian counterpart is the notion of nature as the perfect divine creation. As has been argued above, chemistry is all about understanding radical change, about transformations of substances into one another. If one acknowledges that there are radical changes in nature, understanding and discovering such changes is clearly a scientific endeavor. And because chemical synthesis is the best experimental way we have to study such radical change, it meets all requirements of scientific methods.

Secondly, the distinction between science and technology assumes that the world can clearly be divided up into natural entities and artifacts, which in the Christian (and Platonic) tradition is equivalent to the distinction between entities made by God in the primordial Creation and entities made by humans. In this view, science is about the natural world whereas technology is about producing artifacts from the resources of the natural world. However, pure substances isolated from natural resources are also artifacts because they always result from purification techniques, just as anything produced in an experimental setting counts as an artifact. Moreover, a substance that can be isolated from natural resources through purification can, as a rule, also be synthesized in the laboratory from different compounds, such that there is no scientific way to distinguish between natural and artificial substances, in contrast to artifacts in technologies which can usually be clearly recognized as artifacts. Furthermore, if chemical changes are natural and if nature is essentially process-like, there is no reason to question that the outcomes of such changes are natural, regardless of whether the changes have been experimentally directed or not and whether the outcomes have been known before or not. In sum, the distinction rests on an archaic notion of nature, as something given and static without changeabilities, whereas all modern experimental sciences focus on the study of the dynamics of nature (Schummer 2003b).

While we can therefore reject the idea that chemical synthesis *per se* is a kind of technology rather than science, that does not mean that chemical synthesis is always performed as science. It all depends on the research questions in each case. If the research is performed to study chemical changeabilities, it rather belongs to science. If the synthetic research aims at useful products, it would rather be counted as technological research. However, modern science, in chemistry as well as elsewhere, is a collaborative enterprise that is driven by a variety of motives and intentions that no philosopher is able to identify. One can pursue a specific scientific research question that is also important for a technological goal and integrated in a broader project. And one can pursue at the same time scientific and technological knowledge without much compromising, which some philosophers have recently discovered as the latest move towards "technoscience," although that has been known in chemistry for centuries.

Finally, if we ignore all these complications and take chemical synthesis in the purest sense of science: is it apart from the general level ethically neutral because it is science rather than technology? The answer is no, and the main reason lies again in the fact that chemistry is about radical change. Synthetic chemistry does not only produce knowledge but also actively changes the world, which may affect anybody living in that world.

Assume that in the course of scientific studies on chemical reactivities a chemist has produced a new substance which happens to be extremely toxic, and which, by some means, leaves the laboratory and causes severe human poisoning or an environmental disaster. We would rightly hold the chemist responsible for that harm, not only because of the lack of security measures, but also because the chemist was the original creator of the agent that caused the harm. In such a case, the chemist might insist that he had no intention to cause the harm, which would hardly excuse him because the lack of intention might just be negligence. Also, the argument that he could not foresee the toxic properties of his creation would not count much, because chemists know well that any new substance is unique and has infinitely many properties which, by all scientific standards, bear surprises, such that harmful effects are not unlikely. After all, that is expected from radical change unlike from gradual or marginal change. In sum, even if chemical synthesis is not technology but science, it is beyond the general level ethically relevant because it performs radical changes on the world.

6. Conclusion

Although each of the issues discussed above branches out into various specific issues, there are two running threads throughout this essay, which are radical change and dealing with real-world complexity. First, chemistry is essentially about radical change that cannot adequately be captured by physics; and because radical change enables unlimited synthesis, chemical knowledge is fundamentally incomplete and chemical research ethically relevant in a particular sense. Second, chemistry deals with real-world complexity by adjusting the material world in the laboratory to its classificatory concepts, which are not reducible to physics, and by following methodological pluralism, both of which pose limits to understanding the world outside the laboratory, including predictions of how its synthetic products behave in that world.

The two running threads, presented in this chapter for introductory reasons, might give a too homogeneous impression of current philosophy of chemistry, though. Indeed the field is extremely rich in topics that cover all branches of philosophy, including epistemology, methodology, metaphysics, ontology, ethics, aesthetics, and semiotics. Moreover, there are many important philosophical studies that analyze specific chemical concepts and issues in their particular historical and cultural contexts. From that diversity one might conclude that philosophy of chemistry hardly exists yet as a clearly defined and homogenous field, because it lacks paradigmatic issues and a focused methodology and borrows instead as much from philosophy as from history of science and science studies.

It is certainly true that much of current philosophy of chemistry is still in a process of defining itself anew and that the contemporary zeitgeist is not without impact on that process. However, there are other, perhaps more important, reasons for the diversity. Remember that chemistry follows methodological pluralism rather than universalism, which produces a kind of patchwork knowledge diversified by relevance aspects. Because most of today's philosophers of chemistry have a background, if not a former career, in chemistry, it is likely that their philosophical work is influenced by the epistemological style of chemistry, which deeply distrusts the big pictures and simplifications of universalism. If chemistry also in this way inspires philosophy, the better for philosophy.

References

Baird, D., E. Scerri, and L. MacIntyre (eds.). 2006. *Philosophy of Chemistry: Synthesis of a New Discipline*. Dordrecht: Springer.

Bensaude-Vincent, B. 1998. *Eloge du mixte. Matériaux nouveaux et philosophie ancienne*. Paris: Hachette.

Bhushan, N. and S. Rosenfeld (eds.). 2000. *Of Minds and Molecules: New Philosophical Perspectives on Chemistry*. New York: Oxford University Press.

Earley, J.E. (ed.). 2003. *Chemical Explanation: Characteristics, Development, Autonomy*. Annals of the New York Academy of Science, vol. 988. New York: New York Academy of Sciences.

Janich, P. and N. Psarros (eds.). 1998. *The Autonomy of Chemistry*. Würzburg: Königshausen and Neumann.

Laszlo, P. 1993. *La parole des choses ou le langage de la chimie*. Paris: Hermann.

Primas, H. 1981. *Chemistry, Quantum Mechanics and Reductionism. Perspectives in Theoretical Chemistry*. Berlin: Springer.

Psarros, N. 1999. *Die Chemie und ihre Methoden. Ein philosophische Betrachtung*. Weinheim: Wiley-VCH.

Psarros, N., K. Ruthenberg, and J. Schummer (eds.). 1996. *Philosophie der Chemie*. Würzburg: Königshausen & Neumann.

Ruthenberg K. and J. van Brakel (eds.). 2008. *Stuff: The Nature of Chemical Substances*. Würzburg: Königshausen & Neumann.

Scerri, E.R. 2007. *The Periodic Table: Its Story and Its Significance.* Oxford: Oxford University Press.

Schummer, J. 1996. *Realismus und Chemie. Philosophische Untersuchungen der Wissenschaft von den Stoffen.* Würzburg: Königshausen & Neumann.

Schummer, J. 1998. "The Chemical Core of Chemistry I: A Conceptual Approach." *Hyle: International Journal for Philosophy of Chemistry* 4: 129–62.

Schummer, J. 2001. "Ethics of Chemical Synthesis." *Hyle: International Journal for Philosophy of Chemistry* 7: 103–24.

Schummer, J. 2003a. "The Philosophy of Chemistry." *Endeavour* 27: 37–41.

Schummer, J. 2003b. "The Notion of Nature in Chemistry." *Studies in History and Philosophy of Science* 34: 705–36.

Schummer, J. 2006. "The Philosophy of Chemistry: From Infancy Towards Maturity." In D. Baird, E. Scerri, and L. MacIntyre (eds.), *Philosophy of Chemistry: Synthesis of a New Discipline.* Dordrecht: Springer, 19-39.

Sobczynska, D., P. Zeidler, and E. Zielonacka-Lis (eds.). 2004. *Chemistry in the Philosophical Melting Pot.* Frankfurt: Peter Lang.

van Brakel, J. 2000. *Philosophy of Chemistry. Between the Manifest and the Scientific Image.* Leuven: Leuven University Press.

8 Philosophy of Biology

Matthew H. Haber, Andrew Hamilton,
Samir Okasha, and Jay Odenbaugh

This essay provides sketches of recent and ongoing work in systematics, ecology, and natural selection theory as a way of illustrating: (i) what kind of science biology is and is not; and (ii) an approach to philosophy of biology that is engaged with the scientific details, while still maintaining a conceptual focus. Because biological science is generally not well known among philosophers, we begin by identifying some common misconceptions, and then move on to particular questions: How are inferences made about the deep past? What, if anything, are species? How do ecology and evolution interrelate? Upon what does natural selection operate? These questions and the issues surrounding them do not add up to a survey of the field. We have chosen instead to focus on interesting problems and to exemplify a way of taking them up.

1. Introduction

Philosophy of biology is a vibrant and growing field. From initial roots in the metaphysics of species (Ghiselin 1966, 1974, Hull 1976), questions about whether biology has laws of nature akin to those of physics (Ruse 1970, Hull 1977), and discussions of teleology and function (Grene 1974, Cummins 1977, Brandon 1981), the field has developed since the 1970s to include a vast range of topics. Over the last few decades, philosophy has had an important impact on biology, partly through following the model of engagement with science that was set by first-wave philosophers of biology like Morton Beckner, David Hull, Marjorie Grene, William Wimsatt and others. Today some parts of philosophy of biology are indistinguishable from theoretical biology. This is due in part to the impetus provided by second-wave philosophers of biology like John Beatty, William Bechtel,

Robert Brandon, James Griesemer, Elisabeth Lloyd, and Elliott Sober. Indeed, philosophers have been instrumental in establishing theoretical biology as a field by collaborating with scientists, publishing in science journals, and taking up conceptual questions at the heart of the biological enterprise.

Third-wave philosophers of biology now have a wide array of biological and philosophical topics open to them. Three of these are surveyed here in an effort to show what kinds of questions philosophers of biology are asking, and how these questions are being addressed. While these topics show something of the range of current issues in the field, we have not attempted to provide a survey of philosophy of biology. Lest these selections seem unnatural, we point out that we chose depth over breadth because the method is part of the message. The best contemporary philosophy of biology is characterized by close study of biological details and by asking and answering questions that make contact with the efforts of working scientists.

Before moving on to sketches of recent thinking about systematics, ecology, and natural selection, it will be helpful to provide a bit of the philosophically relevant background on contemporary approaches to the biological sciences, particularly because philosophy of biology and philosophy of science have diverged in important ways over the past couple of decades. One reason for this divergence is that while most philosophers of science have at least some background understanding of the subject matter of physics, very few non-specialists have a similar understanding of the biological world and its study.

2. What Are the Biological Sciences (Not)?

The biological sciences are as diverse as the physical sciences in the kinds of systems they study, in the methods employed, in their standards of evidence, and in what counts as explanations of their phenomena. For example, the daily work and theoretical contexts of molecular genetics are very different from those of comparative morphology or developmental biology. One reason we present a set of sketches of particular topics in the sections that follow is that it is very difficult to say anything that is both substantive and accurate about all of biology.

While it is difficult to characterize all of biology in a meaningful way, it is easy to point to widespread misconceptions among researchers in the humanities about what biology is and what biologists do. These misconceptions have several sources, one of the most important of which is the kinds of biology that are standardly taught in American high schools. Because of the emphasis on cell biology on the one hand and on Mendelian and molecular genetics on the other, one who does not study biology at the university level could be forgiven for concluding that the biological sciences are neither as quantitative nor as richly theoretical as the physical sciences, and therefore that there is no interesting work for philosophers of science to do. As the discussions of phylogenetic inference (§3), evolution and population growth (§4), and kin and group selection (§5) below show, this notion is mistaken.

The evolutionary synthesis that brought together nineteenth-century thinking about phenotypic variation, biogeography, and speciation with twentieth-century efforts to understand genetic inheritance and genotype–phenotype relations was at once highly formal and mechanistic. While some researchers set about the "wet" work of characterizing genes, alleles, chromosomes, and their products and interactions, others built mathematical models that illuminated Darwinian concepts like "fitness" and "selection" against a genetic backdrop. Some of this work is presented in the discussion of evolution and ecology below (§4). It is outside the scope of this essay to discuss the development of theoretical biology in any detail, but it is important to note that several of the biological sciences, including population genetics, epidemiology, and ecosystem ecology, enjoy rich traditions of formal modeling while others are comparative and still others are experimental. Often more than one of these broad approaches is practiced within a single biological science, or even to take up a single research question.

Quantitative approaches to biological systems are now widespread. In addition to the population genetics tradition begun by R.A. Fisher, Sewall Wright, and J.B.S. Haldane in the early part of the twentieth century, game-theoretical (Maynard Smith 1982) and multi-level (Wilson 1975) approaches to natural selection are in wide use, as are differential equations that describe population dynamics as a result of predator–prey interactions (McLaughlin and Roughgarden 1991), and covariance techniques that model the contribution of selection to change over generations (Price 1970, Wade 1985, Queller 1992). There are also very new quantitative approaches to ecosystem ecology (Sterner and Elser 2002, Elser and Hamilton 2007) and to allometry and metabolic scaling (West, Brown, and Enquist 1997). While it is true that Charles Darwin's *On the Origin of Species* (1859) contains no mathematics at all, it is not true that even basic evolutionary biology contains no highly articulated theories, nor theories that are informed by mathematical models.

Formal models have been crucial to biology's development, and have issued in some simple, general statements about the biological world. Biologists and philosophers of biology have therefore continued to ask whether there are laws of biology, and how these might compare with laws of physics (Rosenberg 1994, Mitchell 2003, Brandon 2006, Hamilton 2007). Others have asked what the models *mean*: do abstract and highly idealized models connect with the biological world? If so, what is the appropriate relationship between the model and the world? (Levins 1966, Wimsatt 1981, Odenbaugh 2006). Not all biology is paper-and-pencil theorizing, of course, and this has led other researchers to think about the relationship between the epistemic and social constraints on the "wetter" aspects of biology and other sciences in contrast to formal approaches (Knorr Cetina 1999, Winther 2006).

A second common misconception is that biological kinds – particularly species – are unproblematic *natural* kinds. As we see below in the discussion of species (§3) and levels of selection (§5), some of the most important and interesting conceptual problems biologists are facing arise because it is unclear precisely what the object of study is, how it relates to other objects, and where the edges of biological objects are to be located. This holds for studies of everything from genes to global biodiversity, partly because evolutionary theory requires that there are populations and that they vary, but is silent about what sorts of populations there are, precisely which ones are subject to natural selection and other evolutionary processes, and how they are bounded as they grade into one another (Maynard Smith 1988, Okasha 2006).

With these considerations in mind, we turn to systematics, one of the sciences most responsible for describing the diversity of the living world, understanding its patterns, and discovering the historical and hierarchical relations between organisms and taxa.

3. Systematics

> Systematics can be considered to have two major goals: (1) to discover and describe species and (2) to determine the phylogenetic relationships of these species. (Wiens 2007, 875)

Evolutionary biologists study both the *process* and *pattern* of evolution. Systematists primarily focus on the latter. Studying the pattern of evolution cannot be done without staking out positions that are inherently philosophical in nature. This section is a brief survey of three core issues that must be addressed in some way or another for systematics work to move forward, and are the kinds of questions that represent the overlapping interests of philosophers and biologists.

In order to discover species, systematists must have some idea of what it is to be a species. This turns out to be not simply a biological question, but a deeply philosophical one as well. Furthermore, providing a coherent and useful description means conveying the relevant criteria that suggest that something is, indeed, a species. Determining the phylogenetic, or genealogical, relationships between groups of species requires making an inference about the distant past that is not directly observable. Some justification of inferences that places these claims within the proper theoretical (evolutionary) framework is needed if they are to be scientifically plausible. These challenges are both metaphysical and epistemological in nature. Let us look at each a bit more carefully.

3.1   The species category

For a long time, species were taken to be exemplars of natural kinds (i.e., sets; see below) by philosophers. Now, however, things are not so simple. What this means, and why this matters for both philosophy and biology, is a classic example of a philosophy of biology problem.

Sets are abstract entities defined by their membership. Traditionally understood, sets may be characterized by necessary and sufficient properties, possession of which determines membership in a set. In other words, something is a member of a set by virtue of instantiating the defining properties of that set. If species are sets, then belonging to a species means satisfying conditions for membership in that set, that is, possessing the defining characters of that species. It also means, at least in a traditional sense, that species are abstract entities, and good candidates for being natural kinds.

This characterization of species is at odds with Darwinian thinking, because as Mayr (1959) and Sober (1980) have pointed out, an important feature of Darwinian theory is that populations vary in ways that affect fitness (see §5 below). That is, evolutionary theory requires dynamics that the natural-kind view denies. In response, Michael Ghiselin (1966, 1974) and David Hull (1976, 1978) have forcefully argued that particular species are best understood as individuals, as opposed to sets or classes. The *individuality thesis* is straightforward: the species category (or taxon) is a class, with individual species as members.[1] A species is an individual, as opposed to a set. This means that species are best understood as having parts, as opposed to members, and that species are historical entities, which is to say that they exist in space and time, rather than abstractly. Individual organisms are parts of particular species rather than members of them.

Belonging to a species, then, means being a part of some historical entity, albeit one that is more scattered than other, perhaps more familiar, individuals like organisms. Notice that "being an individual" is not co-extensive with "being an organism" though the two notions are often used interchangeably (Hull 1978, Wilson and Sober 1989, Hamilton, Haber, and Smith 2009). Certainly being an organism entails being an individual, but the reverse is not true.

The timing of the individuality thesis was good. Biologists were generally receptive to it, as it accorded well with shifts in biological practice, namely a general move towards phylogenetic thinking occasioned by the work of German entomologist and theorist Willi Hennig (1966). Taxonomists placed less emphasis on traits and more emphasis on history (or other extrinsic properties) as the theoretically relevant feature of what it meant to be a particular taxon, meaning that taxa are defined by ancestry, not possession of any particular features. The individuality thesis provides a conceptual basis for this position – rather than looking for some essential property, taxonomists instead can specify what it means to be a part of a species in the relevant way (more on this below).

Philosophers have been more resistant to the individuality thesis than most biologists, though generally receptive. Critics of the individuality thesis have complained, among other things, of a conflation of thinking of species as sets with the thesis that species are natural kinds, or that either entails some sort of essentialism (Boyd 1999, Griffiths 1999, Wilson 1999b, Winsor 2006). Other philosophers have argued that thinking of species as sets is not at odds with evolutionary theory, as proponents of the individuality thesis have suggested (Kitcher 1984). Part of the resistance is due to historical inertia – philosophers have often characterized species as exemplars of natural kinds, and the individuality thesis presents a serious challenge to this useful characterization. Either species are not natural kinds, or philosophers need to radically revise their theories of natural kinds.

---

[1] We will come back to what it means for an individual species to be a member of the species category.

3.2   What kinds of individuals are species?

Let us accept the individuality thesis as a working perspective, as it is useful in setting up the next core issue facing systematists.[2] If the species category is the set of all individual species, what are the conditions for membership in this category? What kinds of individuals are species? Answering this question helps answer related questions: what are the relevant parts of a species, and when does speciation occur? The sheer number of answers provided to these questions is overwhelming, and constitutes what is typically referred to as the *species problem*, or debates over *species concepts*.

The species problem concerns what kinds of groups of organisms ought to count as being species. Even this characterization, though, presumes too much. For it might be that species are not groups of organisms, *per se*, but groups of populations, or parts of time-extended lineages (of populations, organisms, or some other genealogical group). Primary concerns for a species concept may range from discovery of some unit of evolution, to epistemological matters of specification (or both). Given space constraints, it will be most useful to take a broad look at two competing species concepts: the biological species concept (BSC) and the phylogenetic species concept (PSC).

The BSC (Mayr 1942, Coyne and Orr 2004) holds that species are interbreeding groups of populations, and that for an organism to belong to a species simply means it belongs to one of those interbreeding groups. Notice that there is some discrepancy as to whether or not the BSC implies actual or potential interbreeding is sufficient. Speciation occurs when a population becomes reproductively isolated from other parts of the species. Species are treated as units of evolution, bound together by a causal process that is likely to produce a unique group moving forward. The BSC is widely known and used, particularly by population biologists – in large part because looking at interbreeding groups coheres well with population biology models of evolution.

Systematists, on the other hand, more often prefer PSCs, of which there are several varieties. Unlike the BSC, PSCs are primarily concerned with groups of organisms with an exclusive shared history. Species, on a PSC account, are composed of organisms that are more closely related to each other than to any organism in a different species. Notably, interbreeding groups of organisms might not meet this criterion, leading to conflicts

---

[2]   The same issues, albeit in a different form, confront systematists even if we were to reject the individuality thesis.

between PSC and BSC advocates, because the species boundaries given by one concept do not necessarily map on the boundaries given by another. Constraining species to groups with unique and exclusive histories is useful for describing the pattern of evolution of taxa, and the distribution of characters across those taxa, both important inferences to be drawn from work in systematics (Baum 1992). *When* to mark a unique history is a matter of some debate among PSC advocates, ranging from very early (initial diagnosability) in a lineage split (Cracraft 1983) to very late (reciprocal monophyly) (Baum and Donoghue 1995) (see Wheeler and Meier 2000 for more on this and other debates).

Many more species concepts have been proposed, and they have been widely discussed (e.g., Ereshefsky 1992a, Howard and Berlocher 1998, Wilson 1999b). Among the issues at stake is whether one ought to be a monist about species concepts, or instead accept a plurality of concepts as legitimate for biological research and theory (Sober 1984a, Ereshefsky 1992b) – or whether to simply reject the notion of species altogether (see especially Mishler 1999)! More recently, Kevin de Queiroz has argued that conflation of the problem of species concepts with the problem of species delimitation underlies much of the confusion around these debates (de Queiroz 2007; see also de Queiroz 1998 and Wilson 1999b). De Queiroz proposes that there is much more agreement surrounding species concepts than previously recognized, though much disagreement persists over criteria of speciation. It remains to be seen whether this position helps resolve disagreements, for either philosophers or biologists.

Notice that the species concepts described above reflect underlying theoretical commitments and the particular research interests of the biologists involved. This is not at all unique to species, but is a feature common to most concepts in science – particularly theoretical entities. More controversial is the philosophical significance of this fact. Kyle Stanford (1995), for instance, has asked whether we ought to adopt an anti-realist stance towards species.

3.3   *Phylogenetic inference*

The task of discovering and describing species turns on what kinds of individuals species are. What of the other major goal of systematics: determining and describing the phylogenetic relationships of species? Phylogeny is the pattern of common descent and is usually represented by phylogenetic trees

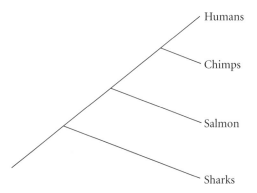

Figure 8.1   A phylogenetic tree displaying the evolutionary relationships between taxa in terms of descent from shared ancestors. Tracing the lines from right to left traces the phylogenetic history of a taxonomic group, with more closely related groups sharing a more recent common ancestor. Thus, salmon are more closely related to chimps and humans than to sharks, by virtue of sharing a more recent common ancestor

(Figure 8.1). Following the branching lines in Figure 8.1 traces the pattern of evolution via common ancestry, and displays the phylogenetic relationships between these groups of species. Phylogenetic studies provide good grounds for evolutionary explanations and inferences – for example, we can appeal to Figure 8.1 to explain why humans and chimps share so many characters (they were passed down from a common ancestor), or to understand the claim that salmon are more closely related to chimps than to sharks (they share a more recent common ancestor).

Discovering phylogenies presents a problem familiar to philosophers of science – a special case of the problem of underdetermination of theory by evidence. For n number of taxa, there are exactly $(2n - 3)!/2^{n-2}(n - 2)!$ possible phylogenetic trees (Felsenstein 2004). In Figure 8.1, there are four taxa and, thus, 15 possible (rooted) trees, though only one that corresponds to the actual historical lineage. The number of possible trees increases exponentially with the number of taxa, such that for 10 taxa there are over 34 million possible trees! Worse, deep evolutionary history cannot be directly observed, and all of these possible trees are *consistent* with the data used to infer this history (where data are simply the distribution of characters across taxa). The challenge facing systematists is twofold: whether phylogenetic inference may be justified in light of such epistemic challenges, and which methods of phylogeny reconstruction allow such justification. As evidenced by the Wiens quotation above, contemporary systematists generally agree that inferring phylogeny is a legitimate (indeed, central) task in modern systematics, and philosophers and biologists have chronicled the emergence of this consensus (see especially Hull 1988).

More relevant to contemporary systematics are debates over which inferential methods provide a justified account of phylogeny. Initially, the phylogenetic technique of choice was parsimony analysis (Hull 1988, Sober 1988). Roughly, parsimony techniques select from among the possible trees the one proposing the fewest number of evolutionary events (Kitching, Forey, Humphries, and Williams 1998). Many leading cladistic theorists initially justified parsimony techniques by explicitly appealing to Karl Popper's falsificationism as a means of solving the problem of phylogenetic inference (Wiley 1975, Farris 1983). The most parsimonious phylogenetic tree was held to be a bold hypothesis of phylogeny, subject to falsification or corroboration depending on future analyses including new or additional data.

In the late 1970s, Joseph Felsenstein demonstrated that lineages of a certain shape were subject to a systematic error in parsimony analysis (Felsenstein 1978, 2004). One example of this is called long-branch attraction (named after the shape of the lineages; see Figure 8.2), in which taxa at the end of long branches are mistakenly grouped together by parsimony analysis, instead of with taxa with which they share a more recent common ancestor. This is not merely an operational problem, but a conceptual challenge to the falsificationist underpinnings of parsimony (Haber 2008). Felsenstein showed that as more data are included, the chance of long-branch attraction increases. This means that parsimony techniques are prone to rejecting hypotheses that correctly capture phylogenetic relationships while corroborating more parsimonious phylogenetic hypotheses that incorrectly group taxa – seemingly producing a systematic inferential error! Felsenstein proposed using statistical techniques (in particular Maximum Likelihood (or ML) methods) to avoid this problem.

Felsenstein's proposal was welcomed by some systematists (*statistical phylogeneticists*), but rejected by others, now typically called *Cladists*.[3] ML techniques were not seen as falling within a falsificationist framework, but instead were viewed as confirmationist. Cladists argued that the cost

---

[3] Here, as is often the case in systematics, the terminology is tricky for the uninitiated. The term "cladist" can be used broadly and narrowly. Broadly, it refers to systematists in the tradition of Willi Hennig who base their work on the foundational principle of monophyly, i.e., proper groups should contain all and only the descendants of a given or hypothesized ancestor. Such groups are called "monophyletic groups," or "clades" (hence the term "cladist"). The Cladists form a narrower group who are cladists, but who also subscribe to a particular methodology (Haber 2008, Williams and Ebach 2008).

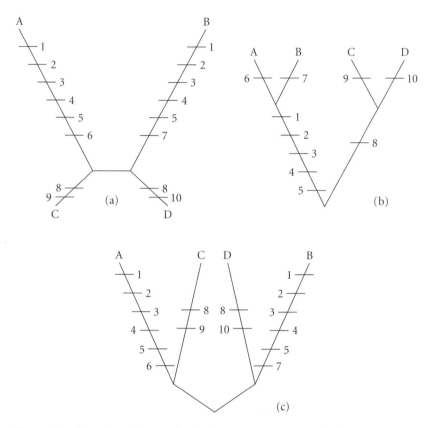

Figure 8.2 Long-branch attraction is a systematic error to which Maximum Parsimony Analysis is subject (see text). (a) is the actual historical pattern of evolution of taxa A, B, C, and D. The length of the branches corresponds to the amount of evolutionary change that has occurred along those branches, with marks representing evolutionary events. (b) is a parsimony analysis of the data from (a), whereas (c) is an ML analysis of that same data. Note that the phylogenetic relationships hypothesized in (b) are not isomorphic with the actual relationships in (a). The long-branch mistake occurs in (b) because parsimony analysis ranks possible phylogenetic hypotheses based on the number of proposed evolutionary events needed to account for all the characters exhibited by the taxa. For example, hypothesis (b) would receive a parsimony score of 10, whereas (c) would receive a parsimony score of 16

of adopting statistical techniques was to abandon good scientific protocol and a return of the problem of inference in systematics; only falsification-ism provided justified methods of surmounting the epistemic challenges facing systematists (e.g., Farris 1983). Thus the persisting debate between Cladists and statistical phylogeneticists is a philosophical disagreement, and not likely to be solved by data in particular cases. What counts as a success in such cases will itself be a matter of dispute. For example, it has been demonstrated that parsimony analysis can be fully construed using ML techniques (Tuffly and Steel 1997, Sober 2004), yet this remains unconvincing to cladists, who reject such characterizations of parsimony as illegitimate due to reliance on statistical approaches (Farris 1983).

4. Ecology and Evolution

While philosophers have paid attention to systematics since the 1960s, they have largely neglected ecology, despite its historical and conceptual richness. This is beginning to change, as works by Cooper (2003), Ginzberg and Colyvan (2004) and Sarkar (2005) attest. One developing area of interest concerns the intersection of ecology with evolution. Two questions can be distinguished:

- What are the relationships between evolutionary and ecological *processes*?
- What are the relationships between evolutionary and ecological *theory*?

In this section, we will explore the relationships between ecology and evolution with regard to both of these questions. The way we answer the former will inform how we approach the latter. We begin by sketching out a common and widely held view about how evolution and ecology are related, then move on to some important concerns about this view and what they might mean for theory in evolution and ecology.

4.1 A two-scale approach

One common way of thinking about how these two disciplines and their systems interact comes from a metaphor first articulated by the eminent ecologist and limnologist G.E. Hutchinson in 1957, who claimed that the "ecological play occurred in the evolutionary theatre." By this he meant that ecological processes occur at rates and scales "below that" of evolutionary processes. To see the contrast, consider two common ecological phenomena: competition and speciation. We say that two species compete when each reduces the rate of growth of the other.[4] More specifically, biologists identify two types of interspecific competition. *Interference competition* occurs when individuals of a species fight with individuals of a different species. By contrast, *exploitative competition* occurs when conspecifics use a common limited resource. Ever since Darwin (1859), biologists of many disciplines have studied competition as it takes place, on scales from a few square meters between two organisms to much larger areas with many more organisms at the ecosystem level.

Now consider speciation, the set of processes by which new species arise. Speciation generally takes thousands of generations to occur. For example, in *allopatric speciation*, a formerly contiguous population of organisms is geographically divided into two or more populations. Over many generations these populations will begin to vary differentially, often because of differences in the selection process as it acts on each population. If and when these two populations meet, they may well find themselves reproductively isolated from each other; that is, members of the two populations may not mate, or they may mate but fail to produce fertile offspring, for any number of reasons.[5]

In the present context what is salient about these two processes – interspecific competition and speciation – is that they are generally understood to occur on different spatial and temporal scales. Competition can occur in a few square meters (look out your window and notice the small birds and mammals foraging away or fighting and you probably see instances of interspecific competition), and within one generation, but the speciation process is generally "splayed out" in space and time.[6]

---

[4]   If one adds an individual of one species, is there a negative, positive, or neutral effect on the per capita rate of growth of the other?

[5]   In the most obvious of cases, the reason why no fertile offspring can be produced will be due to mechanics; however, there are often other more subtle reasons. For example, suppose the one population is diurnal and the other is nocturnal. In this case, they do not encounter one another.

[6]   Not every speciation event need be widely distributed over space and time. Plants have been shown to double in the number of sets of chromosomes in a single generation with the application of any of several ploidy-doubling agents. In addition to phenotypic consequences, ploidy-doubling results in a reproductive barrier between ploidy-doubled plants and their wild-type relatives.

The philosopher Elliott Sober gave a philosophical account of evolutionary theory that fits Hutchinson's two-scale approach rather well. Sober (1984b) articulated and defended the view that evolutionary theory is a "theory of forces."[7] This view is committed to several claims. First, single evolutionary processes like natural selection, mutation, migration, and drift are *forces*, or better yet, *causes* of changes in gene frequencies. These causes are customarily modeled as *singleton laws*. This is a way of describing how natural selection operates when no other cause is at work.[8] For example, consider a single locus on a chromosome with two alleles, *A* and *a*. There are three possible genotypes *AA*, *Aa*, and *aa* and we can associate with each a fitness w_{AA}, w_{Aa}, and w_{aa} respectively. Likewise, let us assume that the frequency of *A* is *p* and the frequency of *a* is *q* (where $p + q = 1$). If we further assume that the only force at work is natural selection, then we have the following dynamical equation for a change in *p*:

$$p' = \frac{p^2 w_{AA} + pq w_{Aa}}{\bar{w}}$$

where $\bar{w} = p^2 w_{AA} + pq w_{Aa} + q^2 w_{aa}$.

Second, these component forces can be combined or "vector summed" together into net forces. These are called *laws of composition*. For example, we can include mutation in our model. Let μ be the fraction of *A* alleles that mutate into *a* alleles. Thus, $(1 - \mu)$ is the fraction of *A* alleles that do not mutate. Our new composition law for natural selection and mutation is:

$$p' = \frac{p^2 w_{AA} + pq w_{Aa} (1 - \mu)}{\bar{w}}$$

According to Sober, the models of evolutionary genetics are *consequence laws* because they predict how gene frequencies will change *given* the laws and the relevant inputs into the equations. However, the inputs – the values of the parameters (genotypic fitnesses and mutation rate) and variables (allele frequencies) – are specified *exogenously*. Sober claims that it is the business

---

[7] It should be noted as well that this view of evolutionary theory has recently come under fire by others, see Sterelny and Kitcher 1988 and Matthen and Ariew 2002 for similar though importantly different criticisms. Likewise, for a more recent defense of Sober's view see Stephens 2004.

[8] Technically speaking, we will be assuming normal Mendelian inheritance which itself is a causal structure. So, we really are modeling natural selection independently of set of circumscribed forces.

of ecology to provide those inputs or explanations of their values. Hence, in ecology, we find *source laws*. On this view, ecology explains how organisms come to differ in their fitnesses, while evolution determines what these fitness differences mean over the long term.

Here is another example of source laws in ecology from what is called "life history theory." Let us say that an organism is *semelparous* if it breeds only once and an organism is *iteroparous* if it breeds more than once. An instance of the difference between these two reproductive strategies might be familiar from the distinction between annual and perennial plants. Given that these strategies are quite different, we might want to know whether it is better from an evolutionary perspective to be an annual or a perennial plant.[9] Consider an annual species and a perennial species and let B_a and B_p be the number of seeds that germinate in the next year respectively. Likewise, let N_a and N_p be the number of annual and perennial plants respectively. Let the survival rate of the perennial species be s and the mortality rate for seeds and seedlings be s_j. For the annuals and perennials, respectively, the equations that describe population growth with time are:

$$N_a(t + 1) = s_j B_a N_a(t)$$
$$N_p(t + 1) = s_j B_p N_p(t) + s N_p(t) = N_p(t)(s_j B_p + s)$$

Suppose the growth rates of annuals and perennials are equal, or[10]

$$s_j B_a = (s_j B_p + s)$$

In effect, we are assuming that being an annual or perennial makes no difference. Rearranging, we have:

$$B_a = B_p + \frac{s}{s_j}$$

However, if we compare the survival rate of mature adults to that of seedlings, the ratio s/s_j is likely to be very large (i.e., $s/s_j \gg 1$). Hence, for

---

[9] For classic discussions of this problem, see Cole 1954, and for amendments and elaborations see Charnov and Shaffer 1973. Our discussion differs in the mathematical formalism of Cole's original discussion and follows Hastings 1997, which is more tractable.

[10] The rate of growth of population i is the ratio of $N_i(t + 1)/N_i(t)$. If this ratio is greater than one, then the population increases; if it is negative, the population declines. So, the annual's rate of growth is $s_j B_A$ and the perennial's rate of growth is $s + s_j B$.

the "annual strategy" to be better evolutionarily than the "perennial strategy", $B_a \gg B_p$. That is, the annuals must produce many, many more seeds than the perennials. Thus, under these circumstances, being a perennial will likely be favored over being an annual.

4.2 Messy interactions: concerns about the two-scale approach

While these models are useful and have been important in the development of evolutionary and ecological thinking, the approach articulated by Hutchinson and elaborated implicitly by Sober is not so tidy empirically. Biologists have found increasing evidence that ecological and evolutionary processes are commensurate in scale. Consider evolutionary processes first. It has been demonstrated that changes in beak size in Darwin's finches can occur remarkably quickly due to changes in rainfall (Grant 1986). Rapid adaptation in insects to insecticides like DDT has also been well documented. By the same token, ecological processes can occur on very long time scales. Consider a mature forest where there has been a forest fire or other disturbance. Some species will go locally extinct and others will migrate into the area. There will be competition for nutrients and light and it may take centuries for the different frequencies of tree species to settle down into something like the previous non-disturbed distribution.

Even a view on which ecological and evolutionary processes occur on the same spatiotemporal scales may still be too simplistic because it is possible for ecological and evolutionary processes to *interact* or *causally affect one another*. *Niche construction* is one example of this kind of interaction. According to niche construction theory, organisms can radically alter their environments in a variety of ways.[11] They can alter how matter and energy move through ecosystems (these organisms are "ecosystem engineers"). For example, there are several species of fungus-growing ants of the tribe Attini, the best known of which are leaf-cutter ants in the genera *Atta* and *Acromyrmex*. Species of these genera cut and move vegetation into nests in order to grow fungus. This provides them with a source of food, and these colonies can reach enormous sizes.

---

[11] Here we focus only on certain aspects of niche construction and not others such as ecological inheritance. For a full discussion of niche discussion one can do no better than Odling-Smee, Laland, and Feldman (2003). For a rich philosophical discussion, see the papers in the January 2005 issue of *Biology and Philosophy*. The examples given above are derived from the Odling-Smee, Laland, and Feldman monograph.

Other organisms can radically change their selective environments. The common cuckoo, *Cuculus canorus*, for example, places its newborns into nests of other species. These newborns then parasitize the brood in the new nest. This behavior has led to important traits such as fast incubation times and the instinct in hatchlings to push "native" eggs from the nest. The cuckoo chicks even exhibit calls that mimic the rate and sound of an entire brood, thereby causing the "mother" to feed only them. As another example, consider the evolution of photosynthesis in bacteria. Before the appearance of these organisms, the earth's atmosphere contained much less oxygen. Oxygen increase as a result of bacteria photosynthesis radically affected biological evolution thereafter.

If these examples are representative of evolutionary and ecological dynamics, then we cannot separate the "evolutionary play" from the "ecological theatre" in the customary way. These concerns about process bring us to the question about theory. Traditionally, theorists constructed models as if Hutchinson's slogan was correct. Ecologists describe their populations of specific species in terms of their demographic properties such as birth and death rates, per capita rates of growth, interaction coefficients, and species densities. In these models, the environment of a population or community is allowed to vary with time. Theoretical ecologists took Hutchinson's slogan to heart and ignored what evolutionary geneticists understand as evolution – change in the frequencies of genes. These geneticists did something very similar as well: they wrote models that predicted the frequency of genotypes as a function of evolutionary "forces" such as natural selection, inbreeding, mutation, migration, genetic drift, etc. Moreover, they have tended to focus on single species and to assume that the environment in which a species finds itself is relatively unchanging. If Hutchinson's slogan is misleading of at least many important biological systems, then we must somehow deal with ecological *and* evolutionary processes occurring in multispecies assemblages in varying environments.

As an illustration of how to model such assemblages, consider a very simple mathematical model of Richard Lewontin's (1983). According to Lewontin, traditional evolutionary theory can be understood in terms of the following pair of differential equations:

$$\frac{dO}{dt} = f(O, E)$$

$$\frac{dE}{dt} = g(E)$$

These equations say that organisms change over time through some functional relationship between themselves and their external environment. However, the environment changes only as some function of itself. In Lewontin's terms, the environment provides "problems" for organisms to "solve"; however, those problems remain unaffected by the organisms. Lewontin argues that this model is wholly inadequate for some of the same reasons as those mentioned above. As an alternative, he offers the following model:

$$\frac{dO}{dt} = f(O, E)$$

$$\frac{dE}{dt} = g(O, E)$$

Organisms and environments *codetermine* each other if processes like niche construction occur frequently. Thus, the phenomena of adaptation – the fit between organism and environment – can be accomplished in more than one way. Through heritable variations in fitness, organisms can "fit" their environments better than alternative types of organisms. However, organisms can also alter their environments, thereby reversing the direction of fit, and forcing it to fit them.

In the end, the processes of evolution and ecology are not necessarily separate either in their spatiotemporal scale or interactively. This suggests that current evolutionary and ecological theory will have to be revised, possibly in radical ways. How radical these revisions should be remains to be seen.

5. Levels of Selection

So far we have presented a brief look at systematics, highlighting some issues of what systematics studies and what issues systematists face when making inferences about the distant past. We have also considered a view that is commonly held by biologists and philosophers of how ecology and evolution are interrelated. In this section we take up several questions around how natural selection works and what it works on. This topic has received rich attention by biologists and philosophers alike, and continues to reward study because of conceptual and empirical advances. This section serves as an introduction to the major issues.

Traditionally, Darwinians have understood selection as acting primarily at the level of the organism. For them, it is the differential survival and reproduction of individual organisms that drives the evolutionary process. There are alternative views, however. Advocates of group selection argue that groups of organisms, rather than individual organisms, may sometimes function as levels of selection; "genic selectionists" such as Richard Dawkins (1976) argue that the true level of selection is in fact the gene; while proponents of multi-level selection (Wilson 1975, Wilson and Wilson 2007) argue that natural selection can occur simultaneously at more than one hierarchical level. Does natural selection act on organisms, genes, groups, colonies, demes, species, or some combination of these?

The levels-of-selection question arises because, in principle, the process of natural selection can operate on any population of entities that satisfies three fundamental requirements, first articulated by Darwin in *On the Origin of Species* and later elaborated more formally by others, including Richard Lewontin (1983) (see §4 above). The first is that the entities should vary in their traits – they must not all be alike. The second is that the variants should enjoy differential reproductive success – some must have more offspring than others. The third requirement is that the traits in question should be heritable, or passed down from parental entities to offspring. If these requirements are satisfied, then the population's composition will change over time in response to selection: fitter entities will gradually supplant the less fit. In this abstract description of how natural selection works, the "entities" are often assumed to be individual organisms. With the rise of molecular genetics, population biology, and ecology in the twentieth century, and accompanying shifts in attention to levels of biological organization, many researchers noticed that biological entities at other hierarchical levels, above and below that of the organism, could satisfy the three key requirements, and hence could form populations that evolve by natural selection. Possibility is not fact, however, and there have long been questions about whether group selection is an empirical reality, and, if so, how important it has been in evolutionary history (Maynard Smith 1964, Wilson and Wilson, 2007).

5.1   The problem of altruism

Debates over the empirical facts and what they might mean has been intimately bound up with the problem of altruism, because altruism is a very clear case in which the level of selection really matters for understanding and explaining the biological world and for evaluating the quality of present evolutionary theory. In evolutionary biology, "altruism" refers to any behavior that is costly to the individual performing the behavior, but benefits others, where the costs and benefits are measured in number of offspring, the units of reproductive fitness. Altruism in this sense is common in nature, particularly among animals living in social groups, but *prima facie*, it is hard to see how it could have evolved by natural selection acting on organisms. By definition, an animal that behaves altruistically will secure fewer resources and have fewer offspring than its selfish counterparts, and so will be selected against. How, then, could altruistic behavior have evolved by a selective process that should eliminate it?

One solution to this puzzle, first suggested by Darwin (1871) himself, is that altruism can evolve by selection at the *group* level. It is possible that groups containing many altruists will out-reproduce groups containing mainly selfish organisms, even though within any group, altruists do worse. In principle, altruism and other group-beneficial behaviors might evolve by natural selection acting on groups, rather than organisms.

Cogent though this argument is, it has been regarded with skepticism by biologists, especially since the publication of G.C. Williams' *Adaptation and Natural Selection* (1966), in which Williams was very critical of what now is sometimes referred to as "naïve group selection" thinking. One reason for doubt about the fact and importance of group selection is that many have argued that the puzzle of altruism can be solved in ways that need not invoke group-level phenomena. According to one influential view, the inclusive fitness or *kin selection* approach first developed by W.D. Hamilton (1964) provides a superior explanation of how altruism evolved.

The basic idea behind kin selection is straightforward. Consider an animal that behaves altruistically, for example by sharing food with others. This behavior is individually disadvantageous, so cannot easily evolve by selection. However, if the animal shares food mainly with its close relatives, rather than with unrelated members of the population, then the behavior can evolve because relatives share genes. In this scenario, there is a certain probability that the recipient of the food will also possess the gene that "causes" the sharing behavior. In other words, if altruistic actions are directed toward kin, the beneficiaries of the actions will themselves be altruists with greater than random chance, and so the altruistic behavior will spread.

Hamilton described these relationships formally in what has come to be known as "Hamilton's rule." The simplest statement of this is $b > c/r$, where b is benefit conferred by the altruist, c is the cost incurred to the altruist, and r is the coefficient of relatedness between the entities. Costs and benefits are calculated in terms of reproductive fitness. This inequality gives the specific conditions under which altruism can be expected to spread, and highlights the importance of genetic relatedness to this way of understanding the evolution of altruism. Hamilton stated the relatedness idea memorably: "To express the matter more vividly . . . we expect to find that no one is prepared to sacrifice his life for any single person but that everyone will sacrifice it when he can thereby save more than two brothers, or four half-brothers or eight first cousins" (Hamilton 1964).

Despite Hamilton's important contributions, the issue of group selection has not been fully settled. Some theorists hold that kin selection, far from being an alternative to group selection, is in fact a *version* of group selection, expressed in different language and using different mathematical models. This issue is partially (though only partially) semantic (cf. Sober and Wilson 1998). Moreover, kin selection can only explain the existence of altruism directed towards relatives, but there are well studied cases of unrelated organisms (those in which r is very low) forming cooperative groups of varying degrees of integration.[12] Finally, some recent theorists have stressed that individual organisms are *themselves* groups of cooperating cells, while each cell is a group of cooperating sub-units, including organelles, chromosomes and genes (Michod 1999). Since cells and multi-celled organisms clearly have evolved, with sub-units that work for the good of the group, it cannot be true that group selection is of negligible importance in the history of life, according to this argument.

From what has been said so far, the levels question may sound purely empirical. Given that natural selection *can* operate at many different levels of the biological hierarchy, surely it is just a matter of finding out the levels(s) at which it *does* act? Surely this is a matter of ordinary empirical enquiry? In fact matters are not quite so simple. The debate over the levels of selection comprises an intriguing mix of empirical, conceptual, and methodological issues, often closely intertwined with each other. This is why the debate is so interesting for philosophers of science.

---

[12]   The range of sociality in insects is one case that philosophers have paid attention to lately and biologists have studied for years. See, for instance, Hamilton, Haber, and Smith (in press) and references therein.

As an example, consider that there is a good deal of debate over how to understand certain biological entities. Are eusocial colonies of insects best understood as groups or as individuals? If they are individuals made up of organisms (individual bees, ants, termites, or wasps) that are parts, it would seem that the individual–group distinction is misleading, and perhaps that models of selection use ontologies that do not properly reflect the organization of the biological world (Wheeler 1911, Wilson and Sober 1989). This comes to a puzzle about colony concepts, akin to the puzzle about species concepts in §3.2 above, but is not peculiar to either of these entities. Similar conceptual problems have come to light in debates about entities at other levels of organization, including groups of related species (Haber and Hamilton 2005, Hamilton and Haber 2006, Okasha 2006) and theoretical early systems of interacting molecules called "hyper-cycles" (Eigen and Schuster 1979, Maynard Smith and Szathmáry 1998). The debate in the latter case is precisely about whether hypercycles are best understood as single entities or as collections of interacting individual biological molecules.

This is not to say that there has not been empirical work to test group, kin, and multi-level selection theory. On the contrary, tests of these theories have been conducted (Wade 1976, 1977, Craig and Muir 1995, Muir 1995, Goodnight and Stevens 1997). However, it is often not clear what these studies *mean* for the debate. As Dawkins demonstrated in *The Selfish Gene* (1976), it is often possible to recast what appears to be altruism at the level of the organism as selfishness at the level of the gene, and it is hard to see what facts might establish that one or the other interpretation is correct. Dawkins' argument is, in part, that evolution operates only at the gene level, and that what appears to be altruism on the part of mothers toward their offspring or of honeybees toward their hive-mates is really behavior that propagates genes that are "ruthlessly selfish." On this view, organisms behave as they do to ensure the survival and differential reproduction of selfish genes. Many have thought that instead of explaining the altruism problem away, Dawkins has shown the need for careful conceptual under-standing of the entities, processes, and theoretical models involved in the way we understand natural selection. This work is ongoing, and is bearing fruit, partly because researchers are asking new questions about major evolutionary transitions, how to incorporate developmental biology within an evolutionary framework, and what relationships groups and individuals bear to one another in tightly (and not-so-tightly) integrated biological systems.

6.   Conclusion

We have tried to give a glimpse of the development of philosophy of biology since the "first wave" in the late 1960s and early 1970s, as well as some details about two areas of biological inquiry – systematics and selection – in which philosophers have had an active role in shaping the conceptual landscape. We have also sketched a mounting problem in the foundations of the study of ecology and evolution, where there is interesting philosophical work to be done. We have not tried to survey the entire field, choosing instead to present miniature case studies of the kind of engagement with science that characterizes not only the best current work in philosophy of biology, but also first- and second-wave philosophy of biology. As the field develops into a widely recognized sub-specialization in philosophy of science, and it becomes possible to spend one's time responding to philosophers rather than to philosophers as well as biologists, it will be increasingly important to emphasize engagement with science and scientists.

Having had a look at the past and present of philosophy of biology, one might ask what shape the fourth wave might take. The prospects are exciting. Because of the efforts of their progenitors, third-wave philosophers (and historians) of biology have found wider acceptance among their peers for work that is conducted collaboratively as well as for work that is conducted across disciplinary boundaries. Interdisciplinarity is not new in philosophy of science, of course, but the possibilities and opportunities are widening and may continue to do so. As more interdisciplinary groups and programs are formed, new avenues for intellectual work are opened, as are new channels of communication between scientists and philosophers. Philosophers have long been involved in cognitive science, evolutionary theory, and systematics, but as the case of ecology shows, there are other areas of biology in which there is work to be done that is both philosophically interesting and potentially of value to theorists in those areas.

These new possibilities for intellectual work are not limited to tackling conceptual problems faced by scientists. As it becomes accepted practice that philosophers should do part of their training in laboratories and in science courses, it also becomes clear that science can inform philosophy. Recent work on explanation is one example of this phenomenon. We inherited from philosophy of physics an account of explanation that did not obviously apply well to biology, as Hull (1974) and Wimsatt (1976) noticed a generation ago because they were deeply engaged in understanding science and its practice. Biological explanation and how it works is now one of the more lively topics of discussion in philosophy of science. Perhaps it is not too much to hope that as the fourth wave develops there will be continued interest in conceptual topics raised by scientists as well as in traditional issues in philosophy of science that are understood in new ways as the relevant science becomes a more prominent impetus for philosophical investigation.

References

Baum, D. 1992. "Phylogenetic Species Concepts." *Trends in Evolution and Ecology* 7: 1–2.

Baum, D. and M.J. Donoghue. 1995. "Choosing Among Alternative 'Phylogenetic' Species Concepts." *Systematic Botany* 20.4: 560–73.

Boyd, R. 1999. "Homeostasis, Species, and Higher Taxa." In R. Wilson (ed.), *Species: New Interdisciplinary Essays*. Cambridge, MA: MIT Press.

Brandon, R.N. 1981. "Biological Teleology: Questions and Explanations." *Studies in History and Philosophy of Science* 12: 91–105.

Brandon, R.N. 2006. "The Principle of Drift: Biology's First Law." *Journal of Philosophy* 102: 319–35.

Charnov, E.L. and M. Schaffer. 1973. "Life History Consequences of Natural Selection: Cole's Result Revisited." *American Naturalist* 107: 791–3.

Cole, L.C. 1954. "The Population Consequences of Life History Phenomena." *Quarterly Review of Biology* 25: 103–27.

Cooper, G.J. 2003. *The Science of the Struggle for Existence: On the Foundations of Ecology*. Cambridge: Cambridge University Press.

Coyne, J.A. and H. Allen Orr. 2004. *Speciation*. Sunderland, MA: Sinauer Press.

Cracraft, J. 1983. "Species Concepts and Speciation Analysis." *Current Ornithology* 1: 159–87.

Craig, J.V. and W.M. Muir. 1995. "Group Selection for Adaptation to Multiple Hen-Cages: Beak-Related Mortality, Feathering, and Body Weight Responses." *Poultry Science* 75: 294–302.

Cummins, R. 1977. "Functional Analysis." *Journal of Philosophy* 72.20: 741–65.

Darwin, C. 1859. *On the Origin of Species*. Cambridge, MA: Harvard University Press, 1975.

Darwin, C. 1871. *The Descent of Man*. London: John Murray.

Dawkins, R. 1976. *The Selfish Gene*. Oxford: Oxford University Press.

de Queiroz, K. 1998. "The General Lineage Concept of Species, Species Criteria, and the Process of Speciation: A Conceptual Unification and Terminological

Recommendations." In Daniel J. Howard and Stewart H. Berlocher (eds.), *Endless Forms: Species and Speciation*. Oxford: Oxford University Press.

de Queiroz, K. 2007. "Species Concepts and Species Delimitation." *Systematic Biology* 56.6: 879–86.

Eigen, M. and P. Schuster. 1979. *The Hypercycle: A Principle of Natural Self-Organization*. Berlin: Springer.

Elser, J.J. and A. Hamilton. 2007. "Stoichiometry and the New Biology: The Future is Now." *PLoS Biology* 5: 1403–5.

Ereshefsky, M. (ed.). 1992a. *The Units of Evolution: Essays on the Nature of Species*. Cambridge, MA: MIT Press.

Ereshefsky, M. 1992b. "Eliminative Pluralism." *Philosophy of Science* 59.4: 671–90.

Farris, J.S. 1983. "The Logical Basis of Phylogenetic Analysis." In N.I. Platnick and V.A. Funk (eds.), *Advances in Cladistics, Volume 2: Proceedings of the Second Meeting of the Willi Hennig Society*. New York: Columbia University Press, 7–36.

Felsenstein, J. 1978. "The Number of Evolutionary Trees." *Systematic Zoology* 27: 27–33.

Felsenstein, J. 2004. *Inferring Phylogenies*. Sunderland, MA: Sinauer Associates, Inc.

Ghiselin, M. 1966. "On Psychologism in the Logic of Taxonomic Controversies." *Systematic Zoology* 15.3: 207–15.

Ghiselin, M. 1974. "A Radical Solution to the Species Problem." *Systematic Zoology* 23.4: 536–44.

Ginzberg, L. and M. Colyvan. 2004. *How Planets Move and Populations Grow*. Oxford: Oxford University Press.

Grant, P.R. 1986. *Ecology and Evolution of Darwin's Finches*. Princeton: Princeton University Press.

Grene, M. 1974. *The Understanding of Nature: Essays in the Philosophy of Biology*. R.S. Cohen (ed.). Dordrecht: Springer.

Griffiths, P. 1999. "Squaring the Circle: Natural Kinds with Historical Essences." In R. Wilson (ed.), *Species: New Interdisciplinary Essays*. Cambridge, MA: MIT Press.

Goodnight, C.J. and L. Stevens. 1997. "Experimental Studies of Group Selection: What Do They Tell Us about Group Selection in Nature?" *American Naturalist* 150: S59–79.

Haber, M.H. 2008. "Phylogenetic Inference." In A. Tucker (ed.), *A Companion to Philosophy of History and Historiography*. Malden, MA: Blackwell Publishing.

Haber, M.H. and A. Hamilton. 2005. "Coherence, Consistency, and Cohesion: Clade Selection in Okasha and Beyond." *Philosophy of Science* 72: 1026–40.

Hamilton, A. 2007. "Laws of Biology, Laws of Nature: Problems and (Dis)solutions." *Philosophy Compass* 2: 592–610.

Hamilton, A. and M.H. Haber. 2006. "Clades are Reproducers." *Biological Theory* 1: 381–91.

Hamilton, A., M.H. Haber, and N.R. Smith. 2009. "Social Insects and the Individuality Thesis: Cohesion and the Colony as a Selectable Individual." In J. Gadau and J. Fewell (eds.), *Organization of Insect Societies: From Genome to Sociocomplexity*. Cambridge, MA: Harvard University Press, 572–89.

Hamilton, W.D. 1964. "The Genetical Evolution of Social Behaviour I and II." *Journal of Theoretical Biology* 7: 1–16 and 17–52.

Hastings, A. 1997. *Population Biology: Models and Concepts*. New York: Springer.

Hennig, W. 1966. *Phylogenetic Systematics*. Urbana, IL: University of Illinois Press.

Howard, D.J. and S.H. Berlocher (eds.). 1998. *Endless Forms: Species and Speciation*. Oxford: Oxford University Press.

Hull, D.L. 1974. *Philosophy of Biological Science*. Englewood Cliffs, NJ: Prentice Hall.

Hull, D.L. 1976. "Are Species Really Individuals?" *Systematic Zoology* 25.2: 174–91.

Hull, D.L. 1977. "A Logical Empiricist Looks at Biology." *British Journal for the Philosophy of Science* 28: 181–94.

Hull, D.L. 1978. "A Matter of Individuality." *Philosophy of Science* 45: 355–60.

Hull, D.L. 1988. *Science as a Process: An Evolutionary Account of the Social and Conceptual Development of Science*. Chicago: University of Chicago Press.

Hutchinson, G.E. 1957. "Concluding Remarks." *Cold Spring Harbor Symposia on Quantitative Biology* 22: 415–27.

Kitcher, P. 1984. "Species." *Philosophy of Science* 51.2: 308–33.

Kitching, I.J., P.L. Forey, C.J. Humphries, and D.M. Williams. 1998. *Cladistics*. 2nd edn. Oxford: Oxford University Press.

Knorr Cetina, K. 1999. *Epistemic Cultures: How the Sciences Make Knowledge*. Cambridge, MA: Harvard University Press.

Levins, R. 1966. "The Strategy of Model Building in Population Biology." *American Scientist* 54: 421–31.

Lewontin, R.C. 1983. "Gene, Organism, and Environment." In D.S. Bendall (ed.), *Evolution from Molecules to Men*. Cambridge: Cambridge University Press.

Matthen, M. and A. Ariew. 2002. "Two Ways of Thinking about Fitness and Natural Selection." *Journal of Philosophy* 49: 55–83.

Maynard Smith, J. 1964. "Group Selection and Kin Selection." *Nature* 201: 1145–7.

Maynard Smith, J. 1982. *Evolution and the Theory of Games*. Cambridge: Cambridge University Press.

Maynard Smith, J. 1988. "Evolutionary Progress and Levels of Selection." In M. Nitecki (ed.), *Evolutionary Progress*. Chicago: University of Chicago Press, 219–30.

Maynard Smith, J. and E. Szathmáry. 1998. *The Major Transitions in Evolution*. Oxford: Oxford University Press.

Mayr, E. 1942. *Systematics and the Origin of Species*. New York: Columbia University Press.

Mayr, E. 1959. "Typological Versus Population Thinking." In *Evolution and Anthropology: A Centennial Appraisal*. Washington, DC: Anthropological Society of Washington, 409–12.

McLaughlin, J. and J. Roughgarden. 1991. "Pattern and Stability in Predator Communities: How Diffusion in Spatially Variable Environments Affects the Lotka-Volterra Model." *Theoretical Population Biology* 40: 148–72.

Michod, R.E. 1999. *Darwinian Dynamics, Evolutionary Transitions in Fitness and Individuality*. Princeton, NJ: Princeton University Press.

Mishler, B. 1999. "Getting Rid of Species?" In R. Wilson (ed.), *Species: New Interdisciplinary Essays*. Cambridge, MA: MIT Press.

Mitchell, S. 2003. *Biological Complexity and Integrative Pluralism*. Cambridge: Cambridge University Press.

Muir, W.M. 1995. "Group Selection for Adaptations to Multiple-Hen Cages: Selection Program and Direct Responses." *Poultry Science* 75: 447–58.

Odenbaugh, J. 2006. "The Strategy of 'The Strategy of Model Building in Population Biology'." *Biology and Philosophy* 21: 607–21.

Odling-Smee, F.J., K.N. Laland, and M.W. Feldman. 2003. *Niche Construction: The Neglected Process in Evolution*. Princeton, NJ: Princeton University Press.

Okasha, S. 2006. *Evolution and the Levels of Selection*. Oxford: Oxford University Press.

Price, G.R. 1970. "Selection and Covariance." *Nature* 227: 520–1.

Queller, D.C. 1992. "A General Model for Kin Selection." *Evolution* 46: 376–80.

Reisman, K. and P. Forber. 2005. "Manipulation and the Causes of Evolution." *Philosophy of Science* 72: 1113–23.

Rosenberg, A. 1994. *Instrumental Biology or the Disunity of Science*. Chicago: University of Chicago Press.

Ruse, M. 1970. "Are There Laws in Biology?" *Australasian Journal of Philosophy* 48: 234–46.

Sarkar, S. 2005. *Biodiversity and Environmental Philosophy: An Introduction to the Issues*. Cambridge: Cambridge University Press.

Sober, E. 1980. "Evolution, Population Thinking, and Essentialism." *Philosophy of Science* 47.3: 350–83.

Sober, E. 1984a. "Discussion: Sets, Species, and Evolution. Comments on Philip Kitcher's 'Species'." *Philosophy of Science* 51: 334–41.

Sober, E. 1984b. *The Nature of Selection: Evolutionary Theory in Philosophical Focus*. Chicago: University of Chicago Press.

Sober, E. 1988. *Reconstructing the Past: Parsimony, Evolution, and Inference*. Cambridge, MA: MIT Press

Sober, E. 2004. "The Contest between Parsimony and Likelihood." *Systematic Biology* 53: 644–53.

Sober, E. and D.S. Wilson. 1998. *Unto Others: The Evolution and Psychology of Unselfish Behavior*. Cambridge, MA: Harvard University Press.

Stanford, K. 1995. "For Pluralism and Against Monism About Species." *Philosophy of Science* 62: 72–90.

Stephens, C. 2004. "Selection, Drift and the 'Forces' of Evolution." *Philosophy of Science* 71: 550–70.

Sterelny, K. and P. Kitcher. 1988. "The Return of the Gene." *Journal of Philosophy* 85: 339–61.

Sterner, R. and J.J. Elser. 2002. *Ecological Stoichiometry: The Biology of Elements from Molecules to the Biosphere.* Princeton, NJ: Princeton University Press.

Tuffly, C. and M. Steel. 1997. "Links between Maximum Likelihood and Maximum Parsimony under a Simple Model of Site Substitution." *Bulletin of Mathematical Biology* 59: 581–607.

Wade, M.J. 1976. "Group Selection among Laboratory Populations of Tribolium." *Proceedings of the National Academy of Sciences* 73: 4604–7.

Wade, M.J. 1977. "An Experimental Study of Group Selection." *Evolution* 31: 134–53.

Wade, M.J. 1985. "Soft Selection, Hard Selection, Kin Selection, and Group Selection." *American Naturalist* 125: 61–73.

West, G.B., J.H. Brown, and B.J. Enquist. 1997. "A General Model for the Origin of Allometric Scaling Models in Biology." *Science* 276: 122–6.

Wheeler, Q.D. and R. Meier (eds.). 2000. *Species Concepts and Phylogenetic Theory: A Debate.* New York: Columbia University Press.

Wheeler, W.M. 1911. "The Ant Colony as Organism." *Journal of Morphology* 22: 307–25.

Wiens, J.J. 2007. "Species Delimitation: New Approaches for Discovering Diversity." *Systematic Biology* 56.6: 875–8.

Wiley, E.O. 1975. "Karl R. Popper, Systematics and Classification: A Reply to Walter Bock and Other Evolutionary Taxonomists." *Systematic Zoology* 24: 233–43.

Williams, D.M. and M.C. Ebach. 2008. *Foundations of Systematics and Biogeography.* New York: Springer.

Williams, G.C. 1966. *Adaptation and Natural Selection.* Oxford: Oxford University Press.

Wilson, D.S. 1975. "A Theory of Group Selection." *Proceedings of the National Academy of Sciences of the United States of America* 72: 143–6.

Wilson, D.S. and E. Sober. 1989. "Reviving the Superorganism." *Journal of Theoretical Biology* 136: 337–56.

Wilson, D.S. and E.O. Wilson. 2007. "Rethinking the Foundation of Sociobiology." *Quarterly Review of Biology* 82: 327–48.

Wilson, R. 1999a. "Realism, Essence, and Kind: Resuscitating Species Essentialism?" In R. Wilson, *Species: New Interdisciplinary Essays.* Cambridge, MA: MIT Press.

Wilson, R. 1999b. *Species: New Interdisciplinary Essays.* Cambridge, MA: MIT Press.

Wimsatt, W. 1976. "Reductive Explanation: A Functional Account." In A.C. Michalos, C.A. Hooker, G. Pearce, and R.S. Cohen (eds.), *PSA 1974*. Dordrecht: Reidel, 671–710.

Wimsatt, W.C. 1981. "Robustness, Reliability, and Overdetermination." In M. Brewer and B. Collins (eds.), *Scientific Inquiry and the Social Sciences*. San Francisco, CA: Jossey-Bass, 124–63.

Winsor, P. 2006. "The Creation of the Essentialism Story: An Exercise in Metahistory." *History and Philosophy of the Life Sciences* 28: 149–74.

Winther, R.G. 2006. "Parts and Theories in Compositional Biology." *Biology and Philosophy* 21: 471–99.

9   Philosophy of Earth Science

Maarten G. Kleinhans, Chris J.J. Buskes, and Henk W. de Regt

 This essay presents a philosophical analysis of earth science, a discipline which has received relatively little attention from philosophers of science. We focus on the question of whether earth science can be reduced to allegedly more fundamental sciences, such as chemistry or physics. In order to answer this question, we investigate the aims and methods of earth science, the laws and theories used by earth scientists, and the nature of earth-scientific explanation. Our analysis leads to the tentative conclusion that there are emergent phenomena in earth science but that these may be reducible to physics. However, earth science does not have irreducible laws, and the theories of earth science are typically hypotheses about unobservable (past) events or generalized – but not universally valid – descriptions of contingent processes. Unlike more fundamental sciences, earth science is characterized by explanatory pluralism: earth scientists employ various forms of narrative explanations in combination with causal explanations. The main reason is that earth-scientific explanations are typically hampered by local underdetermination by the data to such an extent that complete causal explanations are impossible in practice, if not in principle.

1.   Introduction

Earth science has received relatively little attention from philosophers of science. To be sure, plate tectonics and the dinosaur extinction event are frequently cited in the literature, but most of earth science is *terra incognita* to philosophers. One reason for this lack of interest may be found in the widespread idea that earth science is not an autonomous science, but is easily reducible to allegedly more fundamental sciences such as physics and chemistry. While in the past decades serious doubts have been raised about the reductionist program, and disciplines such as biology and psychology have thereby achieved the status of autonomous "special sciences"[1] (Fodor 1974, Dupré 1993), earth science has barely been mentioned in these debates, probably because the phenomena and processes in its domain are assumed to be purely physical and chemical and therefore easily reducible to lower levels.

This essay will make it clear that earth science deserves the attention of philosophers of science. The geosciences have typical features that distinguish them from other natural sciences. These features will be highlighted in the course of an investigation of the nature of earth-scientific explanations. §2 briefly reviews the disciplinary *aims* of earth science. §3 aims to answer the question of whether *reductionism* applies to earth science. First, the nature of earth-scientific *theories* and *laws* is discussed. Subsequently, we investigate whether emergent earth-scientific phenomena exist that are irreducible to phenomena at the chemical and physical level. §4 investigates what kind(s) of *explanation* can be found in earth science. We discuss whether narrative explanation is employed and whether narrative explanation is reducible to causal explanation.[2] The notion of *underdetermination* plays a central role in our argument: earth-scientific theories and hypotheses are usually underdetermined by the available evidence, and therefore complete causal explanations are out of reach. Earth science typically employs combinations of causal and narrative explanations. Finally, we will highlight a methodological strategy that is typical of the geosciences: *abduction* or *inference to the best explanation*.

2.   Object and Aims of Earth Science

Earth scientists study the earth, that is to say, its structure, phenomena, processes, and history. What exactly is it that earth scientists hope to achieve

---

[1]   The term "special sciences" reminds us of the Monty Python movie *Life of Brian*. Imagine a dusty square in Jerusalem, with a mob of scientists from every discipline, chasing philosopher Brian whose statements they accept unconditionally. Then Brian appears at a window and shouts to the crowd: "You are all individuals!" All scientists reply in unison: "We are all individuals!," except the physicist, who replies: "I'm not."

[2]   We use the term "causal explanation" generically for the type of explanation employed in the physical sciences (in contrast to, for example, narrative, functional, and teleological explanation). Of course, we are aware that this type of explanation can further be specified in a variety of ways, but for present purposes this may be ignored (cf. §4).

with their study of the earth? In his *System of the Earth*, James Hutton, one of the founding fathers of earth science, proposed

> to examine the appearances of the earth, in order to be informed of operations which have been transacted in time past. It is thus that, from principles of natural philosophy, we may arrive at some knowledge of order and system in the economy of this globe, and may form a rational opinion with regard to the course of nature, or to events which are in time to happen. (1785, 2)

Apparently, Hutton saw the main aim of earth science as historical description (acquiring knowledge of the "course of nature"), but the statement contains a suggestion that causal explanation and prediction are aims as well. More recently, Rachel Laudan distinguished two aims:

> One is historical: geology should describe the development of the earth from its earliest beginnings to its present form. The other is causal: geology should lay out the causes operating to shape the earth and to produce its distinctive objects. (1987, 2)

This twofold aim arises because some phenomena, like the comet impacts and their effects on the history of life, appear to be understandable only as results of cumulative coincidences, to be inferred and reconstructed from scant evidence, whereas other phenomena, like ocean circulation, appear to have a structure that can be understood on the basis of physical laws.

There are two related issues at stake here (see Kleinhans, Buskes, and de Regt 2005 for more discussion). First is the question of whether earth science is a historical or a nomological (law-formulating) science, or a combination of both. Second is the question of whether earth science aims for description, for explanation, or for both. The questions are related because historical sciences are typically associated with descriptive aims, while nomological sciences are associated with explanatory aims – the traditional distinction between historical description and nomological explanation. However, this is not a necessary connection: one might alternatively claim that historical sciences can explain as well, or that neither historical nor nomological sciences explain. Incidentally, "real" sciences may feature both historical description and nomological explanation, which implies that the distinction is one between "ideal types."

The second question relates to a long-standing debate on the question of whether there exists a separate category of explanations, so-called narrative explanations, that can be provided by historical descriptions. Those who reject this idea either defend that historical sciences do not furnish explanations at all, or believe that their explanations are of the same kind as those in the physical sciences (choosing the latter option still leaves us with a choice between various rival theories of scientific explanation). In §4, we will argue that while earth science is partly a historical science, it does provide explanations because the historical descriptions provided by earth scientists are narrative explanations that integrate causal explanations, sequential reconstructions of the geological past, observations, and background theories.

While earth science has, at least at first sight, a single object (the earth), a deeper look into any textbook shows a bewildering variety of sub-disciplines and approaches. Concepts and techniques are both developed within earth science and borrowed from other disciplines, such as logic, mathematics, physics, chemistry, biology, and computer modeling. A unified body of theory, topics, or techniques seems to be lacking. Earth science is divided into many sub-disciplines with different aims.[3] Sub-disciplines with historical aims ask different questions and use different explanatory strategies than sub-disciplines that focus on causal questions. This can be clarified with the help of Figure 9.1, which visualizes the distinction between deductive, inductive, and abductive explanation.

Scientists are interested in causes, effects, and laws (the three corners of the triangle). Typically, they possess knowledge about two of these and want to infer the third. The three resulting modes of inference are cases of deduction, induction, and adduction, respectively. Employing causes and laws to predict effects is a form of deduction. Combining causes and effects to identify a law or generalization is a form of induction. And inferring causes from knowledge of effects and laws is a case of abduction. For example, geomorphologists claim to be studying general physical processes by deduction and induction, while geologists reconstruct the geological past by abduction. Practitioners associate the former with causal explanation and the latter with historical description, which is believed to be of lesser value by many practitioners of the former. Accordingly, the question of whether historical narratives can be reduced to general process-oriented explanations is a hot topic in institutes which contain both sub-disciplines. (This will be discussed in more detail in §4.1; also see Kleinhans, Buskes, and de Regt 2005 for more discussion.)

---

[3] A non-exhaustive list of disciplines is: geology, geophysics, sedimentology, physical geography, geomorphology, biogeology, biogeography, civil engineering, geodesy, soil science, environmental science, planetology, geochemistry, meteorology, climatology, and oceanography.

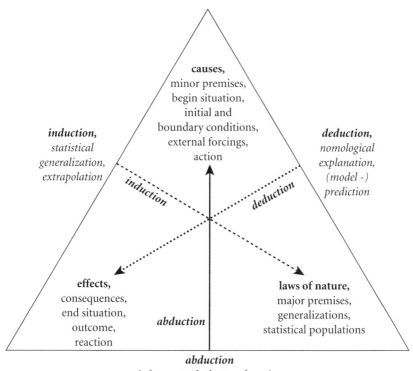

Figure 9.1 Three types of explanation based on causes, effects, and laws, two of which are necessary to arrive at the third. Problems of induction are well known. Abduction is fallible in practice due to problems of underdetermination. Also deductive explanation, particularly in the form of computer modeling, may be hampered by underdetermination problems (see §4.3)

3. The Autonomy of Earth Science

Can earth science be reduced to lower-level sciences like chemistry and physics? According to the traditional model of reduction (Nagel 1961), this requires that the laws and theories of earth science can be deduced from the laws of chemistry and physics, and this in turn requires bridge principles that connect the terms used in the different laws. Nagel-type reduction is "global," in the sense that if a theory is reduced in this fashion all phenomena in the domain of the theory are reduced as well. There are two major obstacles to Nagel's model of reduction: (1) bridge principles often do not exist; and (2) higher-level laws do not conform to the traditional criteria for lawfulness, in particular the requirement that laws are universal and exceptionless generalizations (necessary for global reduction). We first discuss the nature of earth-scientific theories and laws by investigating whether they conform to Nagel's traditional conception of laws or to a recent alternative conception. Subsequently, we consider whether the existence of allegedly emergent phenomena in earth science can be used as an argument against reductionism.

3.1   Theories and laws

What is the nature of the theories and laws that earth scientists actually use? First of all, it should be noted that because of the historical aim of earth science, its practitioners often use the term "theory" in cases of hypothetical historical events. The most famous example is the "impact theory" that is proposed to explain the mass extinction 65 million years ago; this "theory" states that about 65 million years ago a meteorite collided with the earth, causing a radical climate change and a mass extinction of species, among which were the dinosaurs. This hypothesis is a theory in the sense that it postulates an event or chain of events in the past (and therefore not directly observable any more) that explains observed phenomena, but it is not a theory in the sense that it specifies laws or a general model for the explanation of phenomena.

An obvious case of the latter kind is the theory of plate tectonics, often hailed as the grand unifying theory of earth science playing a role comparable to that of natural selection in biology. Plate tectonics provides a mechanistic underpinning of continental drift and was originally conceived by Alfred Wegener as a hypothesis explaining mountain building and the shape of the continents. It explains a host of other phenomena and processes, which cover almost all temporal and spatial scales relevant to earth scientists: continental drift and mountain building are long-term processes, whereas earthquakes and volcanic activity are short-term phenomena which may readily occur within a human lifetime.

But what is the precise nature and status of the generalizations that plate tectonics contains? Do they qualify as laws? The theory describes the formation and movement of plates, and postulates an underlying process in the earth's inner parts (mantle convection) that is responsible for the forces that cause the plates to form and move. There have been attempts to model the mantle dynamics of other planets, which lead to the completely different patterns of rigid-lid mantle convection on Venus and mantle superplume dominance on Mars, agreeing with interpretations of various surface observations. Thus, plate tectonics provides a general model of crust formation and movement and of mantle convection that is valid for very different situations and planets. A simple example of a generalization (a candidate law) within plate tectonics is: "Earthquakes are generated in the rigid plate as it is subducted into the mantle." Is this a law? Not according to the traditional criteria, because it is not universal: it refers to a specific spatiotemporal situation, namely the earth as we know it today. It contains specific earth-scientific terms such as "earthquake," "plate," "subduction," and "mantle." If we specify bridge principles by translating these terms into the language of chemistry and physics, we see that they refer either to contingent distributions of matter in the earth:

- "Mantle" refers to the stony but slightly fluid layer surrounding the core of the planet, made of minerals rich in the elements iron, magnesium, silicon, and oxygen (as opposed to the core, which is the central metallic part of the planet).
- "Plate" refers to a broken piece of the rigid outermost layer of the earth (at present, the earth contains 12 plates).

Or else they refer to processes related to the specific structure of the earth:

- "Earthquake" refers to the failure of the plate when static friction is exceeded and a movement of one block with respect to the other block occurs, giving rise to oscillations or seismic waves.
- "Subduction" refers to the sinking of heavy material of the crust into fluid material, caused by the collision between two plates (which is, in turn, driven by convection in the mantle).

Although plate tectonics is applicable to other earth-like planets as well (see, e.g., van Thienen, Vlaar, and van den Berg 2004), it is clear that its "laws" differ from the laws of physics and chemistry in the sense that their validity is less universal and more tied to the specific constellation of the planet in question. For example, plate-tectonic models would not apply to the gas

giants of the outer solar system. In other words, the model does not contain laws in the traditional sense of universal, exceptionless regularities;
it merely describes contingent phenomena, dependent on the particular
configuration of the earth's structure.

In this respect, plate tectonics fits the analysis that Beatty (1995) provides for the case of biological generalizations. Beatty specifies his "evolutionary contingency thesis" as follows: "All generalizations about the living
world: (a) are just mathematical, physical, or chemical generalizations (or
deductive consequences of mathematical, physical, or chemical generalizations plus initial conditions), or (b) are distinctively biological, in which
case they describe contingent outcomes of evolution" (1995, 46–7). We
claim that Beatty's evolutionary contingency thesis applies equally well to
earth-scientific generalizations such as plate tectonics. This entails that
plate tectonics does not contain autonomous, irreducible laws. Moreover,
there is no reason not to assume that the plate-tectonic model and its laws
can be reduced to lower-level physical or chemical theories. Earth-scientific
generalizations, such as the cited example regarding earthquakes, describe
contingent distributions and processes which can be reduced "locally" because
they can be exhaustively translated in physical and/or chemical terms.

But perhaps there are autonomous, irreducible earth-scientific laws to
be found elsewhere? A candidate might be the set of principles used in the
ordering of rock and sediment layers in historical explanations. The most
important of these are the principles of superposition and of cross-cutting
(Kitts 1966). Superposition means that a layer on top of another layer must
have been formed *after* the lower layer was formed. The lower layer is a
necessary temporal antecedent but not a cause of the formation of the upper
layer. For instance, it is very unlikely that a sediment layer was deposited
below another layer, because it would involve breaking up, eroding, and
redepositing of the upper layer, which obviously destroys it. Likewise, a layer
cross-cutting other layers means that the cross-cutting layer was formed
after the other layers. However, examples can be given in which these
principles lead us astray. A set of layers may have been overturned in severe
folding during mountain building; volcanic activity or fluvial channels
may form a cross-cutting layer at the same time of the deposition of the
planar sediment; and so on. Earth scientists are aware of these pitfalls and
deliberately seek for evidence for such exceptions when applying the
general principles. But the examples show that these geological principles
cannot be regarded as laws in the traditional Nagelian sense: they have many
exceptions. Consequently, these principles cannot be reduced globally,

that is, according to the Nagelian model of reduction. However, this is an argument against Nagel's model rather than an argument against the reducibility of earth-scientific principles. As in the case of plate tectonics, there is a more liberal sense of "local" reduction that applies to generalizations and principles that are less universal than physical laws.

We conclude that earth science does not have irreducible laws, and that the theories of earth science are typically hypotheses about unobservable (past) events, or generalized – but not universally valid – descriptions of contingent processes. In contrast to physics and chemistry but analogously to biology, an important part of theories of earth science consists in descriptions of contingent states of nature (Beatty 1995). The traditional account of reduction (Nagel's model) fails to apply because earth-scientific generalizations do not conform to the traditional criteria for lawfulness. This implies that reductionism is still a viable option (though not in the strict Nagelian sense), because Beatty's account does not entail that higher-level laws are autonomous and is therefore compatible with reductionism.

Incidentally, not only do earth-scientific generalizations resemble biological ones, but there is a close interaction between earth science and biology. Indeed, many earth-scientific phenomena would not exist without interaction with life. Figure 9.2a provides a random selection of examples of earth-scientific phenomena from very small to very large length and time scales. This typical reference to the length and time scales of such phenomena is discussed later. Figure 9.2b gives examples of earth-scientific phenomena which would not exist without biological elements. For example, the composition of the atmosphere of the earth compared to that of Venus and Mars has much more oxygen and much less carbon dioxide. This so-called oxygen revolution was largely caused by photosynthesizing organisms (mostly algae) about 2 billion years ago. The presence of oxygen, in turn, led to increased oxidation of minerals, weathering of rocks, and chemical changes in the oceans. A second example is the variety of effects of life on rivers. Plant roots may stabilize the banks of rivers, which may actually cause rivers to change from a wide, shallow "braided" planform with many mid-channel bars to a narrow, deep "meandering" planform with only one or a few channels. Animals burrowing in or treading on the banks may initiate diversions of the flow and completely new courses of the river. At a larger scale, vegetation in the upstream catchment of a river may strongly damp the surface runoff from rain storms, leading to a much more regular water discharge regime than in unvegetated catchments, with all sorts of consequences for the morphology and geology.

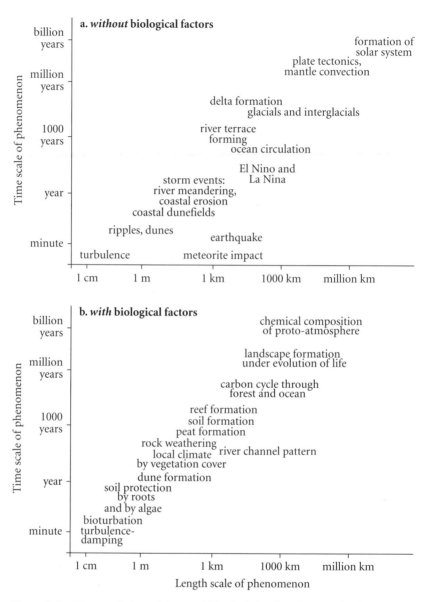

Figure 9.2 The correlation of time and length scale of earth-scientific phenomena, split out between phenomena where biological factors are not important (a) and where these are important (b). The correlation is caused by the limited range of energy available to do work at the earth's surface. Two major exceptions are earthquakes and meteorite impacts; the former because it uses energy from within the planet and the second because it brings extraterrestrial kinetic energy

3.2   The emergent nature of the earth-scientific phenomena

In the previous section, it was argued that reductionism fails in the strong Nagelian sense, but a weaker (local) sense of reduction may still be applicable to earth science. Anti-reductionists might reply by invoking the alleged "emergence" of earth-scientific phenomena as an argument against their reducibility. What precisely is emergence? Humphreys (1997, 341–2) lists the following possible criteria for emergence: emergent properties are novel; are qualitatively different from the properties from which they emerge; could not be possessed at a lower level; have different laws apply to them; result from an essential interaction between constituent properties; and are holistic in the sense of being properties of the entire system (see Kleinhans, Buskes, and de Regt 2005 for more discussion). There is no agreement among philosophers on the question of whether or not emergence is compatible with reduction. Below we investigate whether emergence in earth science provides an argument against reducibility.

A first glance at earth science suggests that it is replete with emergence in one of the senses mentioned above: current ripples, rivers, deltas, volcanoes, mantle plumes, and continents consist of matter that is organized in such a way that novel, qualitatively different properties arise that do not directly follow from physico-chemical laws but seem to comply with higher-level laws. The literature on emergent phenomena with scale-independent characteristics is extensive and focuses mainly on self-organization, self-similarity, and chaos (for examples and details, see Smith 1998, Ball 1999). For earth science, we identify two classes of phenomena.

The first class consists of self-similar, chaotic phenomena. These are characterized by a repeated basic pattern lacking a dominant length scale; the basic pattern occurs at a large range of length or time scales in the phenomenon, which is then called self-similar. The self-similarity at many scales is novel and qualitatively different from the microscopic properties of the constituent properties. Commonly the largest scales occur much less frequently than the smaller scales, which is referred to as $1/f$ ($f =$ frequency) scaling (Bak, Tang, and Weisenfeld 1987). There are many examples of this class in earth science. Clouds and river drainage networks are well known for self-similarity (Ball 1999), as are river discharge records (Mandelbrot and Wallis 1969), sizes of avalanches on sand piles and sizes of forest fires (Bak, Tang, and Weisenfeld 1987), elevation of landscapes and coastlines (Burrough 1981), and iron ore deposits, fault lengths, and fault surfaces (Lam and De Cola 1993). In some cases, self-similarity has been explained on the basis of lower-level physical laws (e.g., sand avalanches, crystal growth, and rock fault lengths), but in many other cases this has not (yet?) been accomplished.

The second class consists of emergent phenomena with a dominant length or time scale. Such a macroscopic regularity emerges from microscopic physical or chemical processes. The microscopic processes may initiate at random or fractal length or time scales, but only one or two frequencies become dominant in the macroscopic pattern. The dominant frequency, or period or length, is in many cases enforced by a boundary condition that is independent of the microscopic processes. Such phenomena are common in earth science. For example, a turbulent water flow in a river provides a chaotic forcing to the sand bed below the current. Out of this chaotic forcing emerges a pattern: a train of large underwater dunes. Despite the enormous variety in current velocity and sand characteristics these dunes (when in equilibrium with the flow) have roughly constant length/height ratios and have a height that is about 20 percent of the water depth. This is an empirical fact for water depths ranging from 0.1 to 100 m and grain sizes from 0.4 to 100 mm. So, in practice, dune height and length are predicted from empirical relations (derived from data by induction) with the water depth as the most important independent variable. Yet, there have been a few promising attempts to predict sizes and behavior of real-world dunes from physics-based models for flow turbulence and sediment transport, so this phenomenon is not emergent in the sense that it cannot be predicted from a lower level.

It should be clear by now that many earth-scientific phenomena possess emergent properties. The question is now whether these properties are also irreducible. At present, this question is hotly debated within earth science. Some earth scientists try to reproduce emergent patterns (e.g., a train of dunes or a braided river pattern) with simple models based only on macroscopic rules. By contrast, others attempt to reproduce these patterns with highly sophisticated mathematico-physical models. The former claim that the phenomena are irreducible, while the latter claim that given enough computing power and detailing of initial conditions the phenomena can be explained (in a mathematical model) on the basis of physical laws only. Accepting emergence as a given fact, debates about reductionism continue among philosophers as well as among earth scientists themselves. Meanwhile, philosophers have generally rejected the traditional Nagelian model of reduction, and rightly so, since in most higher-level sciences, the model is of little use. But this does not imply that in these sciences reduction is completely absent. Accordingly, we need an alternative account of reduction. Wimsatt (1997, S373) suggests that "a reductive explanation of a behaviour or a property of a system is one showing it to be mechanistically explicable in terms of the properties of and interactions among the parts of the system." If one adopts this view of reduction, it appears that many emergent phenomena of earth science may be explained reductively in the future.

4.   Explanation in Earth Science

In §2 we saw that earth scientists have at least two aims, namely to *describe* and to *explain* the (history of) inanimate processes on the earth. Indeed, it seems plausible that all sciences aim at explanation and understanding, over and above mere description, of the phenomena in their domain (see de Regt and Dieks 2005). But we have not yet answered the question of what the nature of earth-scientific explanations is: do they have a special status that distinguishes them from explanations in physics and chemistry? As argued in §3.1, earth-scientific theories are hypotheses about unobservable (past) events or contingent generalizations. How can such theories – which are largely of a descriptive or historical character – provide explanations? In the present section, we address this question. Moreover, we discuss the problem of underdetermination, which is a serious obstacle in the practice of constructing earth-scientific explanations. It will become clear that earth science employs explanatory pluralism: there are several types of explanation, each with its own merits and its appropriate model.[4]

4.1   Do historical narratives explain?

The issue of whether historical narratives can explain, and if so, how they do, is one on which opinions greatly diverge. Some authors claim that

---

[4]   We simplify the discussion of scientific explanation in a way that has no serious consequences for the case of explanation in earth science. We contrast "historical" (or narrative) explanation with "causal explanation," where we take the latter to be exemplified by the physical sciences. Causal explanation can take the form of a deductive-nomological explanation (Hempel 1965) employing causal laws, or of a causal space-time description of events and processes (Salmon 1984). For our purposes, differences between these two modes, as well as alternatives such as the unificationist view of explanation (Kitcher 1989), can be ignored.

narratives explain by integrating an event into a bigger picture (Hull 1989). A narrative thus conveys some greater holistic insight, particularly in history and the social sciences. But others sneeringly call historical sciences varieties of "stamp collecting." Still others see a logical pattern in historical explanation that is fundamentally different from deductive-nomological and inductive-statistical explanation schemes. For human historical narratives it has been argued that the plot, or argumentation structure, of the story (e.g., perspective, ordering in time) conveys an explanation of the events in terms of their necessary (but not sufficient) conditions and their relations (see von Wright 1971).

We will argue that there is more to earth-scientific narratives than merely description. Like evolutionary biology, earth science is partly historiography: it reconstructs and tries to explain past events. The purpose of historical sciences is to provide a correct narrative of the sequence of past events and an account of the causal forces and antecedent conditions that led to that sequence.

The historical narrative approach provides two kinds of explanations relevant for earth science: *robust-process* explanations and *actual-sequence* explanations (Sterelny 1996, 195). While the latter specifies the (causal) chain of events, the former focuses on underlying (robust) causes of the phenomenon. Both kinds of explanation are important in earth science, even for the same event, because they convey distinct breeds of information. Consider two explanations of a catastrophic landslide or mudflow. We may model an observed flow in high detail with physical models. In retrospect, the initiation of the flow that happened at that precise moment in time and at that location can be explained by, for example, heavy rainstorms. This actual-sequence explanation leaves out that the flow was waiting to happen because a certain amount of mud or rubble was on the verge of slope failure. If this specific heavy rainstorm had not initiated the flow, it would very likely have been initiated by another rainstorm, or the spring melting of accumulated snow, or an earthquake. So earth scientists not only model such flows with empirical or physical models but also map areas prone to mudflows as a part of the robust-process explanation for these flows.

In order to be genuinely explanatory, narratives in earth science need not be reduced to physical causal-law explanations. The practice of earth science demonstrates that a domain can be ontologically dependent on another deeper (i.e., physical) domain while at the same time being explanatorily autonomous. In historical narratives, an event is not explained by subsuming it under a generalization. Instead, it is explained by integrating it into an organized whole (Hull 1989). In earth science, the ideas of continental drift and of glacials (ice ages) have played such a role. Mountain formation, earthquakes, volcanism, paleomagnetism, distribution of fossil remains, the form of the continents, and the varying ages of rocks are integrated in an organized whole and all point to one and the same common cause: plate tectonics. Glacial landforms, erratic boulders, lack of vegetation in certain time periods, and evidence for a much lower sea level than today all point to the common cause of glacials. The specific physical processes of plate tectonics and glaciation are captured as contingent regularities with only local reductions to physics and chemistry (see §3.1).

4.2   Underdetermination problems in earth science

While the historical narratives provided by earth scientists can thus be genuinely explanatory, they face a major problem, which we will discuss now, namely the fact that they are usually underdetermined by the available evidence. Philosophers distinguish between weak (practical) underdetermination and strong (logical) underdetermination. In the former case, there is insufficient available evidence to choose one theory over its rivals, while in the latter case theory choice remains impossible no matter how much evidence is gathered. Contrary to physicists, practicing earth scientists very often face situations in which theories are underdetermined by the available evidence. In fact, it is hard to find papers that do not contain at least a paragraph on the way underdetermination was dealt with in practice (although it is usually not explicitly referred to as "underdetermination"). Typical examples of underdetermination problems are the following (see also Turner 2005).

First, the time scale involved in shaping the earth is orders of magnitude larger than the life of human observers or even written history. It is therefore problematic to detect and observe the long-term effects of slow processes that might be extrapolated to the past. Sometimes controlled scale experiments are conducted in the laboratory, but this generates scaling problems that make a direct comparison with the real world more difficult.

Second, many earth-scientific processes and phenomena cannot (yet) directly or even indirectly be observed. Sometimes a phenomenon eludes direct detection by instruments, for instance deep-mantle convection within the earth and other planets. Also, landforms and sediment deposits (with all the clues to processes and events in the past) often have been obliterated by erosion, mountain building, or flooding. An additional practical problem is that current techniques often disturb the observed processes.

Third, many processes are intrinsically random or chaotic and may be very sensitive to initial conditions. A veritable reconstruction of past events from the geologic record becomes extremely arduous because many events and phenomena are so complex that, in theory, an infinite number of possible laws and initial conditions could be involved. This problem has especially received attention in hydrology where it was named "equifinality." Noise and chaos lead to a certain uniqueness of geomorphological phenomena: duplications of events are seldom found and the probability that the long-term river channel development in a particular river delta would take the same course when repeated is near zero.

While complete causal (deductive-nomological) explanations would be highly underdetermined, a narrative explanation does not require an exhaustively detailed set of observations and initial conditions. In many earth-scientific studies, initial and boundary conditions are implicitly (and without the details) given in the reference to the length and time scale of the phenomenon under study or the time period or study area, and in many other studies it is apparent from the context, such as the name of the scientific journal. For example, the Rhine-Meuse delta in the Netherlands has been studied by Henk Berendsen and Esther Stouthamer (2001) for the Holocene time period, which is the past 11,760 years (in 2010). The point of this reference to time is to isolate the phenomenon and its causes from other phenomena by placing the latter in the description of the initial conditions, or, positioning it in the "bigger picture." In this way it becomes unnecessary to specify the whole chain of causes and events from the largest scales (long times past and large areas around the study area) down to the scales of the phenomenon itself.

This approach is not only practical but often also defensible (Schumm and Lichty 1965). First of all, there is a correlation between the relevant length scale and the relevant time scale of most phenomena (see Figure 9.2), so it is not necessary to study the evolution of a large object on a very short time scale or a small object on a very large time scale. For instance, current ripples, which have lengths of about 0.2 m, are formed and destroyed in minutes to hours, so to study ripples over decades is probably useless. Mountain ranges of thousands of kilometers long, on the other hand, are formed in millions of years. In other words, it takes a much longer time to build or break down large things than to build or break down small things.

The reason is that the range of energy available is between two close limits (compared to the extreme energy levels familiar to astronomers): a lower limit necessary to exceed thresholds such as friction or entraining boulders by stream flow, and an upper limit given by the maximum energy available on earth. Consequently, no significant change occurs to the whole mountain range in a period of hours, even though one plot on one slope may have changed significantly by a flood or by mass wasting. Therefore, it is commonly not useful to study large phenomena over very short periods or very small phenomena over large time periods. Notable exceptions are relatively unique events such as the impacts of large meteorites or earthquakes.

4.3   Underdetermination and explanation

How precisely do underdetermination problems lead to the impossibility of finding complete causal explanations and to the use of narrative explanations in earth science? Consider a typical earth-scientific research project that aims to construct a spatiotemporal description *and* an explanation of the course of the river Rhine in the past 10,000 years (see Berendsen and Stouthamer 2001). The hypothesis that the Rhine has been present in the Netherlands is practically unassailable. But for earth scientists this result is only the beginning. What they really want is a description and an explanation of the course of events that is generalizable to other, comparable phenomena in comparable circumstances. For such an explanation much more detailed evidence is needed to distinguish between competing theories. Earth scientists have to infer from present situations to past ones, or from a limited set of observations to a hypothesis or theory. The empirical data gathered by earth scientists often leave room for a wide range of different, incompatible hypotheses. These hypotheses commonly are empirically but not logically equivalent; they cannot be true at the same time. In sum, their inferences are hampered by problems of underdetermination.

The predominant methodology of historical sciences is that of formulating various competing explanations (causes) of present phenomena (effects) and discriminating between them by searching for "smoking guns." Consider a candidate causal explanation for the presently available evidence, for example, regarding the courses of the river Rhine through time. This would take the form of a description of the initial conditions, say, 10,000 years ago, plus a set of causal laws that govern the dynamics of the system. The laws are, as we argued above, derived from physics and chemistry (let us suppose they are given and deterministic). What really matters is the choice of the initial conditions; these are actually referred to as "the cause." But these conditions are not, and indeed can never be, specified *completely* – they merely give an incomplete description of a part of the universe (and this part is never a closed system). In practice, therefore, explanations in earth science are a combination of abductive narratives and causal explanations. The narratives carry most of the explanatory power, for example of how the shifting of the Rhine branches over the delta plane depended on the sea-level rise and other factors.

The causal explanation parts are often applications of computer models to certain aspects of the delta development. Computer modeling based on laws of physics or chemistry is becoming an activity as central in earth science as experimenting and observing. However, the laws in the causal explanations are not simply given and deterministic as assumed above. For instance, for many phenomena it is not *prima facie* clear which physical laws apply. Moreover, the relatively simple laws of physics can seldom be applied directly to the initial conditions to check whether they (deductively) explain the observations under scrutiny. For instance, conservation of momentum and mass are simple physical laws that apply to fluid flow. These simple laws are the basic components of the Navier-Stokes equations that govern fluid flow, which cannot be solved analytically. Therefore these equations are implemented in so-called "mathematical" or "physical computer models." The equations have to be simplified and discrete time steps and grid cells must be used to model the flow in space and time. These simplifications bring a host of necessary numerical techniques to ensure conservation laws and to minimize numerical (computer-intrinsic) error propagation. When initial or boundary conditions are specified for this model, certain laboratory or field conditions can be simulated and compared to the observations. But one can never be certain that a mismatch between model results and observations is not due to the simplifications and numerical techniques. So this does not solve the problems of underdetermination, for as Oreskes, Shrader-Frechette, and Belitz explain:

> Verification and validation of numerical models of natural systems is impossible. This is because natural systems are never closed and because model results are always non-unique. . . . The primary value of models is heuristic. (1994, 641)

Systems that are "never closed" exhibit the underdetermination of initial and boundary conditions and the "non-uniqueness" refers to the equifinality problem. So models cannot be used to hindcast or forecast the systems studied in earth science, unless they are calibrated on data from the past. But in the latter case, one can hardly speak of a true "physical" model because the data carry a part of the explanatory weight.

But there are still great benefits to be gained by modeling. First, the human mind cannot comprehend the results of complex sets of equations in space and time under certain initial conditions, but a model provides comprehensive results. Second, a scientist can manipulate a model in ways that are impossible in nature or even in the laboratory. Various scenarios can be studied and "what-if questions" surveyed under the assumptions of the model, even if the modeled scenario is not what really happened according to the observations. Thus, these scenarios play an important role in robust-process explanations, both to extend the explanation and as counterfactuals in view of the observations. Third, a set of model runs may indicate (though not always prove) whether certain hypotheses are possible at all and whether they conflict with physical laws and mathematical constraints. Modeling in this sense is used as a test for the narrative robust-process explanations. Fourth, the comparison of various different models for the same phenomenon may indicate the robustness of the models: if the modeled phenomenon is similar in the different models, then it is to some extent independent of the model schematizations.

The above considerations make it clear that both historical narratives and causal explanations are needed in earth science. The two kinds of explanation are in fact complementary: causal explanations can be employed on a small spatiotemporal scale and may give further hints for narrative explanations on a much larger scale, such as the scenario modeling approach outlined above, whereas such broadly construed narrative explanations may in turn provide more insight into specific causal processes. The border line between these two types of explanation is often unclear because they gradually merge into one another.

4.4 Inference to the best explanation

Until now we have outlined the problems that earth scientists face. We do not want to suggest that these problems cannot be solved, however. In practice earth scientists *do* come up with explanations, which they regard as more or less corroborated. But how do the scientists arrive at these explanations? An important occupation of earth scientists is formulating hypotheses about possible causes for the phenomena observed. It appears that most earth-scientific explanations are the result of abductive inference, and that earth scientists typically rely on "inference to the best explanation" (Lipton 2004), supported by deductive causal explanation (for instance based on computer modeling). Furthermore, abduction in earth science is extended by a method already described by Thomas Chamberlin (1890) as "the method of multiple working hypotheses." A number of hypotheses are developed which potentially explain the observations. By contrasting and testing a number of (incompatible) hypotheses, a biased attempt at confirmation is prevented. The hypotheses can be processes that are known from observations, or "outrageous hypotheses" of processes that supposedly occurred in past times but are no longer active or that seem to be in conflict with the laws of physics or chemistry. Wegener's hypothesis of "continental drift" is the most famous example of such an outrageous hypothesis: initially rejected as absurd, it is now accepted and supported by plate tectonics. The hypotheses are used to predict testable consequences (deduction), preferably for a wide range of different locations, with the use of different kinds of instruments and often with computer models. When these various data and model scenarios all point to the same (underlying) explanation or common cause, earth scientists accept this explanation as (tentatively) true.

For instance, the occurrence of about 23 glacial–interglacial couplets in the past 2.4 million years ("ice ages") is best explained by a combination of a unique setting of continents and astronomical influences on the global climate. There are cyclic variations in the inclination and direction of the earth's axis and the obliquity of the orbit, of which the periods correspond with those of the glacial–interglacial cycles. These variations cause solar irradiation variations at higher latitudes, which leads to cooling. This, in turn, causes an increase of snow cover which leads to a larger reflection of sunlight and hence causes more cooling. But this orbital forcing has been the case for much of the earth's late history during which glaciation was not observed, so something more is needed. About 2.4 million years ago, Antarctica was detached from the South American continent. Circumpolar currents developed instead of currents circulating between the equator and the pole, which led to thermal isolation and, given the polar position, cooling of Antarctica. The result was increasing ice coverage and hence larger albedo, and consequently the whole planet cooled a few degrees. This cooling, added to the astronomically induced fluctuations, was enough to trigger the glaciation of higher latitudes.

While this set of hypotheses is not without problems, an alternative set of hypotheses that explains these phenomena equally well is not easily conceived. For example, the fluctuation of solar activity due to its stellar dynamics ("solar forcing" on the earth's climate) is a competitive explanation for global temperature oscillations, of which effects have been demonstrated in the geological record. But there is no stellar theory that predicts the 23 rhythmic cycles, whereas the orbital forcing (systematic variations in the earth's orbit) does.

Most earth scientists believe that the former set of (triangulated) hypotheses is the best explanation for the observations, while the latter set of hypotheses is also true but only modifies the effects of orbital forcing. This is an example of inference to the best explanation: the observed ice ages are surprising, but there are hypotheses explaining part of the evidence that would make the observations not surprising. So the set of hypotheses is tentatively accepted as the explanation, and further confronted with new evidence.

Inference to the best explanation is a form of abduction, and, like induction, it has its limitations (Lipton 2004). Foremost, abduction is a method to select *hypotheses* rather than conclusive explanations. We probably will never know whether the best explanation is also the one and only true explanation. In fact, the uniqueness of events and phenomena in the geologic past often requires hypotheses that are at first sight outrageous in view of our present experiences (Davis 1926, Baker 1996). For example, the hypotheses that certain landforms are caused by ice-age glaciers, and that whole continents are drifting and colliding, were once considered outrageous. Over time, however, they were supported with further evidence and robust-process explanations were formulated by extending the hypotheses and by adding causal explanation sketches. The latter are typically developed by using mathematico-physical models, but, as was argued in §4.3, these models are also troubled by underdetermination problems. If the narratives survive tests against an increasing body of evidence, the hypotheses are more generally accepted by consensus as the best explanations.

5. Conclusion

The aim of earth science is to provide descriptions and explanations and, if possible, predictions of phenomena on the earth and earth-like planets.

Earth science can be viewed as a *reductionist* enterprise. However, earth-scientific theories do not contain laws in the traditional sense of universal, exceptionless regularities. In contrast to physics and chemistry but analogously to biology, an important part of theories of earth science consists in descriptions of contingent states of nature. Yet, earth-scientific theories can be reduced locally because they can be exhaustively translated in physical and/or chemical terms. Earth science provides many cases of *emergence*, but recent attempts at reductionistic modeling suggest that many emergent phenomena may be explainable by the laws of physics in the near future.

Earth-scientific explanations are strongly hampered by weak *under-determination*. Underdetermination entails the impossibility of constructing complete causal explanations, but narrative explanations remain possible because they do not require an exhaustively detailed set of observations and initial conditions. Rather, references to time and length scales of interest provide implicit limits for the amount of details needed and refer to relevant background theories.

In general, theories or narratives in earth science provide *explanations* by integrating robust-process explanations and actual-sequence explanations, observations, and background theories. The narratives carry most of the explanatory power in both actual-sequence and robust-process explanations. The causal-explanation parts are, for example, tests whether the hypotheses do not conflict with physical laws or scenarios and counterfactuals derived from computer modeling. A conclusive test of the causal parts is, however, impossible due to the weak underdetermination problems. Two strategies are therefore followed: the method of *multiple working hypotheses* and *inference to the best explanation*. If the narratives survive tests against an increasing body of evidence, the hypotheses are more generally accepted by consensus as the best explanations. When earth scientists reach such a well grounded consensus on a historical narrative, the best explanation is that they possess knowledge of the past.

Acknowledgment

This chapter is based on M.G. Kleinhans, C.J.J. Buskes, and H.W. de Regt, 2005, "*Terra Incognita*: Explanation and Reduction in Earth Science," *International Studies in the Philosophy of Science* 19: 289–317. We thank the editor, James W. McAllister, for his kind permission to include parts of this publication in the present essay.

References

Bak, P., C. Tang, and K. Weisenfeld. 1987. "Self-organised Criticality: An Explanation of 1/f Noise." *Physical Review Letters* 59: 381–5.

Baker, V.R. 1996. "Hypotheses and Geomorphological Reasoning." In B.L. Rhoads and C.E. Thorn (eds.), *The Scientific Nature of Geomorphology*. Chichester: Wiley and Sons, 57–86.

Ball, P. 1999. *The Self-made Tapestry: Pattern Formation in Nature*. Oxford: Oxford University Press.

Beatty, J. 1995. "The Evolutionary Contingency Thesis." In G. Wolters and J.G. Lennox (eds.), *Concepts, Theories, and Rationality in the Biological Sciences*. Pittsburgh: Pittsburgh University Press: 45–81.

Berendsen, H.J.A. and E. Stouthamer. 2001. *Palaeogeographic Development of the Rhine-Meuse Delta, the Netherlands*. Assen: Van Gorcum.

Burrough, P.A. 1981. "Fractal Dimensions of Landscapes and Other Environmental Data." *Nature* 294: 240–2.

Chamberlin, T.C. 1890. "The Method of Multiple Working Hypotheses." *Science* 15: 92–6.

Davis, W.M. 1926. "The Value of Outrageous Geological Hypotheses." *Science* 63: 463–8.

De Regt, H.W. and D. Dieks. 2005. "A Contextual Approach to Scientific Understanding." *Synthese* 144: 137–70.

Dupré, J. 1993. *The Disorder of Things: Metaphysical Foundations of the Disunity of Science*. Cambridge, MA: Harvard University Press.

Fodor, J.A. 1974. "Special Sciences (or: The Disunity of Science as a Working Hypothesis)." *Synthese* 28: 97–115.

Hempel, C.G. 1965. *Aspects of Scientific Explanation*. New York: Free Press.

Hull, D. 1989. *The Metaphysics of Evolution*. New York: SUNY Press.

Humphreys, P. 1997. "Emergence, Not Supervenience." *Philosophy of Science* 64: S337–45.

Hutton, J. 1785. "The System of the Earth, Its Duration, and Stability." In C.C. Albritton (ed.), *Philosophy of Geohistory: 1785–1970*. Stroudsburg: Dowden, Hutchinson and Ross, 24–52.

Kitcher, P. 1989. "Explanatory Unification and the Causal Structure of the World." In P. Kitcher and W. Salmon (eds.), *Scientific Explanation*. Minneapolis: University of Minnesota Press, 410–505.

Kitts, D.B. 1966. "Geologic Time." *Journal of Geology* 74: 127–46.

Kleinhans, M.G., C.J.J. Buskes, and H.W. de Regt. 2005. "*Terra Incognita*: Explanation and Reduction in Earth Science." *International Studies in the Philosophy of Science* 19: 289–317.

Lam, N.S.-N. and L. De Cola. 1993. *Fractals in Geography*. Englewood Cliffs, NJ: Prentice Hall.

Laudan, R. 1987. *From Mineralogy to Geology: The Foundations of a Science, 1650–1830*. Chicago: University of Chicago Press.

Lipton, P. 2004. *Inference to the Best Explanation*. 2nd edn. London: Routledge.

Mandelbrot, B.B. and J.R. Wallis. 1969. "Some Long-run Properties of Geophysical Records." *Water Resources Research* 5: 321–40.

Nagel, E. 1961. *The Structure of Science*. New York: Harcourt, Brace and World.

Oreskes, N., K. Shrader-Frechette, and K. Belitz. 1994. "Verification, Validation and Confirmation of Numerical Models in the Earth Sciences." *Science* 263: 641–2.

Salmon, W.C. 1984. *Scientific Explanation and the Causal Structure of the World*. Princeton, NJ: Princeton University Press.

Schumm, S.A. and R.W. Lichty. 1965. "Time, Space and Causality in Geomorphology." *American Journal of Science* 263: 110–19.

Smith, P. 1998. *Explaining Chaos*. Cambridge: Cambridge University Press.

Sterelny, K. 1996. "Explanatory Pluralism in Evolutionary Biology." *Biology and Philosophy* 11: 193–214.

Turner, D. 2005. Local Underdetermination in Historical Science." *Philosophy of Science* 72: 209–30.

van Thienen, P., N.J. Vlaar, and A.P. van den Berg. 2004. "Plate Tectonics on Terrestrial Planets." *Physics of the Earth and Planetary Interiors* 142: 61–74.

von Wright, G.H. 1971. *Explanation and Understanding*. Ithaca, NY: Cornell University Press.

Wimsatt, W. 1997. "Aggregativity: Reductive Heuristics for Finding Emergence." *Philosophy of Science* 64: S372–84.

Unit 4
Philosophy of the Behavioral and Social Sciences

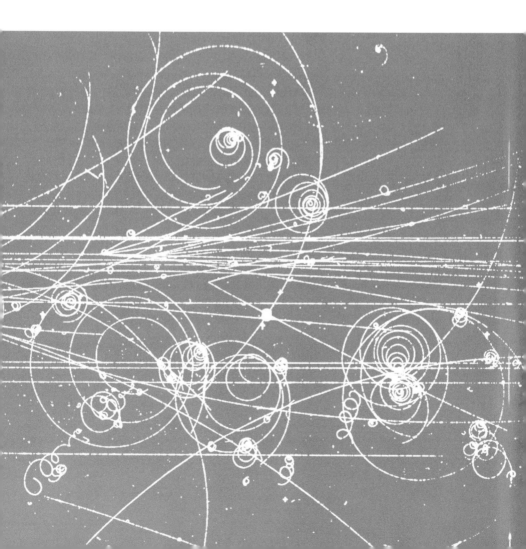

10   Philosophy of
the Cognitive Sciences

William Bechtel and Mitchell Herschbach

 Cognitive science is an interdisciplinary research
endeavor focusing on human cognitive phenomena
such as memory, language use, and reasoning. It
emerged in the second half of the twentieth century
and is charting new directions at the beginning of
the twenty-first century. This essay begins by identifying the disciplines that contribute to cognitive
science and reviewing the history of the interdisciplinary engagements
that characterize it. The second section examines the role that mechanistic explanation plays in cognitive science, while the third focuses
on the importance of mental representations in specifically cognitive
explanations. The fourth section considers the interdisciplinary nature
of cognitive science and explores how multiple disciplines can contribute
to explanations that exceed what any single discipline might accomplish.
The conclusion sketches some recent developments in cognitive science
and their implications for philosophers.

1.   What is Cognitive Science?

Cognitive science comprises a cluster of disciplines including portions of
psychology, linguistics, computer science, neuroscience, philosophy, sociology, and anthropology. Its roots lie in the 1950s; it acquired an academic
identity in the 1970s, and it continues to thrive in the twenty-first century.
It seeks to explain mental activities such as reasoning, remembering,
language use, and problem solving, and the explanations it advances commonly involve descriptions of the mechanisms responsible for these activities.
Cognitive mechanisms are distinguished from the mechanisms invoked in
other domains of biology by involving the processing of information.
Many of the philosophical issues discussed in the context of cognitive

science involve the nature of information processing, especially the central notion of representation. One of the distinctive features of cognitive science is that it is not a discipline but a multi-disciplinary research cluster. It draws upon the contributing disciplines for the problems it investigates, the tools it uses to investigate them, and the explanatory strategies invoked, but the results transcend what is typically achieved in any of the contributing disciplines. This gives rise to philosophical questions about the nature of interdisciplinary research.

The term "cognitive science" was only coined in the mid-1970s. In 1975 it was employed in two books. *Explorations in Cognition*, the product of a collaborative research group at the University of California at San Diego (UCSD), concludes with the suggestion: the "concerted efforts of a number of people from . . . linguistics, artificial intelligence, and psychology may be creating a new field: *cognitive science*" (Norman and Rumelhart 1975, 409). Although situated in psychology, the group employed computational models of semantic networks to explain word recognition, analogy, memory, and semantic interpretation of verbs, sentences, and even brief stories. The collaborative book by computer scientist Daniel Bobrow and cognitive psychologist Allan Collins, *Representation and Understanding: Studies in Cognitive Science* (1975), invoked the term in its subtitle. Two years later, the Alfred P. Sloan Foundation announced its Particular Program in Cognitive Science and, over 10 years, provided $17.4 million to establish and foster interdisciplinary centers at selected research universities. UCSD received one of the early Sloan grants and used a portion of its funding to sponsor the 1979 La Jolla Conference on Cognition, which became the first meeting of the now international Cognitive Science Society. In 1980 the Cognitive Science Society assumed ownership of the journal *Cognitive Science*, which itself had begun publication in 1977.

While it was during the 1970s that cognitive science began to acquire an institutional identity, its roots go back to the middle of the century when a new intellectual perspective began to inspire researchers in psychology and linguistics to reject the strictures that behaviorism had placed on most research in these disciplines in North America since John Watson (1913) issued his manifesto "Psychology as the Behaviorist Views it" and urged a focus on behavior, not hypothetical mental activities. Although B.F. Skinner, who advocated a radical behaviorism that eschewed mental entities, is perhaps the best known behaviorist, the behaviorist tradition was relatively diverse. Not all behaviorists were opposed to positing events inside the head: Clark Hull (1943) appealed to intervening variables such as drive, but stopped short of overtly mentalistic concepts. Edward Tolman (1948) was exceptional among behaviorists in postulating *cognitive maps* to explain navigational abilities of rats. Leonard Bloomfield (1939) carried behaviorism to linguistics where he advanced a strongly empiricist approach to cataloging and analyzing linguistic forms and rejected mentalistic accounts of these forms.

While behaviorism cast a broad shadow over psychology and linguistics in North America, in Europe a variety of alternative perspectives more favorable to mentalistic characteristics of human beings prospered and would come to influence the development of cognitive science. For example, Jean Piaget proposed cognitive operations in his genetic epistemology; Frederick Bartlett introduced schemas (organizing structures) to account for memory distortions; Gestalt psychology recast perception in terms of self-organizing forms; and Lev Vygotsky and Alexander Luria initiated studies demonstrating cultural influences on language and thought. Even in North America, psychophysics and parts of developmental and social psychology were pursued outside of behaviorism's shadow. But for much of psychology, a revolution was required to reverse behaviorism's proscription on appeals to mental phenomena in explaining behavior.

The cognitivism that emerged in the 1950s maintained behaviorism's emphasis on explaining behavioral phenomena and invoking only behavioral evidence. Hence, unlike earlier mentalistic psychology, it rejected introspection as the avenue to the mind. What it required was a way of conceptualizing internal events that construed them as causal processes contributing to the generation of behaviors. This conceptualization was provided by information theory, which developed from formal engineering analyses of communication channels, such as telephones, conducted at Bell Laboratories between the 1920s and 1960s. This endeavor attempted to quantify the capacity to transmit information across channels that are subject to capacity and rate limits and to noise. Construing information as a reduction of uncertainty at the end of the channel about the message at the beginning of the channel, Claude Shannon (1948) introduced the *bit* as the unit of information: the unit of information required to differentiate between two equally likely messages. Shannon also showed that one could determine the redundancy in a message in terms of the reduction of uncertainty. George Miller drew upon this analysis in his Harvard PhD dissertation which used redundancy in a message to explain how messages in spoken English could be understood in noisy environments.

In perhaps his best known research, Miller (1956b) identified comparable capacity limitations in a number of cognitive domains, including short-term memory: humans can hold up to seven, plus or minus two, separate items in memory over a period of minutes (unless they are interrupted earlier). In this research, information is construed as the commodity the mind utilizes and the various tasks it performs (remembering, planning, problem solving) are construed as the processing of information. Donald Broadbent (1958) advanced a model in which information about sensory events is held briefly in a short-term store, and an attentional filter restricts which gets transmitted along a single, limited-capacity channel for further processing. These ideas were incorporated into a general framework for understanding cognitive activity by Ulric Neisser (1967) in his pathbreaking textbook *Cognitive Psychology*.

The idea of the mind as an information processor was further promoted with the introduction of the digital computer as itself an information processor. Shortly after its creation following World War II, some researchers in the new field of computer science began to explore the possibility of programming a computer to behave intelligently (e.g., perform activities that would be judged intelligent if performed by humans). A pivotal conference at Dartmouth College in the summer of 1956 introduced the name *artificial intelligence* and witnessed the first presentation of a program performing intelligently: Alan Newell and Herbert Simon's Logic Theorist, which developed proofs of theorems in symbolic logic.

These contributions were brought together on September 11, 1956, the second day of the second Symposium on Information Theory. Newell and Simon (1956) again reported on Logic Theorist, and George Miller (1956a) presented his research on capacity limitations of short-term memory. In between, a young linguist, Noam Chomsky (1956), presented a paper, "Three Models of Language," in which he argued that various computational systems, such as finite-state automata, were inadequate to model the grammar of human languages and introduced arguments for what he called transformational grammars. These employed procedures for generating core linguistic structures (trees) and transformations to modify these structures. For Miller, on that day, "cognitive science burst from the womb of cybernetics and became a recognizable, interdisciplinary adventure in its own right" (Miller 1979, 4). The interdisciplinary interaction that day between psychology, artificial intelligence, and linguistics became characteristic of cognitive science, although, as discussed above, 20 years were to pass before the name was introduced and the field became institutionalized.

From these beginnings, research in cognitive science has burgeoned. We can here note just a few landmarks that provide an indication of the breadth of the field. Newell and Simon (1972) introduced the idea of a production system, consisting of a working memory and operations (productions) designed to respond and alter the contents of that memory, and employed it to model the strategies humans use to solve problems. Chomsky (1957) developed the first of several grammatical theories (minimalism is the most recent; see Chomsky 1995). Chomsky elicited an opposition movement which rejected the autonomous status Chomsky claimed for syntax and interlaced syntax with semantics (resulting in what Harris (1993) characterizes as the linguistic wars of the 1960s and 1970s). More recently, cognitive linguistics has emphasized how other cognitive processes such as spatial representation (Fauconnier 1994) or metaphors grounded in the body (Lakoff and Johnson 1999) serve as the basis for linguistic structures. Research in the field of memory that started with the distinction between short- and long-term memory has expanded as researchers have distinguished different forms of long-term memory and distinctive features of how each is processed (Schacter 1996, Tulving 2002).

2. Explanation in Cognitive Science

The practitioners of the various cognitive sciences generally construe themselves as engaged in explaining the behavior of human agents. This raises the question of what sort of explanation suffices to explain behavior. Although a number of humanists have contended that the mind must be understood in different terms than other physical systems, cognitive scientists have tended to view their enterprise as contiguous with those of the other natural sciences, especially biology. Traditional philosophical accounts of explanation have construed laws of nature as central, with explanation involving the demonstration that the event to be explained occurred in accordance with laws. On the deductive-nomological (D-N) model, such demonstration involved the derivation of a description of the event to be explained from a statement of the law and initial conditions (Hempel 1965). A law on this account is minimally a true universally quantified conditional statement which supports inferences about counterfactuals (e.g., inferences about what would happen if the conditions specified in the antecedent were true in a given situation).

The D-N account, however, fares poorly in characterizing the explanations biologists and cognitive scientists offer. The central problem with applying the D-N account to research in cognitive science is the paucity of acknowledged laws within the fields of cognitive science. Perhaps, though, there are laws without their being referred to as such. Indeed, as Cummins (2000) has maintained, psychologists often speak of effects where other scientists might refer to laws. Thus, one finds references to the spacing effect (Ebbinghaus 1885), the serial position effect (Glanzer and Cunitz 1966), and the Garcia effect (Garcia, McGowan, Ervin, and Koelling 1968), where each of these provides a generalization about what happens under specified conditions. But, as Cummins also shows, these effects do not provide explanations but rather serve to identify the phenomena in need of explanation. Thus, the spacing effect is the phenomenon that retention is greater when learning is spaced out in multiple learning episodes, rather than compressed (as in cramming for an exam) – a feature of memory encoding that calls out for explanation.

In biology, there is a similar paucity of acknowledged specifically biological laws – a textbook or research report might refer to laws (or equations) from physics and chemistry but not ones specific to biology. A few philosophers have recently followed the lead of biologists themselves, who commonly appeal to mechanisms as providing explanations. These philosophers have attempted to explicate the nature of mechanistic explanation and how it figures in biology. Although they vary in the vocabulary used to characterize mechanisms, the basic idea is that a mechanism consists of component parts which perform different operations and that these parts are so organized and the operations orchestrated that the whole mechanism, in the appropriate context, realizes the phenomenon of interest (Bechtel and Richardson 1993, Machamer, Darden, and Craver 2000, Bechtel 2006, Darden 2006, Craver 2007). Thus, to explain how a cell makes protein, one identifies the various components of the cell that are involved (DNA, mRNA, RNA polymerase, ribosomes, etc.), the operations each of them performs (e.g., RNA polymerase creates an mRNA strand from a DNA template), specifies the organization of the parts, and shows how the various operations are orchestrated to produce a protein.

Appeals to mechanisms to provide explanations are equally ubiquitous in the cognitive sciences, and philosophers have begun to analyze the mechanistic models offered in research on vision and memory (Bechtel 2008) and emotion (Thagard 2006). Memory researchers, for example, have both differentiated memory operations and developed accounts of how they are related. For example, through mental rehearsal an individual can retain for short periods a small number of separate items (e.g., a list of names of people). But humans can also retain for long periods knowledge of facts (e.g., the dates of World War I – a semantic memory) and have the ability to re-experience events in their own lives (e.g., arguing with an officer who gave them a traffic citation – an episodic memory). Explanations of memory processes specify what brain areas and mental operations are involved in, for example, encoding new semantic memories and how they are organized. Thus, on one popular account, for several weeks or months after initial learning, information is encoded in the hippocampus, which then causes changes in regions of the cerebral cortex where the information is maintained for long periods (McClelland, McNaughton, and O'Reilly 1995). A successful mechanistic explanation then explains how it is that humans are able to exhibit the various mental phenomena that they do.

3.   The Distinctive Role of Representations in Cognitive Science Explanations

A difference between many biological mechanisms and cognitive mechanisms is that rather than being concerned with the transformation of materials (e.g., putting amino acids together to constitute proteins), cognitive mechanisms are involved in using information to regulate behavior. Thus, cognitive mechanisms are commonly characterized as *information-processing mechanisms*. The core idea is that states in the head stand in for phenomena outside the head and that by operating on those internal states agents coordinate their behavior with events in the outside world. The states in the head are construed as *re-presentations* of the phenomena outside the head. Consider how you are able to cook a meal from a memorized recipe (or, a bit more challenging, how good cooks can modify memorized recipes to create new dishes). The prototypical cognitive approach treats your knowledge of the recipe as a set of representations in your head and explains your behavior by positing causal processes operating on these representations. The challenge for cognitive science is to characterize these representations more precisely and identify the operations performed on them. There are differing views in cognitive science as to how to meet this challenge.

The idea that the mind trades in representations has roots in the history of philosophy. An innovation of the cognitive revolution was its treatment of the brain, a physical system, as a representational system. One inspiration for the crucial idea that the mind uses representations is that human culture has developed a number of systems used to represent phenomena. The one initially most influential in cognitive science was natural language: we use spoken and written words to communicate with each other because words and the sentences composed from them represent things. But humans operate with a variety of non-linguistic representational systems as well: maps, diagrams, pictures, and so on. Using such external representational systems as models, cognitive scientists posited that states in our heads could similarly be understood as representing things outside the head.

It is important to note, however, that these culturally created external representational systems do not function independently of human beings – if a sandstorm left a tracing in sand on the Martian surface with the shape "Stay out," that would not be a representation, as it was neither constructed by human beings nor processed by them. When "Stay out" is printed on a fence here on earth, it is the fact that it is created and interpreted by human beings that makes it a representation. When cognitive science proposes to incorporate representations in the head as part of the explanation of how we perform cognitive tasks, including the task of interpreting external representations, the question is how states inside the head constitute representations. It would not help to posit a homunculus (i.e., a little person) inside the heads of humans to interpret these internal representations, since that only recreates the problem of explaining the cognitive abilities of the homunculus.

Issues such as this underlie ongoing debates in cognitive science and the philosophy of cognitive science over what makes something a representation and what kinds of representations are required to model cognition. We will present the major accounts of: representational *vehicles*, or the kinds of structures that serve as representations; the types of operations that are performed on these structures; and how the vehicles acquire their *content*, or meaning.

The primary inspiration for one approach to the first two issues emerged from the development of digital computers. As Newell and Simon (1976) put it, computers are "physical symbol systems": they are machines which process information by producing meaningful changes in representations or symbols. The crucial feature of computers that makes this possible is that structures which count as symbols in the computer are composed and transformed via formal or *syntactic* rules – i.e., rules which only concern the physical form of symbols, rather than their meaning or *semantics.* These rules are themselves embodied in physical states in the computer and the manipulations performed on these states mirror the relations among the objects represented. The inspiration, which played a foundational role in the development of artificial intelligence (AI), is that by following purely formal rules, a computer can manipulate symbols in a manner that would count as intelligent reasoning if performed by a human being. Consider a simple addition function: taking the complex input symbol "3+5" (i.e., the concatenation of "3," "+," and "5") and producing the symbol "8" as output. A computer can do this by applying a formal rule indicating that input strings of one physical type should produce outputs of another physical type. The computer need not understand the meaning of the symbols (e.g., that "3" means the number three) or the function being computed (addition); it need only apply the rote procedure characterized by the syntactic rules. By being an information-processing device, the digital computer thus provided a model for how human cognition could be explained in terms of representational processes. The mind was the "program" or "software" running on the "hardware" of the brain.

The physical symbol systems developed by Newell and Simon and other pioneers in AI employed representations modeled on linguistic representations. In applying this model to humans, Fodor (1975) proposed that thinking occurred in a "language of thought" in which, as in natural languages like English, or formal languages like first-order logic, representational vehicles of cognition are sentences constructed from representational atoms (symbols) in accordance with a combinatorial syntax. In these "classical" cognitive architectures, cognitive processes such as planning and reasoning involve the serial manipulation of sentential representations according to syntactic rules, much as how, in formal logic, proofs are constructed through sequential transformations of sentential representations.

The idea that cognitive activities involve formal operations upon symbols was also developed in other domains of cognitive science. To account for the productivity of language with a finite set of principles, Chomsky (1957) advanced transformational grammars in which sentential structures are created using rewrite rules to which transformations are then applied. For a simple illustration, the rewrite rules S→NP+VP (a sentence can consist of a noun phrase and a verb phrase) and VP→V+DO (a verb phrase can consist of a verb and a direct object) could generate "Susie loves Charlie." A transformational rule could then be applied to replace *Charlie* with *whom*, and then move *whom* to the front, to yield the question "Whom does Susie love?"

Psychologists were also attracted to symbol processing models. John Anderson and Gordon Bower (1973), for example, developed a model of human associative memory which provided the basis for Anderson's subsequent attempts to develop a model of the mind that could account for a broad range of cognitive abilities (Anderson 2007).

In relying on the computer as a model of a physical symbol system, symbolic accounts tended to abstract away from the physical details of the brain. Following the computer metaphor for cognition, these accounts are at the "software" or "program" level of description, rather than at the level of physical implementation. Although in the last 20 years a number of symbolic theorists, including both Anderson and Newell (1990), have tried to render their accounts more neurally plausible, other researchers from the very beginnings of cognitive science were attracted to models inspired by the physical structure of the brain. These cognitive scientists investigated how units that send activation signals to each other, modeled loosely on the neurons and neural pathways of the brain, could process information. Warren McCulloch and Walter Pitts (1943), for example, showed how networks of artificial neurons could implement logical relations, while Frank Rosenblatt (1962) explored the capacity of two-layer networks, which he called *perceptrons*, to recognize perceptual patterns. Rosenblatt also introduced a procedure whereby perceptrons could learn to do this. Marvin Minsky and Seymour Papert's (1969) demonstration of the limitations of perceptrons temporarily sidetracked this approach, but the discovery of a learning procedure for multi-layer networks, which do not face the same processing constraints, rejuvenated it in the 1980s. Since it was the weighted connections between artificial neurons that determined the information-processing abilities of such networks, the movement that emerged using such networks to model cognitive processes (Rumelhart and McClelland 1986, Bechtel and Abrahamsen 2002) came to be known as *connectionism*.

Whereas language has made it obvious how to construe symbols as representational vehicles, it is less obvious how to identify representations in connectionist networks. One strategy is to treat each unit as playing a representational role, with its degree of activation serving as a measure of the degree to which it is construed as present in the pattern presented. But a far more interesting approach involves "distributed" representations, in which the representational vehicles are the patterns of activation across a set of units. The same units can figure in multiple vehicles and thereby represent the relations between representations. Cognitive processes are then identified with changes in the network's activation patterns as activity spreads through the network, rather than the application of syntactic rules as on the sentential approach. Learning, as noted above, occurs as the network alters the connections between units, rather than by the acquisition of rules or programs. The distributed nature of connectionist representations accounts for some of the benefits of connectionist networks over classical architectures, such as their ability to generalize and their gracefully degrading performance in response to noisy input or the loss of units (conditions which typically cause catastrophic failure in classical architectures).

Many connectionists view successful connectionist models as providing reason to reject the idea that cognition involves sentential representational vehicles. Critics of connectionism, on the other hand, argue that there are limits to connectionist models that can only be overcome by invoking syntactic rules operating over sentence-like representations. But not everyone sees connectionist networks and sentential accounts as incompatible. Some theorists propose that connectionist networks implement symbolic architectures: that a network can be described at a more abstract level of analysis as a classical architecture operating on sentential representations. This enables researchers to take advantage both of the syntactic operations available in classical architectures and of the generalization and graceful degradation of connectionist networks.

Whatever representational vehicle researchers employ in their cognitive models, an account must also be provided of how these structures come to represent things, how they acquire their content or meaning. Otherwise they are meaningless structures (an objection pressed against classical architectures by Searle (1980) in his Chinese Room argument). This question of how to account for meaning has mainly been addressed by philosophers, rather than cognitive scientists themselves. The major accounts include appeals to causal/informational relations, teleology, functional role, and resemblance.

When talking about representational vehicles, we can distinguish between *types* of vehicles, and concrete instances of a vehicle type, which are called *tokens*. A type of representational vehicle may be defined by, for example, a certain kind of physical structure, so particular entities exhibiting this structure would count as tokens of that representation. The appearance of a representation token in a particular cognitive system is sometimes described as the vehicle type being "tokened" in the system. The type–token distinction applies to all kinds of cognitive architectures. In classical architectures with sentential vehicles, symbol types may be defined by physical shape, so tokens of a symbol would be physical entities with that shape. In connectionist networks, one can distinguish between a type of activation pattern and particular instances of that pattern. Theories of content thus address how tokens of different vehicle types acquire their meaning.

One possibility is that a vehicle represents what it is caused by – e.g., smoke (the vehicle) means fire because smoke is caused by fire. This is the basic idea behind one construal of information (see Dretske 1981): a certain type of representational vehicle would represent or carry information about, say, cats if cats reliably cause vehicles of that type to be tokened in the system. But causal/informational relations alone fail to account for some important features of representations: that we can represent non-existent objects (which could not cause representations to be tokened), and can misrepresent things (as when a representation is caused by something that it does not represent). Further, all kinds of things carry information about their causes without representing those causes (e.g., a gun's firing does not represent its trigger being pulled).

Some theorists have tried to supplement causal/informational accounts with other factors to provide an adequate account of representational content (Cohen 2004). For example, teleological theories propose that something is a representation when it has been selected (by evolution or learning) for the function of carrying information about something in the world (Millikan 1984, Dretske 1995). This means that if a representation is selected to carry information about cats, then it will still represent cats even if on a particular occasion its tokening is caused by a dog. Jerry Fodor's (1987) asymmetric dependence account offers a different way of supplementing causal/informational relations. It claims that vehicles represent only one of the many things they carry information about – namely, the one which is causally responsible for that vehicle's carrying information about the other things. If a symbol carries information about both cats and dogs-seen-at-night, but it does the latter because it does the former, but not the reverse, then the symbol represents cats.

Critics of the above accounts often contend that there is another factor that figures in determining content which these accounts leave out – the functional role of a representation in a cognitive system. In part this role involves the relations a representation has to other representations (Block 1986). Insofar as the functional role of one representation depends on relations to other representations, and these to yet other representations, functional role accounts are *holistic*. This has spurred the objection that since all representations are related to others, one cannot acquire representations one at a time (Fodor and Lepore 1992). In contrast, causal/informational theories are *atomistic*, since each vehicle's content is determined independently of other vehicles, and thus can be learned separately.

So far we have followed the mainstream of the debates, which have treated linguistic representations as the prototypical representational vehicle. Relatively early in the development of cognitive science, however, other theorists focused on mental images as representational structures, where images are viewed as more pictorial in nature. While sentences have a linear order, the spatial properties of the vehicle are not really doing the representational work – this is done by the language's combinatorial syntax. In contrast, pictures are representational vehicles which make use of their spatial properties to represent the spatial layout of objects. Roger Shepard and Jacqueline Metzler (1971) showed that in answering questions about whether one object was a rotated version of another, the time required corresponded to the degree of rotation. This suggested people performed a rotation-like operation on mental images of the objects. Stephen Kosslyn (1980) offered evidence that people can scan, zoom, and rotate their representations just as we do pictures in the world. Since clearly we do not have pictures in our brains, these accounts have explained our mental imagery in terms of the processing mechanisms that our brain uses to process sensory information. Thus, in constructing and reasoning with a visual image, on these accounts, we use our visual system, driven not by visual input but by top-down processes, a proposal that has received support from neural-imaging studies (Kosslyn 1994).

Recently, a number of cognitive scientists have appealed to our capacities for sensory representation to ground an account of our conceptual capacities (Barsalou 1999, Mandler 2004). On these views, language-like representations are not the primary tools of thought, but rather language is a secondary tool for indexing and manipulating those representations. One particularly intriguing way of developing this idea, adumbrated initially by Kenneth Craik (1943) and developed more recently by Jonathan Waskan (2006), is that our representational vehicles are like scale models of things in the world. Just as we use physical models of airplanes in wind tunnels as representations of real airplanes, the brain is thought to operate with scale models structurally isomorphic to what they represent. Whereas the sentential representations used in classical models require separate data structures explicitly indicating how they can be manipulated so as to maintain the semantic relation to what they represent (i.e., syntactic rules), images and scale models are claimed to be structured appropriately such that changes in these representational vehicles automatically mirror changes in the represented system.

Images and scale models introduce a different sort of vehicle than found in classical symbolic models. While there are plausible ways to implement images and scale models in connectionist models, they represent a specific way of employing the connectionist framework – just as in implementing a classical architecture in a connectionist network, researchers need to constrain their networks to implement vehicles that serve as images or scale models. However images or scale models are implemented, they provide a distinctive way of approaching the content issue: resemblance relations or isomorphisms between vehicles and content (Cummins 1996, Waskan 2006). The intuitive appeal of resemblance accounts can be seen in the case of pictures. Pictures seem to represent because they share some of the physical properties of what they picture, such as color. Appealing to such "first-order" isomorphisms between individual objects and individual representations is, however, quite limited: the brain does not share, for example, the color and shape of objects it represents. Appealing to "second-order" isomorphisms – i.e., relations between the relations among various worldly objects and the relations among the associated representations – is a much better option for resemblance theories. Consider maps: although a point on a map bears little resemblance to a location in the world, the distance relations between the points on a map do resemble the distance relations between locations in the world. Such second-order isomorphisms have been found to be a common way brain areas are organized – e.g., the spatial topology of primary visual cortex resembles the spatial topology of the visual field.

Currently there is no consensus about which, if any, of these accounts of vehicles, processing, and content are correct, and vigorous discussions are continuing among both cognitive scientists and philosophers. At the same time, though, a radically alternative perspective has emerged that calls into question the reliance on representations in cognitive science. Some anti-representationalists instead advocate characterizing cognition in terms of the mathematics of dynamical systems theory (Port and van Gelder 1995). Others emphasize the coupling of our brains with our bodies and our world in ways that do not depend upon building up internal models (Clark 1997). This is to view brains not as representing the world, but as dynamically coupled with the body and extra-bodily environment. It is controversial, however, whether dynamical and situated accounts are incompatible with the mental system invoking representations in its engagement with the world (although what is represented may be different when one focuses on how cognitive agents couple with things in their world). Among those advocating a dynamical or situated perspective, some have proposed treating the brain, body, and world as extended cognitive systems, with representations propagating across these various representational media (Hutchins 1995, Clark and Chalmers 1998).

4.   Relations between Disciplines

Insofar as cognitive science presents itself as interdisciplinary, it is important to consider how disciplines can integrate. Much of the philosophical discussion of interdisciplinary relations has focused on the question of reduction and the particular model of reduction advanced by philosophers in the mid-twentieth century. Using this model, philosophers have debated whether the theories of one cognitive discipline, psychology, reduce to those of another, neuroscience (Fodor 1974, Churchland 1986). On the theory reduction model (Nagel 1961), theories are construed as linguistic statements, ideally organized in an axiomatic form, and reduction involves the derivation of the theories of the reduced science from those of the reducing science. One of the requirements of a valid reduction is that common vocabulary is used in the premises and the conclusion. Since this is typically not the case in the relation between neuroscience and psychology, additional premises are required, bridge principles that relate the vocabulary of one discipline to that of the other. Much of the controversy over the reducibility of psychology to neuroscience has turned on the issue of whether appropriate bridge principles can be generated. Fodor (1974) argues that bridge principles are not possible since the concepts used in neuroscience group phenomena in very different ways than those employed in psychology. One version of the argument appeals to the claim that psychological states can be realized in very different types of brains, in which very different states realize the same psychological predicates (Putnam 1967; a claim questioned by Bechtel and Mundale 1999).

The theory reduction account fails, in many ways, to characterize the interactions between disciplines that are characteristic of cognitive science, such as between psychology, AI, linguistics, and philosophy. Although theories are sometimes the point of engagement, many of the engagements go beyond theories, and involve the utilization of techniques of inquiry from different disciplines and the combining of explanatory approaches (e.g., computational modeling from AI in psychology). Finally, it even fails to capture the sort of relations one actually finds between psychology and neuroscience that are characterized as reductionist, since the focus is seldom on deriving one body of theory from another, but rather on working out a mechanism responsible for a phenomenon. Insofar as a mechanism involves both an account of the parts and what they do, and an account of the organization in the whole mechanism and how it confronts its environment, a mechanistic account is inherently an interlevel account to which research on both lower and higher levels contributes.

Lindley Darden and Nancy Maull (1977) introduced the notion of an interfield theory as an alternative account of how theories can relate the results of inquiries in different fields. On this account, the product of interactions between fields or disciplines is not the logical derivation of the theories of one field from those of another, but the integration of information procured in multiple fields to address a common problem. Interfield theories typically develop when investigators in different fields possess different tools which each provide part of the information needed to address the phenomenon of interest. The quest to draw and coordinate resources from contributing disciplines to explain phenomena of shared interest is characteristic of cognitive science. For example, while linguists have focused on developing grammars that account for the productive features of languages, psychologists have been concerned with the mental representations and psychological processes involved in language production and comprehension (Abrahamsen 1987). Different tools are needed and employed to formulate and test grammars than to propose and evaluate psychological mechanisms. In addition, AI researchers contribute to trying to understand language by developing computational models, while neuroscience researchers offer evidence of the brain processes employed and the operations each can perform.

From the 1950s until the 1980s, neuroscience played only a marginal role in cognitive science. Neuroscientists were actively involved in the early interdisciplinary discussions that prefigured cognitive science, but their primary investigatory tools at the time, such as recording from single neurons, could not be employed on human subjects engaged in cognitive tasks. But beginning in the 1980s and especially through the 1990s, new non-invasive neural imaging techniques that could measure brain activity (using blood flow as a proxy) provided an avenue for linking psychological studies of behavior with information about brain activity. This research is often characterized as *cognitive neuroscience* and differentiated from cognitive science proper, but increasingly these tools are being invoked in cognitive science itself.

Much of cognitive science has focused on cognitive operations detached from affect (reflecting a long differentiation in philosophy between reason and emotion). Recent research in various cognitive science disciplines has challenged this segregation. For example, evidence has amassed that effective moral reasoning requires a proper integration of emotion and reason (Damasio 1995). Similarly, pathological conditions such as autism appear to involve deficits in both reasoning and emotion. A brief exploration of this research illustrates some of the most exciting interdisciplinary engagements in contemporary cognitive science.

Autism is a developmental disorder characterized by impairments in social interaction and communication, as well as repetitive and stereotyped patterns of behavior (think of the preoccupation with *The People's Court* displayed by Dustin Hoffman's character in the 1988 movie *Rain Man*). While first under the purview of developmental psychology, autism has become the focus of study for a number of disciplines (see Volkmar, Paul, Klin, and Cohen 2005). One issue has been to characterize more precisely autism's symptoms, particularly its social deficits. Developmental psychologists and neuropsychologists work to construct improved behavioral measures to capture the spectrum of deficits found in people with autism. Given more precise descriptions of the behavioral phenomena, researchers offer theories of the cognitive operations impaired in people with autism. Some propose that autism involves deficits in "executive functions" (skills such as planning, inhibition, and cognitive flexibility); others point to problems of "weak central coherence" (the general ability to integrate pieces of information into coherent or meaningful wholes). One of the most prominent theories explains autism's social deficits in terms of an impaired "theory of mind": that autistic individuals cannot conceive of people's mental states, and thus cannot use this information to guide their social interactions.

While behavioral experiments are used to investigate the predictions of these explanatory proposals, researchers have increasingly turned to neuroscience to determine how the brains of autistic people differ (anatomically, functionally, developmentally) from those of unimpaired people and people with other psychological disorders. A popular neurobiological theory claims autism involves a dysfunctioning "mirror neuron system" (Williams,

Whiten, Suddendorf, and Perrett 2001). Mirror neurons, which were first discovered in primates, fire both when an agent acts and when they observe other agents' actions. An analogous neural system has been found in humans in the pars opercularis of the inferior frontal gyrus, and is proposed to account for, among other things, our understanding of people's intentions, emotions, and other mental states. For example, Mirella Dapretto et al. (2006) argue that we normally understand the emotional significance of people's facial expressions because the mirror neuron system, in concert with the limbic system, causes this emotion to be "mirrored" in us; in this way we "feel" and accordingly understand the perceived person's emotion. Based on an fMRI study showing little activity in the mirror neuron systems of autistic children when imitating or perceiving emotional facial expressions, Dapretto et al. suggest they do not experience the emotions of perceived others in the way unimpaired people do. This proposal thus explains autism's deficits in social understanding and interaction in terms of the inability to automatically experience the emotions and other mental states of social interactants.

As autism research shows, explanations in cognitive science proceed using tools from a variety of disciplines. These tools are brought to characterize the parts, operations, and organization of the cognitive mechanisms underlying our mental abilities.

5.   Conclusion

Cognitive science is not static. The interdisciplinary project continually identifies new problems and develops solutions for solving them. At the same time, the scope of cognitive science has been expanding. We briefly note a few of these developments and some of the implications they have for philosophical discussions of cognitive science.

As we have noted, as new tools for studying brain processing have been developed, cognitive scientists have become increasingly concerned with how cognitive operations are implemented in the brain. How to incorporate information about neural processing poses challenges for both classical and connectionist modeling in cognitive science. Insofar as cognitive models focus on the mental activity, they must to some degree abstract from the neural detail. This frames a philosophical question about how much they can abstract from details of neural processing and still claim to provide accounts of how humans process information. A related issue involves cognitive science's traditional reliance on logical and heuristic techniques to model reasoning. Increasingly, both cognitive and neuroscience researchers are advancing probabilistic models (Chater, Tenenbaum, and Yuille 2006). For philosophers, this raises the question of whether such models should replace more traditional cognitive models, or whether the two kinds of models can be constructively related.

Another major new direction in cognitive science is the concern with the embodied, situated nature of cognition. While traditional cognitive science focused solely on what is going on in the heads of cognizers, recent theorists have argued that the non-neural body and environmental context are not merely inputs to the cognitive system but play a constitutive role in cognition (e.g., Clark 1997). Some of those championing more attention to the organism's body and environment have appealed to a previously untapped philosophical tradition known as phenomenology (comprising writers such as Husserl, Heidegger, and Merleau-Ponty) for insight about these issues (e.g., Wheeler 2005). When the focus is on the real-time responses of organisms to their environment, the temporal dynamics of cognitive processes is obviously important and has been emphasized by those advocating use of dynamical systems tools for understanding cognition (Port and van Gelder 1995). Many advocates of applying dynamical systems theory to cognition have, as we noted, also argued against the reliance on representations in cognitive models. Others, such as Rick Grush (2004), have tried to show how control theory, a dynamical approach, employs neural representations in accounting for motor control, and to extend this approach to other cognitive processes. These and other current debates in cognitive science provide rich opportunities for continued philosophical engagement with cognitive science.

References

Abrahamsen, A.A. 1987. "Bridging Boundaries Versus Breaking Boundaries: Psycholinguistics in Perspective." *Synthese* 72.3: 355–88.

Anderson, J.R. 2007. *How Can the Human Mind Occur in the Physical Universe?* Oxford: Oxford University Press.

Anderson, J.R. and G.H. Bower. 1973. *Human Associative Memory*. New York: John Wiley and Sons.

Barsalou, L.W. 1999. "Perceptual Symbol Systems." *Behavioral and Brain Sciences* 22: 577–660.

Bechtel, W. 2006. *Discovering Cell Mechanisms: The Creation of Modern Cell Biology.* Cambridge: Cambridge University Press.

Bechtel, W. 2008. *Mental Mechanisms: Philosophical Perspectives on Cognitive Neuroscience.* London: Routledge.

Bechtel, W. and A. Abrahamsen. 2002. *Connectionism and the Mind: Parallel Processing, Dynamics, and Evolution in Networks.* 2nd edn. Oxford: Blackwell.

Bechtel, W. and J. Mundale. 1999. "Multiple Realizability Revisited: Linking Cognitive and Neural States." *Philosophy of Science* 66: 175–207.

Bechtel, W. and R.C. Richardson. 1993. *Discovering Complexity: Decomposition and Localization as Strategies in Scientific Research.* Princeton, NJ: Princeton University Press.

Block, N. 1986. "Advertisement for a Semantics for Psychology." In P. French, T. Uehling, and H. Wettstein (eds.), *Midwest Studies in Philosophy* 10. Minneapolis, MN: University of Minnesota Press, 615–78.

Bloomfield, L. 1939. *Linguistic Aspects of Science.* Chicago: University of Chicago Press.

Bobrow, D. and A. Collins (eds.). 1975. *Representation and Understanding: Studies in Cognitive Science.* New York: Academic Press.

Broadbent, D. 1958. *Perception and Communication.* London: Pergamon Press.

Chater, N., J.B. Tenenbaum, and A. Yuille. 2006. "Probabilistic Models of Cognition: Conceptual Foundations." *Trends in Cognitive Sciences* 10.7: 287–91.

Chomsky, N. 1956. "Three Models for the Description of Language. *Transactions on Information Theory* 2.3: 113–24.

Chomsky, N. 1957. *Syntactic Structures.* The Hague: Mouton.

Chomsky, N. 1995. *The Minimalist Program.* Cambridge, MA: MIT Press.

Churchland, P.S. 1986. *Neurophilosophy: Toward a Unified Theory of Mind-brain.* Cambridge, MA: MIT Press.

Clark, A. 1997. *Being There: Putting Brain, Body, and World Together Again.* Cambridge, MA: MIT Press.

Clark, A. and D. Chalmers. 1998. "The Extended Mind." *Analysis* 58: 10–23.

Cohen, J. 2004. "Information and Content." In L. Floridi (ed.), *Blackwell Guide to the Philosophy of Information and Computing.* Oxford: Blackwell, 215–27.

Craik, K. 1943. *The Nature of Explanation.* Cambridge: Cambridge University Press.

Craver, C. 2007. *Explaining the Brain: What a Science of the Mind-brain Could Be.* New York: Oxford University Press.

Cummins, R. 1996. *Representations, Targets, and Attitudes.* Cambridge, MA: MIT Press.

Cummins, R. 2000. "'How Does it Work?' versus 'What Are the Laws?': Two Conceptions of Psychological Explanation." In F. Keil and R. Wilson (eds.), *Explanation and Cognition.* Cambridge, MA: MIT Press, 117–44.

Damasio, A.R. 1995. *Descartes' Error.* New York: G.P. Putnam.

Dapretto, M., M.S. Davies, J.H. Pfeifer, A.A. Scott, M. Sigman, S.Y. Bookheimer, et al. 2006. "Understanding Emotions in Others: Mirror Neuron Dysfunction in Children with Autism Spectrum Disorders." *Nature Neuroscience* 9: 28–30.

Darden, L. 2006. *Reasoning in Biological Discoveries: Essays on Mechanisms, Interfield Relations, and Anomaly Resolution*. Cambridge: Cambridge University Press.

Darden, L. and N. Maull. 1977. "Interfield theories." *Philosophy of Science* 43: 44–64.

Dretske, F.I. 1981. *Knowledge and the Flow of Information*. Cambridge, MA: MIT Press /Bradford Books.

Dretske, F.I. 1995. *Naturalizing the Mind*. Cambridge, MA: MIT Press.

Ebbinghaus, H. 1885. *Über das Gedächtnis: Untersuchungen zur experimentellen Psychologie*. Leipzig: Duncker and Humblot.

Fauconnier, G. 1994. *Mental Spaces: Aspects of Meaning Construction in Natural Language*. Cambridge: Cambridge University Press.

Fodor, J.A. 1974. "Special Sciences (or: The Disunity of Science as a Working Hypothesis)." *Synthese* 28: 97–115.

Fodor, J.A. 1975. *The Language of Thought*. Cambridge, MA: Harvard University Press.

Fodor, J.A. 1987. *Psychosemantics: The Problem of Meaning in the Philosophy of Mind*. Cambridge, MA: MIT Press.

Fodor, J.A. and E. Lepore. 1992. *Holism: A Shopper's Guide*. Oxford: Blackwell.

Garcia, J., B.K. McGowan, F.R. Ervin, and R.A. Koelling. 1968. "Cues: Their Relative Effectiveness as a Function of the Reinforcer." *Science* 160: 794–5.

Glanzer, M. and A.R. Cunitz. 1966. "Two Storage Mechanisms in Free Recall." *Journal of Verbal Learning and Verbal Behavior* 5: 351–60.

Grush, R. 2004. "The Emulation Theory of Representation: Motor Control, Imagery, and Perception." *Behavioral and Brain Sciences* 27: 377–96.

Harris, R.A. 1993. *The Linguistics Wars*. New York: Oxford University Press.

Hempel, C.G. 1965. "Aspects of Scientific Explanation." In C.G. Hempel (ed.), *Aspects of Scientific Explanation and Other Essays in the Philosophy of Science*. New York: Macmillan, 331–496.

Hull, C.L. 1943. *Principles of Behavior*. New York: Appleton-Century-Crofts.

Hutchins, E. 1995. *Cognition in the Wild*. Cambridge, MA: MIT Press.

Kosslyn, S.M. 1980. *Image and Mind*. Cambridge, MA: Harvard University Press.

Kosslyn, S.M. 1994. *Image and Brain: The Resolution of the Imagery Debate*. Cambridge, MA: MIT Press.

Lakoff, G. and M.H. Johnson. 1999. *Philosophy in the Flesh: The Embodied Mind and Its Challenge to Western Thought*. New York: Basic Books.

Machamer, P., L. Darden, and C. Craver. 2000. "Thinking about Mechanisms." *Philosophy of Science* 67: 1–25.

Mandler, J.M. 2004. *The Foundation of Mind: Origins of Conceptual Thought*. Oxford: Oxford University Press.

McClelland, J.L., B. McNaughton, and R.C. O'Reilly. 1995. "Why There Are Complementary Learning Systems in the Hippocampus and Neocortex:

Insights from the Successes and Failures of Connectionist Models of Learning and Memory." *Psychological Review* 102.3: 419–57.

McCulloch, W.S. and W.H. Pitts. 1943. "A Logical Calculus of the Ideas Immanent in Nervous Activity." *Bulletin of Mathematical Biophysics* 7: 115–33.

Miller, G.A. 1956a. "Human Memory and the Storage of Information." *Transactions on Information Theory* 2.3: 129–37.

Miller, G.A. 1956b. "The Magical Number Seven, Plus or Minus Two: Some Limits on Our Capacity for Processing Information." *Psychological Review* 63: 81–97.

Miller, G.A. 1979. *A Very Personal History* (Occasional Paper No. 1). Cambridge, MA: Center for Cognitive Science.

Millikan, R.G. 1984. *Language, Thought, and Other Biological Categories*. Cambridge, MA: MIT Press.

Minsky, M. and S. Papert. 1969. *Perceptrons: An Introduction to Computational Geometry*. Cambridge, MA: MIT Press.

Nagel, E. 1961. *The Structure of Science*. New York: Harcourt, Brace.

Neisser, U. 1967. *Cognitive Psychology*. New York: Appleton-Century-Crofts.

Newell, A. 1990. *Unified Theories of Cognition*. Cambridge, MA: Harvard University Press.

Newell, A. and H.A. Simon. 1956. "The Logic Theory Machine: A Complete Information Processing System." *Transactions on Information Theory 2.3*: 61–79.

Newell, A. and H.A. Simon. 1972. *Human Problem Solving*. Englewood Cliffs, NJ: Prentice Hall.

Newell, A. and H.A. Simon. 1976. "Computer Science as Empirical Inquiry: Symbols and Search." *Communications of the ACM* 19: 113–26.

Norman, D.A. and D.E. Rumelhart. 1975. *Explorations in Cognition*. San Francisco: Freeman.

Port, R. and T. van Gelder (eds.). 1995. *Mind as Motion: Explorations in the Dynamics of Cognition*. Cambridge, MA: MIT Press.

Putnam, H. 1967. "Psychological Predicates." In W.H. Capitan and D.D. Merrill (eds.), *Art, Mind and Religion*. Pittsburgh: University of Pittsburgh Press, 37–48.

Rosenblatt, F. 1962. *Principles of Neurodynamics: Perceptrons and the Theory of Brain Mechanisms*. Washington, DC: Spartan Books.

Rumelhart, D.E. and J.L. McClelland. 1986. *Explorations in the Microstructure of Cognition. Vol. 1. Foundations*. Cambridge, MA: Bradford Books, MIT Press.

Schacter, D.L. 1996. *Searching for Memory: The Brain, the Mind, and the Past*. New York: Basic Books.

Searle, J.R. 1980. "Minds, Brains, and Programs." *Behavioral and Brain Sciences* 3: 417–24.

Shannon, C.E. 1948. "A Mathematical Theory of Communication." *Bell System Technical Journal* 27: 379–423, 623–56.

Shepard, R.N. and J. Metzler. 1971. "Mental Rotation of Three-dimensional Objects." *Science* 171: 701–3.

Thagard, P. 2006. *Hot Thought: Mechanisms and Applications of Emotional Cognition.* Cambridge, MA: MIT Press.

Tolman, E.C. 1948. "Cognitive Maps in Rats and Men." *Psychological Review* 55: 189–208.

Tulving, E. 2002. "Episodic Memory: From Mind to Brain." *Annual Review of Psychology* 53: 1–25.

Volkmar, F.R., R. Paul, A. Klin, and D.J. Cohen (eds.). 2005. *Handbook of Autism and Pervasive Developmental Disorders.* Hoboken, NJ: John Wiley.

Waskan, J. 2006. *Models and Cognition.* Cambridge, MA: MIT Press.

Watson, J.B. 1913. "Psychology as the Behaviorist Views it." *Psychological Review* 20: 158–77.

Wheeler, M. 2005. *Reconstructing the Cognitive World: The Next Step.* Cambridge, MA: MIT Press.

Williams, J.H., A. Whiten, T. Suddendorf, and D.I. Perrett. 2001. "Imitation, Mirror Neurons and Autism." *Neuroscience and Biobehavioral Reviews* 25: 287–95.

11   Philosophy of Psychology

Edouard Machery

Philosophy of psychology takes various forms. Some philosophers of psychology use psychological findings and theories to develop new answers to traditional philosophical questions. A smaller number of philosophers of psychology take their cue from the philosophy of science. They describe and evaluate the discovery heuristics, theories, and explanatory practices endorsed by psychologists. Finally, much philosophy of psychology can be characterized as psychological theorizing. Just like psychologists, philosophers propose empirical theories of specific aspects of our mind, trying to explain relevant psychological phenomena. Focusing mostly on this aspect of the philosophy of psychology, I will consider philosophers' contribution to the theoretical development of psychology in four areas: cognitive architecture and modularity (§2); situated, embodied, and extended cognition (§3); concepts (§4), and mind-reading (§6).[1] Before doing this, however, I will discuss philosophers' and psychologists' views and arguments about the distinctive character of psychology – its mentalistic nature (§1).

---

[1] Philosophers have contributed to the theoretical development of other areas, such as emotions (e.g., Griffiths 1997, Prinz 2004a), consciousness (e.g., Block 1995, Chalmers 1996, Chalmers 2004, Noë 2004, Block 2007), perception (e.g., Jacob and Jeannerod 2003, Noë 2004), psychopathology (e.g., Murphy 2007), moral psychology (e.g., Doris 2002, Nichols 2004, Doris and Stich 2006, Prinz 2007), the relation between language and thought (e.g., Carruthers 2006), and the scientific value of evolutionary psychology (e.g., Buller 2005, Machery and Barrett 2006, Machery forthcoming). For the sake of space, I will not review these contributions here.

1. The Scientific Legitimacy of Mentalism?

1.1 The place of mental states in psychological theories and explanations

It will be useful to start the discussion of the place of mental states in psychology with an example. Everyday experience and experimental evidence suggest that people often reason poorly about probabilistic matters. For instance, Tversky and Kahneman asked participants to read the following story:

> Linda is 31 years old, single, outspoken, and very bright. She majored in philosophy. As a student, she was deeply concerned with issues of discrimination and social justice, and also participated in anti-nuclear demonstrations. (1982, 92)

Participants were then asked to rank various "statements by their probabilities," including the following three:

1. Linda is active in the feminist movement.
2. Linda is a bank teller.
3. Linda is a bank teller and is active in the feminist movement.

Remember that the conjunction axiom of probability theory states that the probability of a conjunction is always smaller than or equal to the probability of one of its conjuncts:

4. $P(p\&q) \leq P(p)$

Thus, participants in Tversky and Kahneman's experiment would be mistaken to answer that it is more probable that Linda is a feminist and a bank teller than simply a bank teller. Nonetheless, 89 percent of participants judged that Linda was more likely a feminist bank teller than a bank teller, a mistake known as "the conjunction fallacy."

Tversky and Kahneman use such mistakes to investigate the psychological mechanisms underlying (correct and incorrect) probabilistic judgments. They propose that people's probabilistic judgments result from simple psychological processes (called "heuristics") that often lead to correct judgments, but occasionally mislead (they are then called "biases") – hence the name of their research program, "heuristics and biases." Thus, according to Tversky and Kahneman, people often use a simple psychological process, called "the representativeness heuristic," to make probability judgments. People evaluate the probability that *a* is an F, according to the similarity between the description of *a* and the stereotype of an F. In the experiment just described, people evaluate the probability that Linda is a bank teller by comparing the description of Linda that is provided in the cover story and the stereotype of a bank teller. Because Linda is less representative of a bank teller than of a feminist bank teller, people rank (3) as more probable than (2), thereby committing the conjunction fallacy.

For present purposes, what matters is that Tversky and Kahneman's account of people's probabilistic judgments is *mentalistic*: it involves ascribing *mental states* to people (viz., internal states that mediate between environmental stimuli and behavior[2]) and *psychological processes* (viz., processes that manipulate mental states). Consider again the representativeness heuristic. When people evaluate the probability that an individual *a* is an F, they retrieve a stereotype of an F (a mental state) from memory. This stereotype is compared with the information about *a*, a psychological process that results in a measure of how representative *a* is of Fs. The probability that *a* is an F is a monotonic function of this measure. Mentalism (viz., the appeal to mental states, psychological processes, and other psychological entities such as personality traits) is a characteristic property of the theories and explanations developed in the various subfields of psychology (e.g., social psychology, cognitive psychology, personality psychology, etc.).

Now, one might wonder whether mentalist theories are legitimate scientific theories. Mental states, psychological processes, and other psychological entities are unobservable entities, which are posited to account for behavior. Like other theoretical entities, claims to their reality should be subject to scrutiny. More importantly, mental states and psychological processes have often been associated with ontological and epistemological properties that are not scientifically correct. Since the seventeenth century, mental states have often been associated with substance dualism – the idea

---

[2]   Because they are diverse, defining what mental states are is a difficult task. Many mental states have semantic properties: they can be true or false (this is the case of, for example, beliefs), or satisfied or unsatisfied (this is the case of, for example, desires). Some mental states also have phenomenal properties: it feels something to have them.

that there are two substances, matter and mind. But, if mental states were distinct from physical states, it would be mysterious how they could causally interact with physical states. Furthermore, it has sometimes been argued that by introspection (the observation of one's own mental states), each of us has a privileged access to our own mental states. But, this first-person privilege seems at odds with the idea that evidence in science is public and accessible from a third-person perspective.

1.2 Methodological behaviorism

In the first decades of the twentieth century, the school of psychology known as "behaviorism" or "methodological behaviorism"[3] formulated the most radical answer to the question "Can psychological entities be legitimately postulated by a scientific theory?" For behaviorists, mental states and other psychological entities had no place in psychology. Behaviorists not only contended that referring to unobservable states between environment and behavior was not required for explaining behavior, but they also argued that it was unscientific. In his influential behaviorist manifesto, John Watson, the father of behaviorism, wrote that psychology could be written without ever using "the terms consciousness, mental states, mind, content, intro-spectively verifiable, imagery, and the like. . . . It can be done in terms of stimulus and response, in terms of habit formation, habit integrations and the like" (1913, 166–7).

Behaviorism was a reaction against the dominant psychology of the time.[4] Much of human psychology in the second half of the nineteenth century was based on introspection. By the end of the nineteenth century, how-ever, the nature of introspection and its value as a scientific method had become a controversial topic among introspective psychologists them-selves (Caldwell 1899, Titchener 1899). By contrast, following the lead of Edward Thorndike and Robert Yerkes, animal psychology had developed controlled and reproducible experimental designs and quantitative measur-ing techniques that allowed for the experimental study of numerous animal

---

[3] "Methodological" is used to distinguish the type of behaviorism discussed here from Gilbert Ryle's (1951) logical behaviorism, according to which mental state predicates pick out behavioral dispositions.

[4] Watson was also influenced by the work of the Russian physiologist Ivan Pavlov. See O'Donnell (1985) for a history of behaviorism.

behaviors (e.g., orientation, problem solving, etc.). The disarray of introspective psychology and the successes of animal psychology paved the way for the reception of behaviorism.

For Watson (1913), the rejection of mental states and other psychological entities as objects of scientific study and as scientific explanatory entities was primarily due to his rejection of introspection. In substance, Watson argues that because introspection is not a proper scientific methodology, the states to which it gives access (viz., the mental states) have no place in a scientific psychology. We will come back to this argument later.

Introspection itself was rejected on the grounds that its products were subjective and unreliable. Watson argued that far from being objective observational reports, introspective reports were influenced by psychologists' theoretical commitments. He also noted that introspection had failed to promote any consensus among psychologists. In addition, introspection prevented the unification of psychology, since it was not used in animal psychology. Rejecting introspection could allow for the transfer of methods from animal to human psychology and for the comparison of results across disciplines.

In addition, one finds in Watson's manifesto the following parsimony argument. Because, for Watson, (animal and human) behavior is an instinctive (i.e., inherited) or habitual (i.e., learned) reaction to measurable aspects of the environment, explaining, predicting, and manipulating behavior merely requires knowledge of the learning history of the agent and of its environment. Introspective data about mental states have no role to play for explaining, predicting, and manipulating behavior.

Although distinct behaviorist theories have been developed (e.g., by Clark Hull, B.F. Skinner, and Edward Tolman), these theories share a common focus on the contingencies between behaviors (called "responses") and measurable environmental conditions (called "stimuli"). Behaviorists attempted to explain why specific behaviors were produced in specific environments by looking at the history of interactions between organisms and their environment. They developed two main accounts of learning: classical conditioning and operant conditioning. Importantly, none of these accounts is mentalist: no reference is made to intervening variables between behavior and the environment.

According to the theory of classical conditioning, inspired by Pavlov's work, organisms have spontaneous responses ("unconditioned responses") caused by environmental stimuli ("unconditioned stimuli"). When a stimulus that is not associated with any response (a "conditioned stimulus") is repeatedly presented in association with an unconditioned stimulus, the organism ends up associating the response with the conditioned stimulus. Thus, in Pavlov's well known experiment, a sound was repeatedly played when food was presented to a dog (causing the dog to salivate) and the dog ended up salivating at the mere hearing of this sound.

The theory of operant or instrumental conditioning, developed by Thorndike and Skinner (e.g., 1938), divides behaviors into two types: responses, which are caused by identified stimuli (e.g., salivating when food is present), and operants, which are not associated with specific stimuli (e.g., pressing a lever for a rat). Focusing on operants, Skinner proposed that organisms tend to repeat operants whose strength is "reinforced."[5] Thus, if a rat receives some food when it presses a lever, the rat will tend to press the lever again. The operant is reinforced and the strength of the operant is measured by how long the organism will press the lever at a rate higher than the base rate (i.e., the rate before reinforcement) under extinction (that is, when no reinforcer follows the operant).

While classical conditioning can explain why an organism (a human or an animal) extends a behavior that is already part of its behavioral repertoire to new contexts, operant conditioning can explain the inclusion of new, originally randomly produced behaviors in the behavioral repertoire of an organism.

Behaviorism has certainly had a lasting and, in many respects, positive influence on psychology. Modern psychology inherited its emphasis on controlled experimental procedures and quantitative, objective measures (rather than introspective reports). Classical and operant conditioning are also important properties of learning (but see Gallistel and Gibbon 2001).

Still, behaviorism was rejected in the second half of the twentieth century for four main reasons.[6] First, the explanatory scope of behaviorist theories turned out to be limited. In his influential review of Skinner's *Verbal Behavior*, Noam Chomsky (1959) noted that while the key notions of

---

[5] The definition of reinforcement was a subject of controversy among behaviorists. I will overlook this difficulty here.

[6] The history of the demise of behaviorism is more complex than is typically acknowledged. Behaviorism remains influential in animal psychology, in educational psychology, and in some fields of neuroscience. Furthermore, although the lore has it that Chomsky's scathing review of Skinner's *Verbal Behavior* was the crucial event in the rejection of behaviorism, efforts to reevaluate mentalism were already on their way in the 1940s (e.g., MacCorquodale and Meehl 1948).

stimulus, reinforcer, and operant were well defined when they were applied to pigeons and rats whose behavior and environment are highly constrained by Skinnerian experimental designs (such as a Skinner box), they were poorly defined outside such a context. He concluded that when used to explain everyday (human and animal) behavior (e.g., language acquisition by children), Skinner's theoretical notions were either misapplied or a misleading paraphrase of mentalistic notions.

Second, psychologists and philosophers have come to realize that to explain people's behavioral competences, it is necessary to postulate intervening states which mediate between environment and behavior. Thus, Chomsky (1959) argued that it was impossible to explain language acquisition without considering both the environment in which learning takes place (the linguistic stimuli) and the contribution of the learner.

Third, we saw above that the rejection of mentalism was principally a consequence of the rejection of introspection. Just like behaviorists, contemporary psychologists typically deny that introspection is a valuable source of evidence about mental states and psychological processes. However, in contrast to behaviorists, they do not conclude that mental states and psychological processes are not proper explanatory entities and objects of scientific study. For contemporary psychologists, the ascription of mental states in psychological explanations is to be justified on explanatory grounds (viz., to account for behavioral competences) rather than on the basis of introspection. As a result, the unreliability and subjectivity of introspective reports do not impugn the justification of mental state ascription.

Finally, as we shall see in the next section, philosophers and psychologists developed a new characterization of mental states and psychological processes that made mentalism scientifically reputable.

1.3   The computational representational theory of mind

The limits of behaviorism show that a purely behavioral psychology is unpromising and that internal states have to be postulated to account for behavioral competences. At this juncture, it seems natural to propose that mental states are the internal states needed for a scientific psychology. This would be premature, however, for it remains to provide a scientifically satisfying account of mental states and psychological processes.

Information theory and the theory of digital computers have provided such an account (Fodor 1975, Newell and Simon 1976, Pylyshyn 1984,

Marcus 2001; also, see Chapter 10 by Bechtel and Herschbach in this volume for additional detail). Mental states are thought to be representations, that is, particulars endowed with a specific content. Written or spoken sentences, maps, paintings, and road signs are representations in that sense: they represent the world as being so and so and can be thereby true or false, accurate or inaccurate. Sentences, maps, paintings, etc., have a derived content, meaning that they have a given content because people use them to represent in a given way. By contrast, mental representations have a non-derived or original content, because, on pain of regress, they cannot intentionally be used to represent. Two mental states of a given type (e.g., the beliefs that Paris is in France and that Berlin is in Germany) are distinguished by the content of their respective representations. Different types of mental states (e.g., beliefs and desires) are distinguished by their functional roles. While the belief that it is noon and the desire that it is noon are both representations that express the proposition that it is noon, they are distinguished by their functional role: beliefs and desires have different causal connections with perceptual stimuli, other mental states, and actions.

Psychological processes consist in transformations of representations. Philosophers and psychologists have proposed that these transformations are computational – hence the name "the computational representational theory of mind." That is, in substance, these transformations are governed by rules that apply to representations in virtue of their formal properties. These rules do not apply to representations in virtue of their content, but in virtue of some non-semantic properties of the representations, in exactly the same way as numerals do not get added by pocket calculators in virtue of their meaning (the numbers they express), but in virtue of their syntactic properties (see Piccinini 2008 for complications).

The computational representational theory of mind assuages worries about the scientific legitimacy of mentalism. Digital computers show that material entities can implement computational processes manipulating representations. They also illustrate how being introspectively accessible is not an essential property of representations. Still, it is important to flag two issues raised by this theory. First, what is the relation between mental representations and brain states? Second, in virtue of what do mental representations have their content? Philosophers of mind and of psychology have extensively discussed these two issues, but for the sake of space, I will not elaborate on them in this essay (see, e.g., Fodor 1987, Stich and Warfield 1994).

2. Cognitive Architecture and Massive Modularity

The organization of the processes that underwrite our perceptual and cognitive architecture – "the cognitive architecture" – is an important topic of debate among philosophers of psychology. Jerry Fodor (1983) has proposed an influential hypothesis about the nature of human cognitive architecture. He distinguishes two types of processes, modules and non-modular processes. A Fodorian module is a psychological process that has most of the following properties: it has a specific type of input and it produces shallow or non-conceptual outputs; its functioning is fast, automatic, cognitively impenetrable (that is, other systems have no access to and no influence upon its internal processing), and informationally encapsulated (that is, it has access to only a subset of the information that is represented in the mind); it is also realized in a discrete brain area, it is innate, and it breaks down in characteristic ways.[7] By contrast, non-modular processes have few (if any) of these properties. For Fodor, modules underwrite a few capacities – particularly, our perceptual capacities and our linguistic faculty. The processes underlying our higher cognitive capacities (e.g., the fixation of our beliefs, the determination of our desires) – what Fodor calls "our central processes" – are supposed to be non-modular.

In contrast to Fodor, many psychologists have argued that the processes underlying some higher cognitive capacities are modular. For instance, Elisabeth Spelke has argued that our capacity to orient ourselves is underwritten by a geometric module (Hermer and Spelke 1996), while Nancy Kanwisher has proposed that a module underwrites our capacity to identify individual faces (Kanwisher, McDermott, and Chun 1997). Going further, some psychologists (e.g., John Tooby, Randy Gallistel) and some philosophers (e.g., Peter Carruthers, Dan Sperber) propose that *all* our psychological processes are modular, a thesis known as "the massive modularity hypothesis."

Various arguments have been proposed in support of the massive modularity hypothesis (Sperber 1994, 2001; for a systematic overview, see Carruthers 2006, ch. 1). I focus here on evolutionary psychologists John Tooby and Leda Cosmides' argument that evolution is unlikely to have selected for non-modular psychological processes (Tooby and Cosmides 1992,

---

[7] In *The Mind Doesn't Work that Way* (2000), Fodor emphasizes the encapsulation of modules.

Cosmides and Tooby 1994).[8] Rather than focusing on the properties that are characteristic of Fodorian modules (see above), Tooby and Cosmides characterize modules in terms of functional specialization (sometimes called "domain-specificity"): modules have been selected for bringing about a specific outcome (that is their function). Evolutionary modules contrast with "domain-general" processes, namely psychological processes that are not functionally specialized. Tooby and Cosmides assume (as I will for the sake of the argument) that psychological processes are adaptations, that is (using evolutionary psychologists' terminology), traits that have been selected because they solved some adaptive problems (e.g., finding food, choosing a fertile mate, avoiding poisons, detecting cheaters, etc.). They argue that a domain-general process would be less efficient than a modular process to solve a given problem, because the latter, but not the former, would have been designed to solve this problem. Thus, natural selection would tend to favor modular processes over non-modular processes. As Cosmides and Tooby famously put it, "as a rule, when two adaptive problems have solutions that are incompatible or simply different, a single general solution will be inferior to two specialized solutions. In such cases, a jack of all trades is necessarily master of none, because generality can be achieved only by sacrificing effectiveness" (1994, 89).

In reply, Richard Samuels (1998) notes that there are two types of modules – computational modules and Chomskyan modules. Computational modules are mechanisms; they are defined by the nature of their processes. The modules hypothesized by Tooby and Cosmides are of this first kind. Chomskyan modules are bodies of knowledge about specific tasks – they are representations, not processes. Chomskyan modules can be used by domain-general reasoning mechanisms. To illustrate this contrast, consider the adaptive problem of avoiding poisonous foods. (Because their diet is not specialized, omnivores have had to solve this problem.) A computational module for solving this problem would be a mechanism for distinguishing safe from unsafe foods. By contrast, a Chomskyan module would be a body of knowledge about safe and unsafe foods, which could be used by a domain-general reasoning mechanism. Having distinguished these two types of module, Samuels notes that natural selection would not prefer

---

[8] Some psychologists and philosophers have also proposed tractability arguments, which state that only modular processes can perform the computations that are required by the tasks defining our cognitive and perceptual capacities (for review and discussion, see Samuels 2005).

a cognitive system made of computational modules to a cognitive system made of Chomskyan modules used by a domain-general reasoning system, because Chomskyan and computational modules are equally specialized for solving adaptive problems. But, if the mind were a cognitive system made of Chomskyan modules used by a domain-general reasoning system, the massive modularity hypothesis would be false. Thus, Samuels concludes, it does not follow from the hypothesis that our cognitive architecture is the product of evolution by natural selection that the mind is massively modular.

The massive modularity hypothesis has been criticized on various grounds (Fodor 2000, Buller 2005). I discuss here only two problems, the input problem (Fodor 2000) and the brain evolution problem (Quartz 2002). Noting that a specific type of input is required to trigger a given module, Fodor (2000) contends that a psychological process (a routing system) is needed to pair each module with the stimuli that trigger it. Because this routing system would have to be activated by all types of stimuli, it could not be modular. Thus, the massive modularity hypothesis is false. Clark Barrett (2005) has convincingly rebutted this argument by drawing an analogy between enzymes and modules. Enzymes come into contact with a large range of substrates. However, because they have specific binding sites, only some of these substrates are bound with enzymes. Similarly, modules could have access to all representations but be activated by only some of them. No non-modular routine process is thus needed in a modular mind.

Barbara Finlay and colleagues' work on brain evolution has also inspired an important objection against the massive modularity hypothesis. They found that across mammals, the volume of the main brain structures is correlated to the volume of the whole brain (Finlay and Darlington 1995). Steve Quartz has argued that these findings show that natural selection did not act on individual brain structures independently of the other brain structures: "These results suggest that neural systems covary highly with one another as a consequence of the restricted range of permissible alterations that evolutionary psychology can act upon. This makes the massive modularity hypothesis of narrow evolutionary psychology untenable" (2002, 189).

Quartz's argument should be resisted (Machery 2007c). A closer look at Finlay and colleagues' data shows that across mammals, the volume of the whole brain does not covary perfectly with the volume of the main brain structures, suggesting that natural selection may have acted upon their volume. Furthermore, there is more to brain evolution than the volume of the brain structures considered by Finlay and colleagues.

Natural selection probably acted upon the nature of brain cells, their organization, or the connectivity between brain areas.

Finally, it is noteworthy that a careless use of the term "module" has muddled the debate about the massive modularity hypothesis. "Module" means different things for different people (Barrett and Kurzban 2006). Particularly, as noted above, evolutionary psychologists define modules as those processes that have a dedicated function. They need not have any of the properties that characterize Fodorian modules. (Similarly, "module" has a distinctive use in neuroscience.) For instance, modules need not be innate, nor need they be automatic or cognitively impenetrable (Machery and Barrett 2006). Rejecting evolutionary psychologists' massive modularity hypothesis on the grounds that our central processes do not possess the properties that characterize Fodorian modules (e.g., they are not automatic, etc.) is thus unsound.

3.   Embodied, Situated, and Extended Cognition

Traditionally, philosophers and psychologists hold that the mind receives some information about its environment through the senses, and uses this information to reason and make decisions, which may lead to action. Some philosophers and psychologists, whose views are often grouped together under the headings "embodied cognition," "situated cognition," and "extended cognition," have criticized this conception of the relation between cognition and the cognizer's environment. Although the views denoted by these headings differ in some respects, for simplicity I will use the expression "extended cognition" in what follows, noting the differences between these views when appropriate. It has often been noted that this new movement combines several distinct positions without clearly marking their differences (e.g., Wilson 2002, Rupert 2004).[9] In this section, I briefly distinguish four threads, before discussing in some detail the idea that mental states and psychological processes are not located in the brain.

A first thread is methodological. Proponents of extended cognition contend that a proper understanding of psychological processes involves examining the environment in which cognition takes place (e.g., Hutchins

[9]   In addition to the works cited in this section, see Clark 1997, Anderson 2003, Shapiro 2004, Wilson 2004.

1995) – a position often referred to by the label "situated cognition."
This methodological claim is sometimes justified on the grounds that
psychological processes are designed (by evolution or by learning) for
specific (physical and social) environments. To illustrate, according to
Gerd Gigerenzer and Ulrich Hoffrage (1995), the processes underlying
probabilistic reasoning are designed to manipulate representations of
natural frequencies, rather than probabilities, consistent with the fact that
for most of human history, probabilistic information was only available in
the form of natural frequencies.

A second thread highlights the importance of agency in understanding
cognition (e.g., Noë 2004, Gallagher 2005). This emphasis is supposed to
stand in contrast to cognitive psychology's traditional focus on situations
that involve no or little action (e.g., chess playing, remembering words on
a list, etc.).

A third, more radical thread takes issue with the idea that cognition involves
manipulating representations (e.g., Brooks 1991, Thelen and Smith 1994).
Anti-representationalists typically focus on some phenomena that proponents of representation-based approaches to cognition explain (or would
explain) by means of representations and representation-based processes.
They then explain these phenomena without positing any process that
manipulates representations. On this basis, they draw the following induction: if postulating representations is not needed to explain these phenomena,
behavior and cognition at large can be explained without representations
(for discussion, see, e.g., Vera and Simon 1993, Clark and Toribio 1994).

A fourth strand of argument, often referred to by the labels "extended
cognition" and "extended mind," focuses on the location of mental states
and psychological processes. Philosophers and psychologists have often
identified token mental states with brain states and psychological processes
with neural processes. In sharp contrast, Mark Rowlands writes that
"[c]ognitive processes are not located exclusively inside the skin of cog-
nizing organisms" (1999, 22); while Andy Clark and David Chalmers
argue: "[W]e will argue that *beliefs* can be constituted partly by features
of the environment, when those features play the right sort of role in driv-
ing cognitive processes. If so, the mind extends into the world" (1998, 12).
According to this view, at least some token mental states are external to
the body or involve extra-corporeal objects as proper parts, while cognition
involves the manipulation of these entities.

Two well known examples might usefully illustrate this view. Consider
first how we perform a complex arithmetical operation by hand, such as

the multiplication of 37 by 23 (Clark and Chalmers 1998, Adams and Aizawa 2001, Noë 2004, Adams and Aizawa 2008). We write down one numeral below the other. Focusing on the rightmost digital of each numeral ("7" and "3"), we multiply the numbers they express. We write down "1" on a third line and write "2" as a carry-over. We then multiply 3 (the number expressed by the leftmost digital of the first numeral) by 3 (for the rightmost digital of the second numeral) and add the carry-over. We write down the numeral "11" left of the numeral "2" (and so on). To perform this multiplication, we create and manipulate objects (viz., numerals) that are external to our body in a rule-governed manner. According to proponents of extended cognition, the numerals are part of our mental states and their rule-governed manipulation counts as psychological processing. As Alva Noë puts it, "[i]f the pencil and paper are necessary for the calculation, why not view them as part of the necessary substrate for the calculating activity?" (2004, 220).

Consider a second example. Clark and Chalmers (1998) propose that in some situations, a notebook can literally be part of someone's memory. They compare a normal woman, Inga, who relies on her memory to determine the address of the Museum of Modern Art, with an Alzheimer patient, Otto, who relies on his constantly available notebook to determine the address of the museum. Clark and Chalmers contend that in spite of the differences between Inga's and Otto's cases, both Otto and Inga believe that the Museum of Modern Art is located on 53rd street:

> To provide substantial resistance, an opponent has to show that Otto's and Inga's cases differ in some important and relevant respect. But in what deep respect are the cases different? To make the case *solely* on the grounds that information is in the head in one case but not in the other would be to beg the question. If this difference is relevant to a difference in belief, it is surely not *primitively* relevant. To justify the different treatment, we must find some more basic difference between the two. (Clark and Chalmers 1998, 6)

This last example is useful to bring to the fore the central argument for the view that mental states and psychological processing extend beyond the skin: there is no significant difference between some states that involve extra-corporeal entities and some brain states, or between the manipulation of extra-corporeal entities, such as consulting a notebook, and the manipulation of mental representations, such as consulting one's memory. If there is really no significant difference between them, then some states that involve extra-corporeal objects are genuine mental states and some processes that involve manipulating these objects are genuine psychological processes.

Unsurprisingly, this fourth thread has caused a fair amount of discussion among philosophers. Most critics grant that if there were no significant differences between states of the brain and states involving extra-corporeal objects as proper parts, or between processes involving only brain states and processes involving extra-corporeal objects, then not all mental states and psychological processes would be in the head, but they deny that the antecedent of this conditional is satisfied. Particularly, endorsing the computational representational theory of mind (§1), Fred Adams and Ken Aizawa (2001) have argued that mental states are representations that are endowed with an original content and that psychological processes are computational processes defined over these representations. For them, these two properties are "the mark of the cognitive." Because the extra-corporeal objects that are manipulated (for instance, the addresses in Otto's notebook) do not have any original content, they do not count as mental states and their manipulations do not count as psychological processes.

There are two main worries with Adams and Aizawa's argument. First, it rests on controversial (though widespread) necessary conditions for something to be a mental state and for something to be a psychological process, ones that might be rejected by proponents of extended cognition. Second, accepting Adams and Aizawa's necessary conditions, proponents of extended cognition might reply that a state counts as mental provided that some of its parts have an original content, and that a process counts as psychological provided that some steps in this process involve states with original content (or states with some parts having an original content).

Robert Rupert's (2004) main argument against extended cognition does not fall prey to these worries, because he does not assume a specific mark of the cognitive. Rather, focusing on memory, he highlights the differences between the properties of memory retrieval on the one hand and the use of extra-corporeal objects to store information on the other. A large number of generalizations have been found about how people store information in memory (e.g., interference effects[10]) and how they retrieve information from memory (e.g., recency effects[11]). He correctly notes that few of these generalizations apply to the gathering of information from physical mnemonic aids, such as notebooks. Furthermore, any generalization that

---

[10]   Associating two objects makes it more difficult to associate one of them with a new object.

[11]   Objects recently memorized are easier to retrieve from memory.

could apply to information retrieval from both memory and mnemonic aids would probably be about a much larger class of systems, which would include, but not be identical to, the class of cognitive systems. Rupert concludes that treating states and processes within the brain and states and processes involving extra-bodily objects as physical parts is not a promising strategy for cognitive science.

4. Concepts

People classify objects into classes, samples into substances, and events into event types. This capacity, typically called "categorization," is a basic capacity of human cognition: without it, we would be unable to acquire any general knowledge. Psychologists assume that when we categorize an object into a class (often called "a category" in psychology), we rely (maybe unconsciously) on some knowledge about this class.[12] Thus, when I classify an object as a table, I use some knowledge about tables. Psychologists call "concepts" those bodies of knowledge that are used by default to categorize (for an overview of the psychology of concepts, see Murphy 2002, Machery 2009; for a history of the psychology of concepts, see Machery 2007a). Importantly, in addition to categorization, concepts are also used by the psychological processes underlying other capacities, such as induction. Thus, a concept of water is a body of knowledge about water that is used by default to categorize samples as being samples of water, to reason inductively about water, and so on. Psychologists interested in concepts attempt to describe the properties of these bodies of knowledge.

Philosophers have long paid attention to the psychology of concepts.[13] Famously, Fodor has argued that all the theories of concepts developed in psychology were incorrect (Fodor 1994, 1998). His favorite target has been the prototype theory of concepts (Rosch and Mervis 1975, Hampton 1979). Prototype theorists argue that a concept is a body of statistical knowledge about a class (or a substance, etc.). The simplest versions propose that a prototype is a body of knowledge about the typical properties of the members of a class. Thus, a prototype of a dog is a body of knowledge

---

[12]   Here, "knowledge" is used as psychologists use it. It roughly means information or misinformation.

[13]   In addition to the issues discussed here, philosophers have also discussed the acquisition of concepts (Fodor 1981, Laurence and Margolis 2002).

about the typical properties of dogs. Fodor's main objection against the prototype theory can be put simply. Concepts compose. Thus, anybody who can think about dogs and about blue things can *ipso facto* think about blue dogs. But prototypes do not compose. Thus, concepts cannot be prototypes. To support the second premise of this argument, Fodor has put forward several considerations. The pet fish argument is the best known of these. Fodor notes that a poodle might be a prototypical pet and that a shark might be a prototypical fish, while a prototypical pet fish is a golden fish. Thus, our prototype of a pet fish is not derived from our prototype of a pet and our prototype of a fish. Rather, it is derived from our experience with pet fish. Thus, prototypes do not compose.

Fodor's pet fish argument is unconvincing. The fact that the prototype of a pet fish does not result from the combination of the prototype of a pet and the prototype of a fish does not show that prototypes do not compose, *when we have no experience with members of the extension of the complex concepts*. That is, we might combine a prototype of an *x* (say, of a spy) and the prototype of a *y* (say, of a grandmother), when we have no experience with objects that are *x* and *y* (spy grandmothers). Experimental evidence does suggest that people produce complex prototypes in these conditions (see Murphy 2002, ch. 12 for a review). Particularly, people seem to assume that the typical properties of an *x* and of a *y* tend to be also typical of an object that is both an *x* and a *y* (Hampton 1987). Thus, a grandmother spy has the typical properties of a grandmother: she might have grey hair and wear out-of-fashion clothes (see, however, Connolly, Fodor, Gleitman, and Gleitman 2007).

Recent philosophical work on concepts has focused on two main issues: whether an empiricist theory of concepts is viable (Prinz 2002) and whether concepts form a natural kind (Machery 2005, 2009). I consider these two issues in turn. Following psychologist Lawrence Barsalou (1999), Jesse Prinz (2002) has argued that recent developments in the psychology of concepts support a view of concepts that has many affinities with David Hume's empiricist theory of ideas. Although there are several differences between Barsalou's, Prinz's, and others' neo-empiricist theories of concepts, they all endorse the two following theses (Machery 2006b):

1. The knowledge that is stored in a concept is encoded in several perceptual representational formats.
2. Conceptual processing involves reenacting some perceptual states and manipulating these perceptual states.

Thesis (1) is about how we encode our conceptual knowledge (Prinz 2002, 109). Neo-empiricists assume that each perceptual system involves a distinct representational format. Thesis (1) asserts that our conceptual knowledge is encoded in these perceptual representational formats. By contrast, amodal theorists argue that our conceptual knowledge is encoded in a representational format that is distinct from our perceptual representational formats. That is, for amodal theorists, we possess a distinct, *sui generis* representational format, which is used to encode our conceptual knowledge, in addition to our perceptual representational formats. This distinct representational format is usually thought of as being language-like. To illustrate this distinction, according to neo-empiricists, Marie's conceptual knowledge of apples consists of the visual, olfactive, tactile, somatosensory, and gustative representations of apples that are stored in her long-term memory. These representations are a subset of the perceptual representations of apples Marie has had in her life. According to amodal theorists, Marie's conceptual knowledge of apples consists of representations encoded in a single, distinct representational format.

Thesis (2) concerns the nature of the psychological processes underlying categorization, induction, deduction, analogy-making, linguistic comprehension, and so forth. The central insight is that retrieving a concept from long-term memory during reasoning, categorization, etc., consists in producing some perceptual representations. For instance, retrieving the concept of dog when we reason about dogs consists in producing some visual, auditory, etc., representations of dogs. This process is called "simulation" or "reenactment." Thinking about dogs during reasoning, thus, consists of simulating seeing, hearing, and smelling dogs. Our psychological processes consist in manipulating these reenacted percepts. Thus, according to Barsalou, when we decide whether some object has a given part, for example whether lions have a mane, we produce a visual representation of a lion and another of a mane and we match these two representations; if they do match, we decide that lions have a mane (Solomon and Barsalou 2001, 135–6).

Two main lines of reply to these neo-empiricist theories of concepts have been developed in the philosophical literature. Some philosophers and psychologists have argued that neo-empiricist theories of concepts suffer from the very same problems that plagued David Hume's theory of ideas. For instance, John Sarnecki (2004) and Arthur Markman and Hunt Stilwell (2004) argue that Prinz's empiricist account of concepts cannot be applied to abstract concepts (see also Machery 2006b), while Markman and

Stilwell argue that it cannot be applied to relational concepts, such as the concept of an uncle (see Prinz 2004b for a rejoinder to Sarnecki and Markman and Stilwell).

Instead of discussing the theoretical problems of the neo-empiricist theories of concepts, I have focused on the evidence for these theories, for it is supposed to favor neo-empiricism over amodal theories of concepts (Machery 2007b). Thus, Barsalou and colleagues write: "Amodal theories have been attractive theoretically because they implement important conceptual functions, such as the type-token distinction, categorical inference, productivity, and propositions. [...] Conversely, *indirect* empirical evidence has accumulated for modality-specific representations in working memory, long-term memory, language, and thought" (Barsalou, Simmons, Barbey, and Wilson 2003, 85–6).

The evidence for neo-empiricism is not as strong as might appear at first. Three problems plague most of the current empirical research inspired by neo-empiricism. First, neo-empiricists erroneously assume that for a given experimental task, a single prediction can be derived on behalf of amodal theorists and can be tested experimentally. In fact, however, there are numerous competing amodal models of the psychological process involved in a given task and these competing models make different predictions about subjects' performance in this task. Because different amodal models make different predictions, it is not the case that Barsalou's and others' experimental findings are inconsistent with an amodal view of concepts in general. Rather, they are inconsistent with specific amodal models.

Second, neo-empiricists have not acknowledged that amodal theorists, such as Fodor or Zenon Pylyshyn, recognize that we can use imagery to solve various problems. Thus, when one is asked to count the number of windows in one's own house, one typically visualizes one's house and counts the number of visualized windows. Because amodal theorists admit the existence and importance of imagery, they do predict that in some tasks people will simulate having perceptual states. But neo-empiricist researchers have often failed to focus on tasks for which amodal theorists would not expect people to use perceptual imagery.

Finally, neo-empiricists have not acknowledged the possibility that in some domains, or for some tasks, or, maybe, in some contexts, people might use perceptual representations, while using amodal representations in other domains, or in other tasks, or in other contexts.

The second debate among philosophers of psychology interested in concepts concerns whether the class of concepts is a natural kind. Most psychologists working on concepts assume that concepts share many scientifically important properties and attempt to describe those properties. They assume thereby that concepts constitute a natural kind, that is, roughly, a class of entities about which numerous non-accidental, scientifically important generalizations can be made – an assumption I have called "the Natural Kind Assumption" (Machery 2005, 2009). Against the Natural Kind Assumption, I have argued that most classes are represented by several concepts that belong to kinds that have little in common. For instance, I propose that the class of dogs is typically represented by several concepts of dog and that these concepts have few (scientifically relevant) properties. To support this proposal, I have shown that for each relevant cognitive capacity (categorization, induction, concept combination), some phenomena are best explained if one posits a first kind of concepts (namely, prototypes), other phenomena are best explained if one posits a second kind of concepts (what psychologists call "exemplars"), and yet other phenomena are best explained if one posits a third kind of concepts (what psychologists call "theories").[14] Because these three kinds of concepts have little in common, I have concluded that concepts are not a natural kind. Furthermore, I have proposed that the notion of concept is ill-suited for a scientific psychology and that the term "concept" should be eliminated from its theoretical vocabulary, exactly as the notion of superlunear objects was eliminated from astronomy.

The claims that concepts are not a natural kind and that the notion of concept should be eliminated from psychology have come under criticism. While agreeing that there are different kinds of concepts, Gualtiero Piccinini and Sam Scott (2006; see Machery 2006a for a reply) have argued that if most classes were represented by several concepts (for instance, if dogs were represented by a prototype, by a set of exemplars, and by a theory), then, contrary to the conclusion I drew, concepts would be a natural kind. For, each concept consists of several parts. For instance, the concept of dog would have three parts (one corresponding to the prototype of a dog, one corresponding to exemplars of particular dogs, one corresponding to a theory about dogs). Dan Weiskopf (Weiskopf 2009) has developed a different

[14] In substance, an *exemplar* is a representation of an individual. Thus, an exemplar of dog is a representation of a particular dog. According to the exemplar view of concepts, a concept is a set of exemplars. In substance, according to the *theory* view of concepts, a concept is similar to a scientific theory. Thus, a concept of dog might consist of some nomological knowledge about dogs.

criticism. While highlighting the diversity of concepts, he argues that numerous generalizations can in fact be made about concepts. He concludes that eliminating "concept" from the theoretical vocabulary of psychology would prevent the formulation of numerous important generalizations.

5. Mindreading

Mindreading is the practice of ascribing mental states, such as beliefs, desires, emotions, and perceptions, to others and to oneself. It is an essential and automatic component of our everyday life. Consider watching Alfred Hitchcock's *Rear Window*. At the end of the movie, L.B. Jeffries, played by James Stewart, hides in the dark in his apartment. We understand this behavior because we know that he *believes* that the killer is coming to his apartment and because he does not *want* to be killed. Stewart's behavior is meaningful because we have ascribed some specific mental states to the character he is playing.

Philosophers of psychology have been involved in an interdisciplinary attempt to characterize the psychological mechanisms underlying mindreading (an attempt that also involves developmental psychologists, psychopathologists, neuropsychologists, and animal psychologists). Two main accounts have been developed: the theory theory and the simulation theory. I consider them in turn.[15]

Although proponents of the theory theory, such as philosophers Peter Carruthers, Fodor, Shaun Nichols, and Steve Stich, and psychologists Simon Baron Cohen, Alison Gopnik, Alan Leslie, and Joseph Perner, disagree on various points, they concur that people have a large and complex body of knowledge about mental states, the relations between mental states, the relations between mental states and stimuli, and the relation between mental states and behaviors (Wellman 1990, Perner 1991, Baron-Cohen 1995, Gopnik and Meltzoff 1997). For instance, we might know that if an individual sees that *p*, she typically believes that *p*. We might also know that for an individual to see an object, she has to stand in some physical

---

[15]   Philosophers have also participated in the debates about whether non-human primates can mindread (e.g., Povinelli and Vonk 2004), about whether the theory of mind is modular (e.g., Fodor 1992), and about the nature of mechanisms underlying self-knowledge (e.g., Nichols and Stich 2003, Carruthers 2006).

relation to this object (e.g., her face has to be turned toward this object, her eyes have to be directed toward this object, etc.). People use this body of knowledge when they ascribe beliefs and desires to others: "The central idea shared by all versions of the 'theory-theory' is that the processes underlying the production of most predictions, intentional descriptions, and intentional explanations of people's behavior exploit an internally represented body of information (or perhaps mis-information) about psychological processes and the ways in which they give rise to behavior" (Stich and Nichols 1995, 87–8). Note that we need not be aware that we possess and use this body of knowledge: it might be subdoxastic, exactly as our linguistic knowledge is supposed to be.

Simulation theorists, such as philosophers Alvin Goldman, Robert Gordon, and Jane Heal, and neuropsychologists Jean Decety and Vittorio Gallese, doubt that we have an extensive body of knowledge about mental states and that mindreading involves using this body of knowledge. They contend that ascribing mental states to someone else consists in simulating her mind, or, to put it metaphorically, putting oneself in her shoes (Gordon 1986, Heal 1986, Goldman 1989, 1992, Meltzoff and Decety 2003). Gallese and Goldman write: "ST [simulation theory] arose partly from doubts about whether folk psychologizers really represent, even tacitly, the sorts of causal/explanatory laws that TT [theory theory] typically posits. ST suggests that attributors use their own mental mechanisms to calculate and predict the mental processes of others" (1998, 496).

In spite of a few disagreements about the nature of simulation, simulation theorists agree that simulating involves mimicking the mental life of the individuals to whom we want to ascribe mental states. Mindreading involves having similar states and similar thought processes to the individuals to whom mental states are ascribed. For Goldman, for instance, simulation typically involves three steps. One first pretends to have some mental states. Thus, while playing chess, if one attempts to predict one's partner's next move, one pretends to have the mental states one's partner might have (e.g., her desire to win the game and her knowledge of chess). Second, these pretend states are used as inputs to one's own reasoning and decision processes. Once one has pretended to have some mental states, one reasons as if these mental states were one's own. Thereby, one mimics the chain of thoughts others might have. For instance, to predict the next move of one's chess partner, one decides what move to make, pretending to have the mental states one's chess partner might have. Finally, one ascribes a mental state to others. For instance, instead of acting on one's decision about what move to make, this decision is "taken off-line" and used to predict what one's partner will do.

Philosophers have spent much energy clarifying these two approaches and contrasting various versions of both approaches. Rather than looking at these details, I will now sketch some of the main arguments developed on behalf of each approach.

Stich and Nichols (1992) have developed an influential argument for the theory theory. They note that specific biases influence the processes involved in reasoning and in decision making. They focus particularly on a phenomenon called "the endowment effect" (Thaler 1980): people are only willing to sell an object that they possess for more money than they paid when they acquired it. For instance, people might be willing to pay $10 to acquire the poster of a movie, but be unwilling to sell it for less than $15. Now, consider a situation where we have to predict what price someone is going to ask for selling an object and what price she would be willing to pay for acquiring this object. Suppose, as Goldman would have it, that we use our own decision processes to make these predictions. Then, because our own decision processes are biased, we should predict that others will fall prey to the endowment effect. The theory theory is not committed to this prediction because it might not be part of our (maybe implicit) theory of mind that people's decision processes are so biased. Thus, Stich and Nichols write: "If there is some quirk in the human decision making system, something quite unknown to most people that leads the system to behave in an unexpected way under certain circumstances, the accuracy of predictions based on simulations should not be adversely affected" (1992, 263). Stich and Nichols note that people are very poor at predicting that they themselves and others would be victims of the endowment effect.

Goldman and other proponents of simulation theories have replied that systematic errors in mindreading are in fact consistent with simulation theory (see, e.g., Goldman 2006; for discussion, see Stich and Nichols 1995). Simulation is accurate only when the pretend states (e.g., in a chess game, pretending to have one's partner's knowledge of chess and her desire to win the game) are accurate. If the pretend states differ in a systematic manner from the states that the target of the simulation actually has, simulation will lead to systematic mistakes. In reply, a theory theorist might concede that simulation theory predicts systematic mistakes, while questioning whether it predicts the very mistakes highlighted by Stich and Nichols.

I now focus on some arguments for the simulation theory. In the 1990s, neuropsychologists Giacomo Rizzolatti, Vittorio Gallese, and colleagues discovered some neurons in the ventral premotor cortex of macaques (area F5) that not only fire when the macaques are doing an action (as was expected), but that also fire when the macaques observe another macaque or the human experimenter do the same action (Gallese, Fadiga, Fogassi, and Rizzolatti 1996). These neurons have been dubbed "mirror neurons" (for recent reviews, see Rizzolatti and Craighero 2004, Gallese 2007). Evidence suggests that humans also have mirror neurons. Remember that according to simulation theorists, mindreaders mimic others: they have the same (or similar) mental states and they go through the same (or similar) chains of thoughts. Gallese and Goldman have proposed that the existence of mirror neurons supports simulation theory, because they are activated both when one decides to act and when one understands others' actions: "MN activity seems to be nature's way of getting the observer into the same 'mental shoes' as the target – exactly what the conjectured simulation heuristic aims to do" (1998, 498).

Goldman has also argued that neuropsychological findings show that the recognition and ascription of emotions, such as fear, disgust, or anger, involves a simulation process (Goldman and Sripada 2005, Goldman 2006). Roughly, when another person expresses an emotion facially and behaviorally, we are supposed to experience the emotion that this person is experiencing. We then ascribe to her the emotion that we are experiencing (see Goldman and Sripada 2005 for a careful discussion of several models of this process). Seeing John make a disgust face (brows narrowed, upper lip raised, lip corners drawn down and back, and nose drawn up and wrinkled) causes disgust in me. Because I recognize that I feel disgust, I ascribe disgust to John. Goldman and Sripada have convincingly argued that the hypothesis that emotion ascription involves a simulation process accounts for the finding that following lesions in the brain areas involved with specific emotions, patients who are unable to experience these emotions are also impaired in recognizing them.

Recent philosophical contributions to the understanding of mindreading have partly moved beyond the original debate between the simulation theory and the theory theory. Researchers are now developing various hybrid accounts of the mechanisms underlying mindreading – that is, accounts that include both simulation- and theory-based psychological processes (Nichols and Stich 2003, Goldman 2006).

6.   Conclusion and Future Directions

In this essay, I have reviewed some important issues that have been at the center of the philosophy of psychology: the legitimacy of mentalism, the modularity of the cognitive architecture, the situated and extended nature of cognition, the nature of concepts, and the mechanisms underlying mind-reading. To conclude, I will sketch what the philosophy of psychology may possibly look like in the forthcoming years. Certainly, philosophers will probably continue contributing to the areas discussed in this essay (see footnote 1 for other areas of active debate), but they will also probably turn to new issues. These might include questions about the methodology of psychology, which has so far attracted little attention among philosophers (but see Trout 1998, Glymour 2001), such as the role of null hypothesis significance testing (Trout 1999, Fidler 2005, Machery, no date) and the relation between group data and the study of the mind of individuals. In addition, philosophers of psychology will probably attempt to improve our understanding of the relation between psychology and neuroscience (e.g., Schouten and Looren de Jong 2007).[16]

References

Adams, F. and K. Aizawa. 2001. "The Bounds of Cognition." *Philosophical Psychology* 14: 43–64.
Adams, F. and K. Aizawa. 2008. *The Bounds of Cognition*. Oxford: Blackwell.
Anderson, M.L. 2003. "Embodied Cognition: A Field Guide." *Artificial Intelligence* 149.1: 91–103.
Baron-Cohen, S. 1995. *Mindblindness: An Essay on Autism and Theory of Mind*. Cambridge, MA: MIT Press.
Barrett, H.C. 2005. "Enzymatic Computation and Cognitive Modularity." *Mind and Language* 20: 259–87.
Barrett, H.C. and R. Kurzban. 2006. "Modularity in Cognition: Framing the Debate." *Psychological Review* 113: 628–47.
Barsalou, L.W. 1999. "Perceptual Symbol Systems." *Behavioral and Brain Sciences* 22: 577–660.

---

[16]   I would like to thank Luc Faucher and my research assistant Kara Cohen for their comments on a previous version of this essay.

Barsalou, L.W., W.K. Simmons, A.K. Barbey, and C.D. Wilson. 2003. "Grounding Conceptual Knowledge in Modality-specific Systems." *Trends in Cognitive Sciences* 7: 84–91.

Block, N. 1995. "On a Confusion about a Function of Consciousness." *Behavioral and Brain Sciences* 18.2: 227–87.

Block, N. 2007. "Consciousness, Accessibility and the Mesh between Psychology and Neuroscience." *Behavioral and Brain Sciences* 30: 481–548.

Brooks, R.A. 1991. "Intelligence without Representation." *Artificial Intelligence* 47: 139–59.

Buller, D.J. 2005. *Adapting Minds: Evolutionary Psychology and the Persistent Quest for Human Nature.* Cambridge, MA: MIT Press.

Caldwell, W. 1899. "The Postulates of a Structural Psychology." *Psychological Review* 6: 187–91.

Carruthers, P. 2006. *The Architecture of the Mind.* New York: Oxford University Press.

Chalmers, D.J. 1996. *The Conscious Mind: In Search of a Fundamental Theory.* New York: Oxford University Press.

Chalmers, D.J. 2004. "How Can We Construct a Science of Consciousness?" In M.S. Gazzaniga (ed.), *The Cognitive Neurosciences III.* Cambridge, MA: MIT Press, 1111–20.

Chomsky, N. 1959. "Review of *Verbal Behavior* by B.F. Skinner." *Language* 35.1: 26–58.

Clark, A. 1997. *Being There: Putting, Brain, Body and World Together Again.* Cambridge, MA: MIT Press.

Clark, A. and D.J. Chalmers. 1998. "The Extended Mind." *Analysis* 58: 7–19.

Clark, A. and J. Toribio. 1994. "Doing without Representing?" *Synthese* 101: 401–31.

Connolly, A.C., J.A. Fodor, L.R. Gleitman, and H. Gleitman. 2007. "Why Stereotypes Don't Even Make Good Defaults." *Cognition* 103: 1–22.

Cosmides, L. and J. Tooby. (1994). "Origins of Domain Specificity: The Evolution of Functional Organization." In L.A. Hirschfeld and S.A. Gelman (eds.), *Mapping the Mind: Domain Specificity in Cognition and Culture.* Cambridge: Cambridge University Press, 85–116.

Doris, J. 2002. *Lack of Character: Personality and Moral Behavior.* New York: Cambridge University Press.

Doris, J. and S. Stich. 2006. "Moral Psychology: Empirical Approaches." In E.N. Zalta (ed.), *Stanford Encyclopedia of Philosophy*, http://plato.stanford.edu/entries/moral-psych-emp/ (accessed February 1, 2008).

Fidler, F. 2005. "From Statistical Significance to Effect Estimation: Statistical Reform in Psychology, Medicine and Ecology." PhD dissertation, Department of History and Philosophy of Science, University of Melbourne.

Finlay, B.L. and R.B. Darlington. 1995. "Linked Regularities in the Development and Evolution of Mammalian Brains." *Science* 268: 1578–84.

Fodor, J.A. 1975. *The Language of Thought.* Cambridge, MA: Harvard University Press.

Fodor, J.A. 1981. "The Present Status of the Innateness Controversy." In J.A. Fodor, *Representations: Philosophical Essays on the Foundations of Cognitive Science.* Cambridge, MA: MIT Press, 257–316.

Fodor, J.A. 1983. *The Modularity of Mind: An Essay on Faculty Psychology.* Cambridge, MA: MIT Press.

Fodor, J.A. 1987. *Psychosemantics: The Problem of Meaning in the Philosophy of Mind.* Cambridge, MA: MIT Press.

Fodor, J.A. 1992. "A Theory of the Child's Theory of Mind." *Cognition* 44: 283–96.

Fodor, J.A. 1994. "Concepts: A Potboiler." *Cognition* 50: 95–113.

Fodor, J.A. 1998. *Concepts: Where Cognitive Science Went Wrong.* Oxford: Oxford University Press.

Fodor, J.A. 2000. *The Mind Doesn't Work that Way.* Cambridge, MA: MIT Press.

Gallagher, S. 2005. *How the Body Shapes the Mind.* Oxford: Oxford University Press.

Gallese, V. 2007. "Before and Below 'Theory of Mind': Embodied Simulation and the Neural Correlates of Social Cognition." *Philosophical Transactions of the Royal Society B: Biological Sciences* 362: 659–69.

Gallese, V. and A.I. Goldman. 1998. "Mirror Neurons and the Simulation Theory of Mindreading." *Trends in Cognitive Sciences* 2.1: 493–501.

Gallese, V., L. Fadiga, L. Fogassi, and G. Rizzolatti. 1996. "Action Recognition in the Premotor Cortex." *Brain* 119: 593–609.

Gallistel, C.R. and J. Gibbon. 2001. "Computational Versus Associative Models of Simple Conditioning." *Current Directions in Psychological Science* 10: 146–50.

Gigerenzer, G. and U. Hoffrage. 1995. "How to Improve Bayesian Reasoning without Instruction: Frequency Formats." *Psychological Review* 102: 684–704.

Glymour, C. 2001. *The Mind's Arrows: Bayes Nets and Graphical Causal Models.* Cambridge, MA: MIT Press.

Goldman, A.I. 1989. "Interpretation Psychologized." *Mind and Language* 4: 161–85.

Goldman, A.I. 1992. "In Defense of the Simulation Theory." *Mind and Language* 7: 104–99.

Goldman, A.I. 2006. *Simulating Minds.* New York: Oxford University Press.

Goldman, A.I. and C. Sripada. 2005. "Simulationist Models of Face-based Emotion Recognition." *Cognition* 94: 193–213.

Gopnik, A. and A.N. Meltzoff. 1997. *Words, Thoughts, and Theories.* Cambridge, MA: MIT Press.

Gordon, R. 1986. "Folk Psychology as Simulation." *Mind and Language* 1: 158–71.

Griffiths, P.E. 1997. *What Emotions Really Are: The Problem of Psychological Categories.* Chicago: University of Chicago Press.

Hampton, J.A. 1979. "Polymorphous Concepts in Semantic Memory." *Journal of Verbal Learning and Verbal Behavior* 18: 441–61.

Hampton, J.A. 1987. "Inheritance of Attributes in Natural Concept Conjunctions." *Memory and Cognition* 15: 55–71.

Heal, J. 1986. "Replication and Functionalism." In J. Butterfield (ed.), *Language, Mind and Logic*. Cambridge: Cambridge University Press, 135–50.

Hermer, L. and E.S. Spelke. 1996. "Modularity and Development: The Case of Spatial Reorientation." *Cognition* 61: 195–232.

Hutchins, E. 1995. *Cognition in the Wild*. Cambridge, MA: MIT Press.

Jacob, P. and M. Jeannerod. 2003. *Ways of Seeing: The Scope and Limits of Visual Cognition*. Oxford: Oxford University Press.

Kanwisher, N., J. McDermott, and M.M. Chun. 1997. "The Fusiform Face Area: A Module in Human Extrastriate Cortex Specialized for Face Perception." *Journal of Neuroscience* 11: 4302–11.

Laurence, S. and E. Margolis. 2002. "Radical Concept Nativism." *Cognition* 86.1: 22–55.

MacCorquodale, K. and P.E. Meehl. 1948. "On a Distinction between Hypothetical Constructs and Intervening Variables." *Psychological Review* 55: 95–107.

Machery, E. 2005. "Concepts Are Not a Natural Kind." *Philosophy of Science* 72: 444–67.

Machery, E. 2006a. "How to Split Concepts. Reply to Piccinini and Scott." *Philosophy of Science* 73: 410–18.

Machery, E. 2006b. "Two Dogmas of Neo-empiricism." *Philosophy Compass* 1.4: 398–412.

Machery, E. 2007a. "100 Years of Psychology of Concepts: The Theoretical Notion of Concept and its Operationalization." *Studies in History and Philosophy of Biological and Biomedical Sciences* 38: 63–84.

Machery, E. 2007b. "Concept Empiricism: A Methodological Critique." *Cognition* 104: 19–46.

Machery, E. 2007c. "Massive Modularity and Brain Evolution." *Philosophy of Science* 74: 825–38.

Machery, E. 2009. *Doing Without Concepts*. New York: Oxford University Press.

Machery, E. Forthcoming. "Discovery and Confirmation in Evolutionary Psychology." In J.J. Prinz (ed.), *Oxford Handbook of Philosophy of Psychology*. Oxford: Oxford University Press.

Machery, E. No date. "Null Hypothesis Significance Testing: Reports of Its Death Have Been Greatly Exaggerated." Unpublished manuscript.

Machery, E. and H.C. Barrett. 2006. "Debunking *Adapting Minds*." *Philosophy of Science* 73: 232–46.

Marcus, G.F. 2001. *The Algebraic Mind*. Cambridge, MA: MIT Press.

Markman, A. and H. Stilwell. 2004. "Concepts à la Modal: An Extended Review of Prinz's *Furnishing the Mind.*" *Philosophical Psychology* 17.3: 391–401.

Meltzoff, A.N. and J. Decety. 2003. "What Imitation Tells Us about Social Cognition: A Rapprochement between Developmental Psychology and Cognitive Neuroscience." *Philosophical Transactions of the Royal Society: Biological Sciences* 358: 491–500.

Murphy, D. 2007. *Psychiatry and the Scientific Image.* Cambridge, MA: MIT Press.

Murphy, G.L. 2002. *The Big Book of Concepts.* Cambridge, MA: MIT Press.

Newell, A. and H.A. Simon. 1976. "Computer Science as Empirical Enquiry." *Communications of the Association for Computing Machinery* 19: 113–26.

Nichols, S. 2004. *Sentimental Rules. On the Natural Foundations of Moral Judgment.* New York: Oxford University Press.

Nichols, S. and S. Stich. 2003. *Mindreading: An Integrated Account of Pretense, Self-awareness and Understanding Other Minds.* Oxford: Oxford University Press.

Noë, A. 2004. *Action in Perception.* Cambridge, MA: MIT Press.

O'Donnell, J.M. 1985. *The Origins of Behaviorism: American Psychology, 1870–1920.* New York: New York University Press.

Perner, J. 1991. *Understanding the Representational Mind.* Cambridge, MA: MIT Press.

Piccinini, G. 2008. "Computation without Representation." *Philosophical Studies* 137: 205–41.

Piccinini, G. and S. Scott. 2006. "Splitting Concepts." *Philosophy of Science* 73.4: 390–409.

Povinelli, D.J. and J. Vonk. 2004. "We Don't Need a Microscope to Explore the Chimpanzee Mind." *Mind and Language* 19: 1–28.

Prinz, J.J. 2002. *Furnishing the Mind: Concepts and their Perceptual Basis.* Cambridge, MA: MIT Press.

Prinz, J.J. 2004a. *Gut Reactions: A Perceptual Theory of Emotions.* New York: Oxford University Press.

Prinz, J.J. 2004b. "Sensible Ideas: A Reply to Sarnecki and Markman and Stilwell." *Philosophical Psychology* 17.3: 419–30.

Prinz, J.J. 2007. *The Emotional Construction of Morals.* New York: Oxford University Press.

Pylyshyn, Z.W. 1984. *Computation and Cognition.* Cambridge, MA: MIT Press.

Quartz, S.R. 2002. "Toward a Developmental Evolutionary Psychology: Genes, Development, and the Evolution of the Human Cognitive Architecture." In S.J. Scher and F. Rauscher (eds.), *Evolutionary Psychology: Alternative Approaches.* Dordrecht: Kluwer, 185–206.

Rizzolatti, G. and L. Craighero. 2004. "The Mirror-neuron System." *Annual Review of Neuroscience* 27: 169–92.

Rosch, E. and C. Mervis. 1975. "Family Resemblances: Studies in the Internal Structure of Categories." *Cognitive Psychology* 7: 573–604.

Rowlands, M. 1999. *The Body in Mind: Understanding Cognitive Processes.* Cambridge: Cambridge University Press.

Rupert, R. 2004. "Challenges to the Hypothesis of Extended Cognition." *Journal of Philosophy* 101: 389–428.

Ryle, G. 1951. *The Concept of Mind.* London: Hutchinson.

Samuels, R. 1998. "Evolutionary Psychology and the Massive Modularity Hypothesis." *British Journal for the Philosophy of Science* 49: 575–602.

Samuels, R. 2005. "The Complexity of Cognition: Tractability Arguments for Massive Modularity." In P. Carruthers, S. Laurence, and S. Stich (eds.), *The Innate Mind: Structure and Content.* Oxford: Oxford University Press, 107–21.

Sarnecki, J. 2004. "The Multimedia Mind: An Analysis of Prinz on Concepts." *Philosophical Psychology* 17.3: 403–18.

Schouten, M. and H. Looren de Jong (eds.). 2007. *The Matter of the Mind: Philosophical Essays on Psychology, Neuroscience and Reduction.* Oxford: Wiley Blackwell.

Shapiro, L.A. 2004. *The Mind Incarnate.* Cambridge, MA: MIT Press.

Skinner, B.F. 1938. *The Behavior of Organisms: An Experimental Analysis.* New York: Appleton-Century.

Solomon, K.O. and L.W. Barsalou. 2001. "Representing Properties Locally." *Cognitive Psychology* 43: 129–69.

Sperber, D. 1994. "The Modularity of Thought and the Epidemiology of Representations." In L.A. Hirschfeld and S.A. Gelman (eds.), *Domain Specificity in Cognition and Culture.* Cambridge: Cambridge University Press, 39–67.

Sperber, D. 2001. "In Defense of Massive Modularity." In E. Dupoux (ed.), *Language, Brain and Cognitive Development: Essays in Honor of Jacques Mehler.* Cambridge, MA: MIT Press, 47–58.

Stich, S. and S. Nichols. 1992. "Folk Psychology: Simulation or Tacit Theory." *Mind and Language* 7.1: 35–71.

Stich, S. and S. Nichols. 1995. "Second Thoughts on Simulation." In M. Davies and T. Stone (eds.), *Mental Simulation: Philosophical and Psychological Essays.* Oxford: Blackwell, 87–108.

Stich, S. and T.A. Warfield (eds.). 1994. *Mental Representation.* Oxford: Blackwell.

Thaler, R.H. 1980. "Toward a Positive Theory of Consumer Choice." *Journal of Economic Behavior and Organization* 1: 39–60.

Thelen, E. and L.B. Smith. 1994. *A Dynamic Systems Approach to the Development of Cognition and Action.* Cambridge, MA: MIT Press.

Titchener, E.B. 1899. "Structural and Functional Psychology." *Philosophical Review* 8: 290–9.

Tooby, J. and L. Cosmides. 1992. "The Psychological Foundations of Culture." In J.H. Barkow, L. Cosmides, and J. Tooby (eds.), *The Adapted Mind: Evolutionary Psychology and the Generation of Culture.* New York: Oxford University Press, 19–136.

Trout, J.D. 1998. *Measuring the Intentional World: Realism, Naturalism, and Quantitative Methods in the Behavioral Sciences.* New York: Oxford University Press.

Trout, J.D. 1999. "Measured Realism and Statistical Inference: An Explanation for the Fast Progress of 'Hard' Psychology." *Philosophy of Science* 66: S260–72.

Tversky, A. and D. Kahneman. 1982. "Judgments of and by Representativeness." In D. Kahneman, P. Slovic, and A. Tversky (eds.), *Judgment Under Uncertainty: Heuristics and Biases.* Cambridge: Cambridge University Press, 84–98.

Vera, A.H. and H.A. Simon. 1993. "Situated Action: A Symbolic Interpretation." *Cognitive Science* 17: 7–48.

Watson, J.B. 1913. "Psychology as the Behaviorist Views it." *Psychological Review* 20: 158–77.

Weiskopf, D. 2009. "The Plurality of Concepts." *Synthese* 169: 145–73.

Wellman, H.M. 1990. *The Child's Theory of Mind.* Cambridge, MA: MIT Press.

Wilson, M. 2002. "Six Views of Embodied Cognition." *Psychonomic Bulletin and Review* 9.4: 625–36.

Wilson, R. 2004. *Boundaries of the Mind: The Individual in the Fragile Sciences.* Cambridge: Cambridge University Press.

12   Philosophy of Sociology

Daniel Little

This essay provides a basis for examining the basic ontological, explanatory, and theoretical characteristics of social scientific knowledge, with special application to sociology. The philosophy of the social sciences is the field within philosophy that thinks critically about the nature and scope of social scientific knowledge and explanation. The essay considers some of the ways that philosophers and social scientists have conceptualized the nature of social phenomena, and argues for an ontology based on socially situated individuals in interaction. The essay examines several features of social explanation, focusing on the idea of a causal mechanism. It argues that social explanations come down to a claim about social causation, and social causation in turn should be understood in terms of a hypothesized causal mechanism connecting one set of social facts with another. The essay turns finally to several issues of epistemology. How are social science hypotheses and theories to be tested empirically? And what are some of the limitations of positivism and naturalism as theories of social science knowledge? The essay closes by returning to ontology in a consideration of methodological individualism and holism.

1.   What Is the Philosophy of the Social Sciences?

The philosophy of social science is a group of research traditions that are intended to shed light on various aspects of the intellectual effort of understanding and explaining social phenomena.[1] In brief, it is the study

---

[1]   Noteworthy contributions to the philosophy of social science in the past two decades include Braybrooke 1987, Rosenberg 1988, Elster 1989, Bohman 1991, Little 1991, Martin and McIntyre 1994, Bunge 1996, Kincaid 1996, Turner and Roth 2003, Sherratt 2006, and the continuing debates and discussions contained in the journal edited by Ian C. Jarvie, *The Philosophy of Social Science*.

of the social sciences from the point of view of the quality of knowledge they offer, the types of explanations they advance, and the important conceptual problems that are raised in the course of social science research. Core questions include: What are the scope and limits of scientific knowledge of society? What is involved in arriving at a scientific understanding of society? What are the most appropriate standards for judging proposed social explanations? Is there such a thing as social causation? How are social theories and assertions to be empirically tested? How do social facts relate to facts about individuals?

The philosophy of social science is in one sense a "meta" discipline – it reviews and analyzes the research of other, more empirical researchers. But in another sense the philosopher is a direct contributor to social science research; by discussing and reflecting upon the methods, assumptions, concepts, and theories of working sociologists, the philosopher is also contributing to the improvement of sociology research. On this perspective it is arbitrary to draw a line between the theoretical and conceptual inquiry of the applied social scientist, and the similar studies of the philosopher.

A philosophy can guide us as we construct a field of knowledge, and it can serve as a set of regulative standards as we conduct and extend that field of knowledge. Philosophy has served both intellectual functions for the social sciences in the past century and a half. Philosophical ideas about the nature of knowledge and the nature of the social world guided or influenced the founding efforts by such early sociologists as Max Weber, Émile Durkheim, or Herbert Spencer in the formulation of their highest-level assumptions about social processes and their most general assumptions about what a scientific treatment of society ought to look like. So there has been an important back-and-forth relationship between philosophy and the social sciences from the start. John Stuart Mill and William Whewell framed many of the assumptions about the social sciences that would govern the development of many areas of the social sciences in the English-speaking world; whereas European philosophers such as Heinrich Rickert and Wilhelm Dilthey articulated a vision of the "human sciences" based on the idea of meaningful action that would have great influence on European social science development. At its best, philosophy can function as an equal collaborator with the creators and practitioners of the social sciences, helping to arrive at more durable and insightful theories and methods. At its worst, philosophical doctrines can blind social researchers to more fertile and innovative avenues of theory development and explanation.

The importance of the philosophy of social science derives from two things: first, the urgency and complexity of the challenges posed by the poorly understood social processes that surround us in twenty-first-century society; and second, the unsettled status of our current understanding of the logic of social science knowledge and explanation. We need the best possible research and explanation to be conducted in the social sciences, and current social science inquiry falls short. We need a better-grounded understanding of the social, political, and behavioral phenomena that make up the modern social world. Moreover, the goals and primary characteristics of a successful social science are still only partially understood. What do we want from the social sciences? And how can we best achieve these cognitive and practical goals? There are large and unresolved philosophical questions about the logic of social science knowledge and theory on the basis of which to arrive at that understanding. And philosophy can help articulate better answers to these questions. So philosophy can play an important role in the development of the next generation of social science disciplines.

2.   What Are the "Social Sciences"?

The social sciences, and sociology in particular, are largely the product of the nineteenth century – and the European nineteenth century at that. Inspired by the successes of the natural sciences, including especially chemistry, physics, and biology, thinkers such as Auguste Comte, John Stuart Mill, and Karl Marx set the goal of providing a "scientific" theory of society. At bottom were several goals: rigorous factual study of social phenomena, explanation of social outcomes and patterns on the basis of more fundamental theories and laws, and understandings of social processes that would permit more effective interventions (e.g., policies, reforms, revolutions) to address social problems. These goals highlight several epistemic values: rigorous empirical inquiry, explanation, and a basis for effective intervention. But they also borrow some features of the natural sciences that prove to be ill-suited to the study of the social world. By this, I mean what can be referred to as "naturalism" (Thomas 1979, McIntyre 1996): an assumption that nature is governed by laws, a quest for unification under a single comprehensive theory, and a quest for systematization and prediction that proves to be unattainable with respect to social phenomena.

The social sciences today encompass a much wider terrain, and they have proliferated into a handful of core disciplines, including economics, sociology, political science, and anthropology. Each discipline further subdivides into a diverse set of methods and theories, each inspiring separate approaches to the study of social phenomena. There are cross-disciplinary combinations of these core disciplines (e.g., ethnographic sociology, political economy), and there are new interdisciplinary configurations that draw on methods and topics of the core disciplines but align themselves separately (e.g., Asian studies, women's and gender studies, and globalization studies). The definition and evolution of the disciplines is itself an understudied topic in the philosophy of social science. Several sociologists have provided thoughtful analysis of the logic and development of the disciplines of sociology (Zald 1992, 1995, Abbott 1999, 2001). Broadly, we might say that the disciplines of the social sciences are defined and differentiated by topic, method, and theory; under each of these headings are specific and differentiating assumptions about the way in which the sociologist or the economist will investigate and explain a body of social phenomena. But, finally, we have no reason to think that the current arrangement of the disciplines provides an ideal form of coverage for all social inquiries.

It should be emphasized that the social sciences today are "works in progress." We have no basis for thinking that we have arrived at the best ways of breaking down the central problems of social life into areas of study, or the best ways of investigating social processes, or the best theories and explanatory ideas that will provide a basis for better understanding of social phenomena. So the definition of the "core" disciplines and primary methods needs to be understood in a very provisional way; we hope we will have a better sociology, political science, or economics in the future. And the division of labor across specific disciplines should be understood as provisional and shifting.

Consider one discipline in particular, sociology. We might break the research task of sociology into several component types of intellectual work. First, sociology involves description. Social phenomena are observable, and it is straightforward to design rigorous research efforts aimed at establishing the facts about a particular domain. This aspect of sociology involves rigorous empirical study of social phenomena. Examples of descriptive research include ethnographic research and micro-sociology along the lines of the Chicago School of sociology. But large-scale description is feasible as well, including empirical description of large social patterns and institutions. Descriptive findings often take the form of statistical estimates of the frequency of a feature within a group – for example, rates of suicide among Protestants (Durkheim), or rates of diabetes among various racial groups. Properties may be correlated with one another within a given population; variation in one variable may be associated with variation in another variable. Descriptive research can thus sometimes reveal patterns of behavior or social outcomes – for example, patterns of habitation and health status. And patterns such as these invite efforts to find causal relationships among the characteristics enumerated.

Second, sociology involves discovery of social causation and mechanisms. What are the conditions or processes that lead to variation across social groups? What are the causes of such phenomena as the "demographic transition" from high-fertility to low-fertility behavior? What accounts for the rapid growth of cities during certain periods? Sociology can provide explanations of some social outcomes as a causal consequence of proposed social mechanisms. Once we have a generic idea of what social causal mechanisms look like, we can turn to specifics and try to discover the processes through which behavior is created and constrained. So we can try to discover or hypothesize the mechanisms through which governments in countries based on tropical agriculture tend to under-serve farmers (Bates 1988). Social theories are hypotheses about social causal mechanisms; so theories provide a basis for explanation of social phenomena.

Third, there is the dimension of "theory" formation. Marx offered theories of capitalism, Weber of bureaucracy, Charles Tilly of state formation, and Mayer Zald and John McCarthy of resource mobilization. What is a theory in sociology? I suggest that "theory" functions in two ways in sociology, and each is different from the use of theory in the natural sciences. First, there is "grand" theory – capitalism, social cohesion, state formation. These concepts should be understood as articulating an organizing mental framework within which to organize many of the empirical details and characteristics of a given domain of social phenomena. This is analogous to Weber's concept of an ideal type (Weber and Shils 1949). The second use is what Robert Merton calls "theories of the middle range" (Merton 1967), and what I believe can be paraphrased as an exposition of a concrete social causal mechanism. The theory of free-riding; the theory of charismatic leadership; the theory of the flat organization – each involves an exposition of a social context and a characteristic social mechanism that can take place in that context. When we explain the low productivity of collective farming as a result of "free-riding" or "easy-riding," we are highlighting features of individual decision making and the context of supervision that leads to low levels of productivity on the part of individual workers.

What none of these examples provides – and what I think cannot be provided in the social sciences – is a theory that can serve as the unifying, deductively articulated foundation of all explanation in a given domain. Whereas Maxwell's electromagnetic theory was capable of deductively explaining a wide range of physical phenomena, there are no such comprehensive theories in the social sciences. And this is easily understood, given the multiplicity of motives, structures, institutions, and mentalities through which any given social phenomenon results. Once we look closely at the ontology of social phenomena, there emerge major and visible limitations on the degree of systematicity, interconnection, and determination that should be expected of social phenomena. The social world is highly contingent, the product of many independent actors. So we should only expect a weak degree of systematic variation among social phenomena.

So sociology consists of descriptive research, the search for social mechanisms that would explain empirical findings, and the formulation of mid-level theories that can provide a basis for better understanding the causes thus identified. Finally, the epistemic setting provided by the disciplinary institutions offers a basis for estimating the rational credibility of social science knowledge: journals, peer review, tenure evaluation. Social research and explanation remain fairly close to the level of the facts. Researchers in the disciplines and sub-disciplines are charged to test and explore the empirical and theoretical claims of their peers.

The results of a science including these components will be empirically disciplined, theoretically eclectic, and systemically modest. The goal of providing an over-arching theory that demonstrates the systematic integration of the social is abandoned. (It is worth observing that there was a period in the history of sociology when the epistemic values of the discipline were most consistent with this view. That was the period of the Chicago School (Abbott 1999).)

3. How Do We Define the Social?

Ontology is the study of what exists. Philosophers have devoted most of their attention to epistemology of scientific research; but it is perhaps even more important to think carefully about the ontology of the domain to be studied.

The social sciences study "social phenomena" – but what are social phenomena? Philosophers have paid less attention than is needed to the question of social ontology; and yet good social science theorizing requires good thinking about the fundamental nature of the social. So an important contribution that philosophers of social science can make is to return to the most basic questions about the nature of social entities. (Ian Hacking has done some important thinking about some of these questions of social ontology; see Hacking 1999, 2002. Other original contributions include Gould (1978) and Gilbert (1989).)

How have social scientists defined the nature of the subject matter of their investigations? Social scientists have singled out a wide variety of types of social phenomena for study. Consider this range: comparison of social types; crime and deviancy; urbanization; social movements and collective action; the workings of legislatures; voter behavior; and consumer behavior. And, clearly enough, it is not necessary to have a comprehensive definition of all social phenomena in order to define a domain for description and analysis.

Some social theorists have sometimes treated social constructs as unified macro-entities with their own causal powers. Structuralist theories maintain for example that "capitalism causes people to value consuming more than family time" or "democracy causes social cohesion." Likewise, some theorists have held that moral systems and cultures cause distinctive patterns of behavior – "Confucian societies produce cohesive families." Each of these claims places a large social entity in the role of a causal factor. A variant of this ontology is the idea of "systems theory" treating social entities as interconnected and self-regulating systems.

Is this a coherent way of talking? Can large structures and value systems exercise causal influence? The problem here is that statements like these look a lot like "action at a distance." We are led to ask: *How* do capitalism, democracy, or Confucianism influence social outcomes? In other words, we want to know something about the lower-level mechanisms through which large social facts impact upon behavior, thereby producing a change in social outcomes. We want to know something about the "microfoundations" of social causation (Little 1998).

One point seems obvious – and yet it is often overlooked or denied. Social behaviors are carried out by individuals, and individuals are influenced only by factors that directly impinge upon them (currently or in the past). Consider a particular voter's process of deciding to support a particular candidate. This person experienced a particular history of personality formation – a particular family, a specific city, a work history, an education. So the person's current political identity and values are the product of a sequence of direct influences. And at the moment, this socially constructed person is now exposed to another set of direct influences about the election race – newspapers, internet, co-workers' comments, attendance at political events, etc. In other words, his or her current political judgments and preferences are caused or influenced by a past and current set of experiences and contexts.

This story brings in social factors at every stage – the family was Catholic, the city was Chicago, the work was a UAW-organized factory. So the individual is socially influenced and formed at every stage. But here is the important point: every bit of that social influence is mediated by locally experienced actions and behaviors of other socially formed individuals. "Catholicism," "Chicago culture," and "union movement" have no independent reality over and above the behaviors and actions of people who embody those social labels.

Consider this formulation of a better ontology for the social sciences. Social phenomena are the manifestation of the actions, choices, thoughts, and behaviors of individuals. Individuals interact with each other in a variety of ways. Social outcomes are the aggregate result of these interactions and choices. Social outcomes can be observed and described, so we can provide a set of empirical social facts (e.g., demography, rates of suicide, or urban conditions). With a body of empirical observations in hand, the social scientist can ask a variety of questions: Are there patterns among these data? Are there causes that can be identified that lead to variations in the data? What sorts of causal mechanisms can be identified at work among social phenomena? And the social scientist can consider a variety of ways of further investigating these data: surveys of participants, ethnographic research, collection of data such as prices and quantities of commodities, formulation of ideas about social structures, and so on. A social science might then be loosely defined as a research tradition or paradigm that takes a particular definition of the empirical scope of the research; the methods that appear most illuminating; some core theoretical and explanatory ideas; and some specific criteria of empirical evaluation of claims in this area.

4.   What Is a Social Explanation?

To explain an outcome is to demonstrate why it occurred; what conditions combined to bring it about; what caused the outcome in the circumstances, or caused it to be more likely to occur (Salmon 1984, Miller 1987, Kitcher and Salmon 1989, Knowles 1990, Little 1991). The most fundamental aspect of an explanation of an outcome or pattern is a hypothesis about the causal processes through which it came about. So social explanation requires that we provide accounts of the social causes of social outcomes. This approach begins in a different place from the central theory of explanation offered in the philosophy of science, the covering-law model of explanation (Hempel 1965). This deviation is deliberate: the underlying conception of a domain consisting of law-governed regularities is not well suited to the real nature of social phenomena. Instead, the central component of this theory of explanation is the idea of a causal social mechanism.

There are other theories of social explanation that have been advanced: functional explanation (X occurs because of the beneficial consequences it produces for system S), hermeneutic explanation (X expresses such-and-such meaning within the cultural system), rational-intentional explanation (X occurs because circumstances C give agents reason to behave B, aggregating to X), and materialist explanation (X exists because it serves the material interests of G). (These and other models of explanation are more fully discussed in Little 1991.) Functional explanations in the social sciences have been greatly discredited because they generally fail to provide an account of the sustaining causal mechanism, from future benefits to current social arrangements (Elster 1982, 1983, 1990). And materialist and rational-intentional explanations are really just varieties of causal explanations. So causal relationships among social phenomena (and environmental circumstances) continue to serve as the most plausible foundation for social explanations.

The next section will look more closely at the idea of social causation. But it is worth asking whether all causal explanations work in roughly the same way; and the evidence of social science practice suggests that they do not. Instead, there are some large categories of "types" of causal explanations that can be discovered when we consider a large number of social and historical explanations of outcomes and patterns as provided by sociologists, economists, anthropologists, and historians. (Robert Nozick approached the theory of explanation from this point of view, particularly in his theory of "filtering" explanations; see Nozick 1974, 1981.) For example, consider:

Selection mechanisms. Why are passengers on commercial aircraft better educated than the general population? Because most airline passengers are business travelers, and high-level and mid-level business employees tend to have a higher level of education than the general population.

Evolutionary explanations. Why does the level of efficiency of a firm tend to rise over time? Because the net efficiency of a firm is the product of many small factors. These small factors sometimes change, with an effect on the efficiency of the firm. Low efficiency firms tend ultimately to lose market share and decline into bankruptcy. Surviving firms will have features that produce higher efficiency.

Imitation mechanisms. Why did the no-huddle offense become so common in the NFL in the 1980s? Because it was successful for a few teams, and other coaches copied the offense in the hope that they too would win more games.

Rational-intentional explanations. Why do boycotts often fizzle out? Because participants are rational agents with private goals, and they make calculating decisions about participation.

Aggregative explanations. Why does technological innovation occur continuously within a market-based society? Because firms are constantly looking for lower-cost and higher-value-added methods of manufacturing, and this search leads them to innovations in products and technologies. These individual efforts aggregate to a social pattern of technology change.

Conspiracy explanations. Why did the United States move away from passenger railroads as the primary form of intercity transportation? Because powerful industries took political actions to assure that private automobiles would be encouraged as the primary form of transport.

Culture explanations. Why were so many Quaker men conscientious objectors at great personal cost during World War II? Because their religious beliefs categorically rejected the violence of war and they refused to participate in this immoral activity.

Path-dependency explanations. Why do we still use the very inefficient QWERTY keyboard arrangement that was devised in 1874? Because this arrangement, designed to keep typists from typing faster than the mechanical keyboard would permit, was so deeply embodied in the typing skills of a large population and the existing typewriter inventory by 1940 that no other keyboard arrangement could be introduced without incurring massive marketing and training costs.

Upon analysis, it is clear that these are all different aspects of causal processes in social life: how certain social arrangements combine to bring about other arrangements or outcomes. So let us now turn to the topic of social causation.

5.   What Is Social Causation?

It was noted above that social explanations most typically are causal explanations: specification of the causal conditions that brought about the outcome or regularity of interest. So we need to raise two sorts of questions. First, what kind of thing is a social cause – how do social facts cause other social facts? Is there such a thing as social causation? What does social causation derive from? What is the ontology of "social necessity" (analogous to natural necessity) – the way in which one set of circumstances "brings about" another set of outcomes? And second, what kinds of social research can allow us to identify the causes of a social outcome or pattern? In general, we can begin with an ontology grounded in purposive social action by agents within institutional settings and environments. Social causation derives from the patterns of behavior that are produced in this setting. (For example, we can explain the degradation of environmental quality of a common resource as the consequence of free-riding behavior.)[2]

Generally speaking, a cause is a condition that either necessitates or renders more probable its effect, in a given environment of conditions. It is a circumstance that has the power or capacity to bring about the effect (Cartwright 1989). (For the philosophers, this means that "C is sufficient in the circumstances to bring about E or to increase the likelihood of E." See Mackie 1974 for a detailed exposition.) Normally a cause is also necessary for the production of its effect – "if C had not occurred, E would not have occurred." The probabilistic version can be formulated this way: "If C had not occurred, the likelihood of E would have been lower." (Wesley Salmon explores the intricacies in much greater detail; see Salmon 1984.) These features of necessity, sufficiency, and differential conditional probabilities provide the basis for developing specific empirical methods for testing causal relationships.

This account depends upon something that David Hume abhorred: the idea of causal necessity. For natural causes we have a suitable candidate in the form of natural necessity deriving from the laws of nature: "When C

---

[2]   For an extensive and fair view of the salience of causal explanation in sociology, see Abbott 1998. Abbott provides an admirable analysis of the itinerary of "causalism" through the development of sociology and the historical social sciences. He offers an interpretation of "causal explanation" for the social sciences that does not reduce to correlation, association, general laws, inductivism, or positivism.

occurs, given the laws of nature, E necessarily occurs." However, there are no "laws of society" that function ontologically like laws of nature. So how can there be "social necessity"? Fortunately, there is an alternative to law-based necessity, in the form of a causal mechanism. A mechanism is a particular configuration of conditions and processes that predictably leads from one set of conditions to an outcome. Mechanisms bring about specific effects. For example, "over-grazing of the commons" is a mechanism of resource depletion. Moreover, we can reconstruct precisely why this mechanism works for rationally self-interested actors in the presence of a public good. So we can properly understand a claim for social causation along these lines: "C causes E" means "there is a set of causal mechanisms that convey circumstances including C to circumstances including E."

There is an important recent body of work in the philosophy of social science on social mechanisms that converges with original and useful work on methodology of comparative research coming from within the historical social sciences.[3] This work makes the case for placing the discovery of concrete causal mechanisms at the center of our conception of historical and social inquiry. Social mechanisms are *concrete social processes in which a set of social conditions, constraints, or circumstances combine to bring about a given outcome.*[4] On this approach, social explanation does not take the form of inductive discovery of laws; rather, the generalizations that are discovered in the course of social science research are subordinate to the more fundamental search for causal mechanisms and pathways in individual outcomes and sets of outcomes.[5] This approach casts doubt

[3]   Important recent exponents of the centrality of causal mechanisms in social explanation include Hedström and Swedberg (1998), McAdam, Tarrow, and Tilly (2001), and George and Bennett (2005). Volume 34, numbers 2 and 3 of *Philosophy of the Social Sciences* (2004) contain a handful of articles devoted to the logic of social causal mechanisms, focused on the writings of Mario Bunge.

[4]   The recent literature on causal mechanisms provides a number of related definitions: "We define causal mechanisms as ultimately unobservable physical, social, or psychological processes through which agents with causal capacities operate, but only in specific contexts or conditions, to transfer energy, information, or matter to other entities" (George and Bennett 2005). "Mechanisms are a delimited class of events that alter relations among specified sets of elements in identical or closely similar ways over a variety of situations" (McAdam, Tarrow, and Tilly 2001, 24). "Mechanisms . . . are analytical constructs that provide hypothetical links between observable events" (Hedström and Swedberg 1998, 13).

[5]   Authors who have urged the centrality of causal mechanisms or powers for social explanation include Sørensen (2001), Harré and Secord (1972), Varela and Harré (1996), Cartwright (1983, 1989), Salmon (1984), and Dessler (1991).

on the search for generalizable theories across numerous societies. It looks instead for specific causal influence and variation. The approach emphasizes variety, contingency, and the availability of alternative pathways leading to an outcome, rather than expecting to find a small number of common patterns of development or change.[6] The contingency of particular pathways derives from several factors, including the local circumstances of individual agency and the across-case variation in the specifics of institutional arrangements – giving rise to significant variation in higher-level processes and outcomes.[7]

This approach places central focus on the idea of a causal mechanism: to identify a causal relation between two kinds of events or conditions, we need to identify the typical causal mechanisms through which the first kind brings about the second kind. What, though, is the nature of the social linkages that constitute causal mechanisms among social phenomena? I argue for a *microfoundational* approach to social causation: the causal properties of social entities derive from the structured circumstances of agency of the individuals who make up social entities – institutions, organizations, states, economies, and the like (Little 1998). So this approach advances a general ontological stance and research strategy: the causal mechanisms that create causal relations among social phenomena are compounded from the structured circumstances of choice and behavior of socially constructed and socially situated agents.

Now let us turn to inquiry. How would we detect social causation? Fundamentally there are three ways. We can exploit the mechanism requirement and seek out particular or common social mechanisms. Both social theory and process-tracing can serve us here (Goertz and Starr

---

[6] An important expression of this approach to social and historical explanation is offered by Charles Tilly: "Analysts of large-scale political processes frequently invoke invariant models that feature self-contained and self-motivating social units. Few actual political processes conform to such models. Revolutions provide an important example of such reasoning and of its pitfalls. Better models rest on plausible ontologies, specify fields of variation for the phenomena in question, reconstruct causal sequences, and concentrate explanation on links within those sequences" (Tilly 1995, 1594).

[7] McAdam et al. describe their approach to the study of social contention in these terms: "We employ mechanisms and processes as our workhorses of explanation, episodes as our workhorses of description. We therefore make a bet on how the social world works: that big structures and sequences never repeat themselves, but result from differing combinations and sequences of mechanisms with very general scope. Even within a single episode, we will find multiform, changing, and self-constructing actors, identities, forms of action and interaction" (McAdam, Tarrow, and Tilly 2001, 30).

2003, George and Bennett 2005). Second, we can exploit the "necessary and sufficient condition" feature by using comparative methods like Mill's methods (Ragin 1987). And third, we can exploit the probabilistic and statistical implications of a causal assertion by looking for correlations and conditional probabilities among the conditions associated with hypothesized causal mechanisms (Woodward 2003). This feature underpins standard "large-n" quantitative research methods in social science. In each case, we must keep fully in mind the centrality of causal mechanisms. A discovery of a statistical association between X and Y is suggestive of causation, but we need to be able to hypothesize the mechanism that would underlie the association if we are to attribute causation. Likewise, the discovery that a study of multiple cases suggests that A is necessary for E and A&B are sufficient for E requires us to consider the question, what is the concrete social mechanism that links A, B, and E?

6.   How Are Social Science Hypotheses and Theories Given Empirical Justification?

What makes sociology "scientific"? An important component of a reply is that assertions, hypotheses, and theories are subject to the test of empirical evidence. Hypotheses need to be evaluated in terms of observations of how the real world behaves. We should evaluate our assertions in terms of their fit with the empirical facts. This is the "empiricist" constraint. So we might begin by saying that social science assertions are tested and evaluated on the basis of disciplined observation of social phenomena; careful articulation of the logical consequences of theoretical assertions; and evaluation of theoretical assertions on the basis of their consistency with a wide range of empirical observation. And in fact, much social science research takes the form of painstaking discovery of social facts in a range of levels of scale.

Post-positivist philosophers of science have noticed that these simple ideas raise many puzzles, however (Hanson 1958, Glymour 1980, Brown 1987). Consider these points:

•   No set of observable facts guarantees the truth of a scientific assertion.
•   There is no sharp distinction between observation and theory; our observations of the empirical facts commonly depend upon the assumption of some elements of scientific theory. Observations are "theory-laden."

- Even the empirical "facts" are subject to multiple interpretations; it is often possible to redescribe a set of observations in a way that appears to support contradictory hypotheses.

In the social sciences there are additional complexities about how to arrive at empirical observations and measurements.

- Social observations require us to "operationalize" the empirical facts we want to observe. For example, we may want to observe the standard of living of the working class. But we cannot achieve this directly. Instead, we need to arrive at "proxies" that are plausibly indicative of the property in question. So the wage basket that can be purchased with a given average money wage may be the index we use for measuring the standard of living. But there are other defensible ways of operationalizing the standard of living, and the various criteria may yield results that behave differently in given times and places.
- Social observation requires aggregation of measurements over a diverse group of individuals. We have to make judgments and choices when we arrive at a process for aggregating social data – for example, the choice of using the Gini coefficient rather than the share of income flowing to the bottom 40 percent as a measure of income inequality, or using the median rather than the mean to observe changes in income distribution. These choices must be made – and there are no decisive empirical reasons that would decide the issue.
- Social concepts are needed to allow us to break down the social world into a set of facts. But there are plausible alternative conceptual schemes through which we can understand the nature and varieties of social phenomena. So, once again, we cannot hold that "observation" determines "theory."

These are familiar logical difficulties with the basic requirement of empiricism within the philosophy of science. However, they are not fatal difficulties. At bottom, it remains true that there is such a thing as social observation. It is necessary to accept that observations are theory-laden, that no observation is incontrovertible, and that empirical evaluation depends upon judgment. All this accepted, there is a range of social observation that is relatively close to the ground and to which we can attribute some degree of epistemic warrant. Finally, there is available to us a coherence epistemology that permits a holistic and many-sided process of conveying warrant.

My view, then, is that the situation of the social sciences is less like physics (highly dependent on long chains of reasoning in order to assess empirical warrant) and more like journalism (grounded in careful and reasoned constructions of observations of the social world). The social world is reasonably transparent. We can arrive at reasonably confident observations of a wide range of social facts. And we can provide a logical analysis of the degree of credibility a given sociological theory has, given a fixed set of (corrigible) observations. Much of sociology is closely tied to descriptive inquiry, and the epistemic challenges come in at the stage of building our observations rather than our theories.

Moreover, the common views that natural science theories are "underdetermined" by all available evidence (so that multiple theories can be equally well supported) and that scientific theories can only be supported or undermined as wholes (with no separate confirmation for parts of theories) appear to be largely inapplicable to the social sciences. Rather, social theories are more commonly of the "middle range," permitting piecemeal observation, testing, and empirical evaluation.

This also means that the hypothetico-deductive model of confirmation is less crucial in the social sciences than in the natural sciences. The key explanatory challenge is to discover a set of causal processes that might explain the observed social world. And sophisticated observation is often the bulk of what we need. (Ian Shapiro's recent book, *The Flight from Reality in the Human Sciences* (2005), is a tough critique of excessive formalism and theorism in the social sciences.)

7. Positivism, Naturalism, and General Social Laws

There is a strong current of positivism in sociology. Other paradigms exist – feminism, Marxism, comparative historical sociology, and ethnographic sociology, to name several. But the claim of science is generally couched in terms of a positivist theory of science and inquiry. The core assumptions of positivism include these central ideas, which we can refer to as naturalism – that social science is identical in its logic to natural science, and that science involves the search for general laws about empirical phenomena. The historical dominance of positivism in sociology is unsurprising, in that several of the founders of sociology (Comte, Mill, and Durkheim in particular) were most emphatic in asserting the necessary connection between the two ideas, and Comte invented both "positivism" and "sociology" as modern terms. The positivism that entered the social sciences through Mill and Comte received additional momentum as a result of the influence of the logical positivism of the Vienna Circle in the 1910s and 1920s and American funding strategies in the 1950s. Some philosophers of social science have often embraced the naturalistic program for the social sciences. Some have done so explicitly – for example, Thomas (1979). Others have embraced the idea that social science knowledge needs to be expressed in the form of law-like generalizations (Kincaid 1990, McIntyre 1996).

The grip of positivism on sociology has been at the center of much recent analysis and debate within the social sciences themselves in recent years (Wallerstein 1999, Adams, Clemens, and Orloff 2005, Steinmetz 2005). Some of the most valuable methodological and epistemological writing that is currently appearing in the social sciences is focused on the goal of broadening the palette of tools through which social scientists analyze and explain the social world (McDonald 1996, Ortner 1999, Mahoney and Rueschemeyer 2003). Nonetheless, the preference for quantitative findings continues to dominate the premier journals in sociology.

The positivist program in sociology might be summarized in these terms: identify a set of units for study; measure a selected set of properties of these units; use statistical techniques to discover correlations and regularities among the units; and attempt to arrive at causal hypotheses about the relationships that are observed among the variables. The units may be individual persons (voters, criminals, social workers), or they may be higher-level social entities (cities, national governments, types of social or political structures). These methodological presuppositions led to a strong preference for quantitative methods in social science, in which the ideal research result is a large data set upon which the researcher has successfully conducted statistical analysis leading to a discovery of associations among some of the variables. And favored tools of inquiry include survey research and large-n studies based on large data sets constructed by national agencies. (Andrew Abbott (1998, 1999) refers to this as the variables paradigm.)

An associated methodological premise – adopted with admiring glances at physics and economics – is a preference for formal mathematical models of the social situations that are being studied. Here the methodological ideal is to identify a set of abstract axioms that constitute the "theory" of the field. These axioms may be drawn from mathematical decision theory or game theory. Then, the theorist undertakes to construct a model of the empirical phenomena based on these premises with the goal of explaining outcomes as equilibrium solutions to the axioms given the antecedent conditions. But all too often the elegance of the mathematics overshadows the real empirical challenges of the investigation. (Ian Shapiro (2005) encapsulates his criticisms of this formalistic predilection with the title *The Flight from Reality in the Human Sciences*.) Critics such as Shapiro argue that it is important for the disciplines (including his own, that is, political science) to re-emphasize empirical studies that include detailed, careful descriptions of various social phenomena – e.g., legislatures, labor unions, or social service agencies. The point here is not that formal models are inappropriate in the social sciences. It is rather that we need always to be asking the question, how does this formalism help to illuminate the real, underlying social processes? And how can it be given empirical content?

This positivistic ideal for social science research devalues several other approaches to social investigation: qualitative or ethnographic research, comparative "small-n" research, and case-study research (King, Keohane, and Verba 1994). Alternative approaches to sociological research include at least these:

Qualitative research. Considers the individuality and particularity of the actors who are the subject of investigation. Borrows from the methods of anthropology, making use of ethnographic methods and findings. Makes use of field research and participant-observer methods. Makes a rigorous effort to discover and understand features of the lived human experience of the social actors who make up the social unit under examination.

Comparative research. Identifies a small number of complex and similar cases (revolutions, social welfare systems, labor unions). Carefully structures the definitions of the characteristics to be observed. Identifies the outcome to be explained (occurrence of ethnic conflict). Uses a variety of methods of controlled comparison to sort out causal connections (for example, Mill's methods of similarity and difference to sort among the antecedent conditions as "necessary," "sufficient," or "null").

Historical research. Places the social institutions, organizations, or movements into historical context; discovers some of the conjunctural and path-dependent circumstances that shaped their current characteristics (Thelen 2004).

Some of the best work in sociology today combines all of these approaches; for example, C.K. Lee's multi-stranded approach to the task of understanding the roles of gender, class, and management goals within Chinese factories (Lee 1998), or Leslie Salzinger's treatment of similar issues in the maquiladoras in Mexico (Salzinger 2003).

Naturalism is a poor guide for social science inquiry. Instead, we need to approach social science research with greater attention to the ontology of social life – the sorts of things that make up the social world. And when we pay attention to a more realistic ontology, focusing on individuals in interaction with each other as the basic element, we will be led to a readiness to find contingency, heterogeneity, path-dependence, and particularity among the phenomena that we study – corresponding to the plasticity of human institutions and human agency. So let us now ask whether there are social laws. Is there anything like a "law of nature" that governs or describes social phenomena?

My view is that this is a question that needs to be approached very carefully. As a bottom line, I take the view that there are no "social laws" analogous to "laws of nature," even though there are some mid-level regularities that can be discovered across a variety of kinds of social phenomena. But care is needed because of the constant temptation of naturalism – the idea that the social world should be understood in strong analogy with the natural world. If natural phenomena are governed by laws of nature, then social phenomena should be governed by "laws of society." But the analogy is weak.

In fact, there are few law-like generalizations about social entities and processes (Little 1993). The deepest basis for this judgment is ontological. Social phenomena do not fall into fixed and distinct "types," in which the members of the type are homogeneous. We can generalize about "water," but not about "revolution," for the simple reason that all samples of pure water have the same structure and observable characteristics; but not so for all "revolutions." The category of "revolution" is not a "kind," and we should not imagine that we can arrive at a set of necessary and sufficient conditions for membership in this group. Each revolution, for example, proceeds according to a historically specific set of causes and circumstances. And there are no genuinely interesting generalizations across the whole category.

Of course there are observable regularities among social phenomena. Urban geographers have noticed a similar mathematical relationship in the size distribution of cities in a wide range of countries. Durkheim noticed similar suicide rates among Catholic countries – rates that differ consistently from those found in Protestant countries. Political economists notice that there is a negative correlation between state spending on social goods and the infant mortality rate. And we could extend the list indefinitely.

But what does this fact demonstrate? Not that social phenomena are "law-governed." Instead, it results from two related facts: first, there are social causal mechanisms; and second, there is some recurrence of common causes across social settings.

Take the mechanism of "collective action failures in the presence of public goods." Here the heart of the mechanism is the analytical point that rationally self-interested decision makers will take account of private goods but not public goods; so they will tend to avoid investments in activities that produce public goods. They will tend to become "free-riders" or "easy-riders." The social regularity that corresponds to this mechanism is a "soft" generalization – that situations that involve a strong component of collective opportunities for creating public goods will tend to demonstrate low contribution levels from members of affected groups. So public radio fundraising will receive contributions only from a small minority of listeners; boycotts and strikes will be difficult to maintain over time; and fishing resources will tend to be over-fished. And in fact, these regularities can be identified in a range of historical and social settings.

However, the "free-rider" mechanism is only one of several that affect collective action. There are other social mechanisms that have the effect of enhancing collective action rather than undermining it. For example, the presence of competent organizations makes a big difference in eliciting voluntary contributions to public goods; the fact that many decision makers appear to be "conditional altruists" rather than "rationally self-interested maximizers" makes a difference; and the fact that people can be mobilized to exercise sanctions against free-riders affects the level of contribution to public goods. (If your neighbors complain bitterly about your smoky fireplace, you may be incentivized to purchase a cleaner-burning wood or coal.) The result is that the free-rider mechanism rarely operates by itself – so the expected regularities may be diminished or even extinguished.

The conclusion that can be drawn from this is pretty simple. It is that social regularities are "phenomenal" rather than "governing": they emerge as the result of the workings of common social causal mechanisms, and social causation is generally conjectural and contingent. So the regularities that become manifest are weak and exception-laden – and they are descriptive of outcomes rather than expressive of underlying "laws of motion" of social circumstances.

And there is a research heuristic that emerges from this discussion as well. This is the importance of searching out the concrete social causal mechanisms through which social phenomena take shape. We do a poor job of understanding industrial strikes if we simply collect a thousand instances and perform statistical analysis on the features we have measured against the outcome variables. We do a much better job of understanding them if we put together a set of theories about the features of structure and agency through which a strike emerges and through which individuals make decisions about participation. Analysis of the common "agent/structure" factors that are relevant to mobilization will permit us to understand individual instances of mobilization, explain the soft regularities that we discover, and account for the negative instances as well.

8. How Do Social Facts Relate to Facts about Individuals and Facts about Psychology?

Finally, we turn to the topic that many regard as the most fundamental in the philosophy of social science: the question of the relationship between social facts and individual facts. This is the question to which methodological individualism is a famous reply. The question raises issues about ontology – what is the nature of social entities? Are they composed solely of features of individuals? But it also raises questions about inter-theoretical reduction: should we expect that all statements or explanations at the level of the social should be reducible, in principle, to statements at the level of individuals? Or are there social explanations that are autonomous with respect to facts about specific individuals?

What is the relationship between facts about society and facts about individuals? Are there any social facts that are autonomous with respect to facts about individuals, as Durkheim maintained? Or are social facts reducible to some set of facts about individuals? Methodological individualism is the view that holds that social explanations need to be based on facts about individuals and their motivations. The position holds that there are no causal properties at the level of complex social entities; rather, all social causes operate through features of individual behavior. The position of methodological individualism extends back to Mill (1879), Weber (1949), and J.W.N. Watkins (1968). It is primarily opposed to the idea that there are "holistic" social facts (Durkheim) – social facts that are autonomous with respect to facts about individuals.

The contrary position is holism or structuralism. Durkheim was an explicit social holist; he maintained that social facts are autonomous with respect to facts about individuals, and that we need to focus the scientific attention of sociology upon large social facts such as morality, law, or the division of labor. Individual consciousness and behavior are influenced by these large social facts, but individuals do not constitute or determine those facts. (Steven Lukes' biography of Durkheim provides a very astute analysis of his conception of social holism; see Lukes 1972.) Structuralism derives from different intellectual roots; but the view gives similar autonomy to the level of social structures as exercising primacy over individual actions. Structures determine individuals rather than individuals determining structures. Structuralist theorists include Godelier (1972), Althusser and Balibar (1971), and Hindess and Hirst (1975), who attribute explanatory primacy to structures such as the capitalist mode of production, and anthropologists such as Claude Levi-Strauss, who give primacy to structures of cultural realities, such as the kinship system within a culture, over the thoughts and behavior of individuals (Levi-Strauss 1969).

The thesis of methodological individualism can be formulated as a statement about explanation, as a thesis about social ontology, or as a statement about inter-theoretic reduction. The explanatory version holds that "social phenomena are best explained in terms of the motivations and interactions of individual persons" (Steel 2006, 440). The ontological version maintains that social entities and their properties are constituted by individuals and their actions: "This perspective affirms that there are large social structures and facts that influence social outcomes, but it insists that these structures are only possible insofar as they are embodied in the actions and states of socially constructed individuals" (Little 2006, 346). The intertheoretic version holds that it is "possible to reduce theories containing social predicates to theories stated in terms of individuals and their properties only" (Zahle 2003, 77).

Methodological individualism has had a rebirth of interest in recent years (Udehn 2001, Zahle 2007). James Woodward (2000) and Daniel Steel (2006) have explicated the position in terms of a theory of what the claim about explanation amounts to. Woodward's fundamental idea is that one explanation is more fundamental than another if it is "invariant" with respect to a wider range of interventions.

Micro-economics and rational choice theory are the areas of the social sciences that are most explicitly founded in the assumptions of methodological individualism. The goal of these fields is to explain social outcomes as the aggregate result of decisions by rationally self-interested agents.

A recent strategy in approaching the issue of the relationship between social facts and individual facts is to postulate that social-level statements and causal judgments need to be provided with micro-foundations: descriptions of the pathways through which socially situated individuals are led to act in such a way as to bring about the macro-level fact. (See Little 1989, 1998 and Elster 1989 on this position.)

Reducibility means that the statements of one scientific discipline should be logically deducible from the truths of some other "more fundamental" discipline. It is sometimes maintained that the truths of chemistry ought in principle to be derivable from those of quantum mechanics. A field of knowledge that is not reducible to another field R is said to be "autonomous with respect to R." Philosophers sometimes further distinguish "law-to-law" reduction, "type-to-type" reduction, "law-to-singular-fact" reduction, and "type-to-token" reduction.

Are social sciences such as economics, sociology, or political science reducible in principle to some other more fundamental field – perhaps psychology, neurophysiology, or the theory of rationality? To begin to answer this question we must first decide what items might be reduced: statements, truths, laws, facts, categories, or generalizations. Second, we need to distinguish several reasons for failure of reduction: failure in principle, because events, types, and laws at the social level are simply not fixed by states of affairs at "lower" levels; and failure for reasons of limits on computation. (The motions of a five-body system of stars might be determined by the laws of gravitation even though it is practically impossible to perform the calculations necessary to determine future states of the system.)

So now we can consider the question of social reduction in a reasonably clear form. Consider first the "facts" that pertain to a domain of phenomena – whether these facts are known or not. (I choose not to concentrate on laws or generalizations, because I am doubtful about the availability of strong laws of social phenomena.) Do the facts of a hypothetically complete theory of human psychology "fix in principle" the facts of economics or sociology, given appropriate information about boundary conditions?

One important approach to this problem is the theory of supervenience (Kim 2005, 1993). A level of description is said to supervene upon another level just in case there can be no differences of state at the first level without there being a difference of state at the second level. The theory is first applied to mental states and states of neurophysiology: "no differences in mental states without some difference in neurophysiology states." Supervenience theory implies an answer to the question of whether one set of facts "fixes in principle" the second set of facts. (It has been taken as obviously true that social facts supervene upon facts about individuals; how could it be otherwise? What other constitutive or causal factors might influence social facts, beyond the actions and ideas of individuals?) If the facts about social life supervene upon facts about the psychological states of individuals, then it follows that the totality of facts about individual psychology fixes in principle the totality of facts about social life. (Otherwise there would be the situation that there are two total social worlds corresponding to one total "individual psychology" world; so there would be a difference at the social level without a difference at the level of individual psychology.)

So this provides the beginnings of an answer to our question: if we believe that social facts supervene upon facts about individuals, then we are forced to accept that the totality of facts about individuals "fix" the facts about society. However, supervenience does not imply "reducibility in principle," let alone "reducibility in practice" between levels. In order to have reducibility, it is necessary to have a system of statements describing features of the lower level which are sufficient to permit deductive derivation (or perhaps probabilistic inference) of all of the true statements contained in the higher-level domain. If it is a social fact that "collective action tends to fail when groups are large," then there would need to be a set of statements at the level of individual psychology that logically entail this statement.

Two additional logical features would appear to be required for reduction: a satisfactory set of bridge statements (linking the social term to some construction of individual-level terms; "collective action" to some set of features of individual agents, so there is a mapping of concepts and ontologies between the two domains), and at least some statements at the lower level that have the form of general laws or law-like probabilistic statements. (If there are no general statements at the lower level, then deductive inference will be limited to truth-functional deduction.)

Now it is time for a speculative leap: a judgment call on the question of whether we ought to look for reductive links between social facts and individual-level facts. My intuition is that it is not scientifically useful to do so, for several reasons. First is the point about computational limits: even if the outcome of a riot is "fixed" by the full psychological states of participants *ex ante* and their strategic interactions during the event, it is obviously impossible to gather that knowledge and aggregate it into a full and detailed model of the event. So deriving a description of the outcome from a huge set of facts about the participants is unpromising. Second, it is telling that we need to refer to the strategic interactions of participants in order to model the social event; this means that the social event has a dynamic internal structure that is sensitive to sub-events that occur along the way. (Jones negotiates with Smith more effectively than Brown negotiates with Black. The successful and failed negotiations make a difference in the outcome but are unpredictable and contingent.) Third, the facts at the social level rarely aggregate to simple laws or regularities that might have been derived from lower-level laws and regularities; instead, social outcomes are contingent and varied.

So, for a variety of reasons, it is reasonable to take the view that social facts supervene upon facts about individuals, but that social explanations are autonomous from laws of psychology. At the same time, the view that social explanations require micro-foundations appears to be a reasonable one: we need to know what it is about the circumstances and motives of individuals such that their ordinary socially situated choices and behavior result in the social processes and causal connections that we observe. And in field after field it is possible to demonstrate that it is possible to provide such micro-foundations.

9. Conclusion

This essay has taken on a large subject: what is involved in understanding society? What sorts of ontological assumptions do we need to make as we attempt to analyze and explain social processes? What is involved in arriving at an explanation of a social outcome or pattern? How do we provide empirical confirmation for our hypotheses and theories about the social world? And what help or hindrance can be derived from the legacies of positivism and naturalism? How can philosophy contribute to the creation of a better sociology for the twenty-first century?

We began by noticing that this inquiry is not of merely academic concern. Understanding society better is an urgent need for all of us in the twenty-first century. Our quality of life, our physical security, our ability to provide for greater social justice globally and locally, and our ability to achieve the sustainability of our natural environment all depend upon social processes and social behavior. The better we understand these processes and behavior, the better we will be able to shape our futures in ways that serve our needs and values. And currently our understanding of important social processes is highly limited. We need better theories, better research methodologies, and better conceptions of the basic nature of social phenomena, if we are to arrive at a more realistic understanding of the social world. The philosophy of social science can contribute to these important tasks.

Three particularly central ideas emerge from the considerations presented above. First is a point about social ontology. Social research should be based on a realistic understanding of the fact that social phenomena are constituted by socially embedded individuals in interaction with each other. Higher-level social entities – states, organizations, institutions – are real enough, but they must be understood as being composed of individuals in interaction. So social science must avoid the error of reification – the assumption that social entities have some kind of abiding permanence independent of the individuals who constitute them.

This ontology should in turn lead social researchers to expect a substantial degree of contingency and plasticity in the phenomena they study. Given that institutions and organizations are constituted by the social individuals who make up the institution or organization, we should expect that they will mutate over time – that is, we should expect plasticity of social entities. And we should anticipate contingency. Rather than the iron laws of history that Marx hoped to find, we should understand that social outcomes depend upon many independent factors, and that outcomes will be path-dependent and contingent. This means, further, that we should not expect to find law-governed regularities among social phenomena, and we should not define "science" as the discovery of law-governed regularities among a set of phenomena.

Finally, it has been argued here that there is a credible basis for finding a degree of order among social phenomena, in the form of causal relationships between various social facts. The discovery of social causal mechanisms is the foundation of social explanation. Moreover, there is a very consistent relationship between the idea of a social causal mechanism and the social ontology of "socially situated individuals" that was developed above. Social causation flows through the structured social actions of individuals. And empirical social research can inform us about various aspects of the processes of social causation: the social institutions within which individuals act; the historical processes of development through which individuals came to have their mental models; moral ideas, and preferences; and the powers and constraints that are embodied in a set of social relationships at a given time.

References

Abbott, A. 1998. "The Causal Devolution." *Sociological Methods and Research* 27.2: 148–81.

Abbott, A.D. 1999. *Department and Discipline: Chicago Sociology at One Hundred.* Chicago: University of Chicago Press.

Abbott, A.D. 2001. *Chaos of Disciplines.* Chicago: University of Chicago Press.

Adams, J., E.S. Clemens, and A.S. Orloff (eds.). 2005. *Remaking Modernity: Politics, History, and Sociology, Politics, History, and Culture.* Durham, NC: Duke University Press.

Althusser, L. and E. Balibar. 1971. *Reading Capital.* New York: Pantheon Books.

Bates, R.H. (ed.). 1988. *Toward a Political Economy of Development: A Rational Choice Perspective.* California Series on Social Choice and Political Economy 14. Berkeley: University of California Press.

Bohman, J. 1991. *New Philosophy of Social Science: Problems of Indeterminacy.* Cambridge, MA: MIT Press.

Braybrooke, D. 1987. *Philosophy of Social Science.* Englewood Cliffs, NJ: Prentice Hall.

Brown, H.I. 1987. *Observation and Objectivity.* New York: Oxford University Press.

Bunge, M. 1996. *Finding Philosophy in Social Science.* New Haven, CT: Yale University Press.

Cartwright, N. 1983. *How the Laws of Physics Lie.* Oxford: Oxford University Press.

Cartwright, N. 1989. *Nature's Capacities and Their Measurement.* Oxford: Oxford University Press.

Dessler, D. 1991. "Beyond Correlations: Toward a Causal Theory of War." *International Studies Quarterly* 35: 337–55.

Elster, J. 1982. "Marxism, Functionalism, and Game Theory." *Theory and Society* 11: 453–82.

Elster, J. 1983. *Explaining Technical Change: A Case Study in the Philosophy of Science. Studies in Rationality and Social Change.* Cambridge: Cambridge University Press.

Elster, J. 1989. *Nuts and Bolts for the Social Sciences.* Cambridge: Cambridge University Press.

Elster, J. 1990. "Merton's Functionalism and the Unintended Consequences of Action." In J. Clark, C. Modgil, and S. Modgil (eds.), *Consensus and Controversy.* London: Falmer Press.

George, A.L. and A. Bennett. 2005. *Case Studies and Theory Development in the Social Sciences.* BCSIA Studies in International Security. Cambridge, MA: MIT Press.

Gilbert, M. 1989. *On Social Facts.* Princeton, NJ: Princeton University Press.

Glymour, C.N. 1980. *Theory and Evidence.* Princeton, NJ: Princeton University Press.

Godelier, M. 1972. *Rationality and Irrationality in Economics.* London: New Left Books.

Goertz, G. and H. Starr (eds.). 2003. *Necessary Conditions: Theory, Methodology, and Applications.* Boulder, CO: Rowman & Littlefield.

Gould, C.C. 1978. *Marx's Social Ontology: Individuality and Community in Marx's Theory of Social Reality.* Cambridge, MA: MIT Press.

Hacking, I. 1999. *The Social Construction of What?* Cambridge, MA: Harvard University Press.

Hacking, I. 2002. *Historical Ontology.* Cambridge, MA: Harvard University Press.

Hanson, N.R. 1958. *Patterns of Discovery: An Inquiry into the Conceptual Foundations of Science.* Cambridge: Cambridge University Press.

Harré, R. and P.F. Secord. 1972. *The Explanation of Social Behaviour.* Oxford: Blackwell.

Hedström, P. and R. Swedberg (eds.). 1998. *Social Mechanisms: An Analytical Approach to Social Theory. Studies in Rationality and Social Change.* Cambridge: Cambridge University Press.

Hempel, C.G. 1965. *Aspects of Scientific Explanation, and Other Essays in the Philosophy of Science.* New York: Free Press.

Hindess, B. and P.Q. Hirst. 1975. *Pre-capitalist Modes of Production.* London: Routledge & Kegan Paul.

Kim, J. 1993. *Supervenience and Mind: Selected Philosophical Essays.* Cambridge: Cambridge University Press.

Kim, J. 2005. *Physicalism, or Something Near Enough.* Princeton Monographs in Philosophy. Princeton, NJ: Princeton University Press.

Kincaid, H. 1990. "Defending Laws in the Social Sciences." *Philosophy of the Social Sciences* 20.1: 56–83.

Kincaid, H. 1996. *Philosophical Foundations of the Social Sciences: Analyzing Controversies in Social Research.* Cambridge: Cambridge University Press.

King, G., R.O. Keohane, and S. Verba. 1994. *Designing Social Inquiry: Scientific Inference in Qualitative Research.* Princeton, NJ: Princeton University Press.

Kitcher, P. and W.C. Salmon (eds.). 1989. *Scientific Explanation.* Minnesota Studies in the Philosophy of Science 13. Minneapolis: University of Minnesota Press.

Knowles, D. (ed.). 1990. *Explanation and its Limits.* Royal Institute of Philosophy lecture 27. Cambridge: Cambridge University Press.

Lee, C.K. 1998. *Gender and the South China Miracle: Two Worlds of Factory Women.* Berkeley: University of California Press.

Levi-Strauss, C. 1969. *The Elementary Structures of Kinship*. Boston, MA: Beacon Press.

Little, D. 1989. "Marxism and Popular Politics: The Microfoundations of Class Struggle." *Canadian Journal of Philosophy* 15 (suppl.): 163–204.

Little, D. 1991. *Varieties of Social Explanation: An Introduction to the Philosophy of Social Science*. Boulder, CO: Westview Press.

Little, D. 1993. "On the Scope and Limits of Generalizations in the Social Sciences." *Synthese* 97: 183–207.

Little, D. 1998. *Microfoundations, Method and Causation: On the Philosophy of the Social Sciences*. New Brunswick, NJ: Transaction Publishers.

Little, D. 2006. "Levels of the Social." In S. Turner and M. Risjord (eds.), *Handbook for Philosophy of Anthropology and Sociology*. Amsterdam: Elsevier.

Lukes, S. 1972. *Émile Durkheim; His Life and Work: A Historical and Critical study*. New York: Harper & Row.

Mackie, J.L. 1974. *The Cement of the Universe: A Study of Causation*. Oxford: Clarendon Press.

Mahoney, J. and D. Rueschemeyer. 2003. *Comparative Historical Analysis in the Social Sciences*. Cambridge Studies in Comparative Politics. Cambridge: Cambridge University Press.

Martin, M. and L.C. McIntyre (eds.). 1994. *Readings in the Philosophy of Social Science*. Cambridge, MA: MIT Press.

McAdam, D., S.G. Tarrow, and C. Tilly. 2001. *Dynamics of Contention*. Cambridge Studies in Contentious Politics. New York: Cambridge University Press.

McDonald, T.J. (ed.). 1996. *The Historic Turn in the Human Sciences*. Ann Arbor: University of Michigan Press.

McIntyre, L.C. 1996. *Laws and Explanation in the Social Sciences: Defending a Science of Human Behavior*. Boulder, CO: Westview Press.

Merton, R.K. 1967. *On Theoretical Sociology: Five Essays, Old and New*. New York: Free Press.

Mill, J.S. 1879. *A System of Logic, Ratiocinative and Inductive, Being A Connected View of the Principles of Evidence and the Methods of Scientific Investigation*. 10th edn. London: Longmans Green and Co.

Miller, R.W. 1987. *Fact and Method: Explanation, Confirmation and Reality in the Natural and the Social Sciences*. Princeton, NJ: Princeton University Press.

Nozick, R. 1974. *Anarchy, State, and Utopia*. New York: Basic Books.

Nozick, R. 1981. *Philosophical Explanations*. Cambridge, MA: Harvard University Press.

Ortner, S.B. (ed.). 1999. *The Fate of "Culture": Geertz and Beyond*. Berkeley: University of California Press.

Ragin, C.C. 1987. *The Comparative Method: Moving beyond Qualitative and Quantitative Strategies*. Berkeley: University of California Press.

Rosenberg, A. 1988. *Philosophy of Social Science*. Boulder, CO: Westview Press.

Salmon, W.C. 1984. *Scientific Explanation and the Causal Structure of the World*. Princeton, NJ: Princeton University Press.

Salzinger, L. 2003. *Genders in Production: Making Workers in Mexico's Global Factories*. Berkeley: University of California Press.

Shapiro, I. 2005. *The Flight from Reality in the Human Sciences*. Princeton, NJ: Princeton University Press.

Sherratt, Y. 2006. *Continental Philosophy of Social Science: Hermeneutics, Genealogy, Critical Theory*. Cambridge: Cambridge University Press.

Sørensen, A.B. 2001. "Careers, Wealth and Employment Relations." In A.L. Kalleberg and I. Berg (eds.), *Source Book on Labor Markets: Evolving Structures and Processes*. New York: Plenum Press.

Steel, D. 2006. "Methodological Individualism, Explanation, and Invariance." *Philosophy of the Social Sciences* 36.4: 440–63.

Steinmetz, G. (ed.). 2005. *The Politics of Method in the Human Sciences: Positivism and its Epistemological Others, Politics, History, and Culture*. Durham, NC: Duke University Press.

Thelen, K.A. 2004. *How Institutions Evolve: The Political Economy of Skills in Germany, Britain, the United States, and Japan*. Cambridge Studies in Comparative Politics. Cambridge: Cambridge University Press.

Thomas, D. 1979. *Naturalism and Social Science: A Post-Empiricist Philosophy of Social Science*. Themes in the Social Sciences. New York: Cambridge University Press.

Tilly, C. 1995. "To Explain Political Processes." *American Journal of Sociology* 100.6: 1594–610.

Turner, S.P. and P.A. Roth (eds.). 2003. *The Blackwell Guide to the Philosophy of the Social Sciences*. Malden, MA: Blackwell Publishing.

Udehn, L. 2001. *Methodological Individualism: Background, History and Meaning*. London: Routledge.

Varela, C.R. and R. Harré. 1996. "Conflicting Varieties of Realism: Causal Powers and the Problem of Social Structure." *Journal for the Theory of Social Behaviour* 26.3: 313–25.

Wallerstein, I.M. 1999. *The End of the World as We Know it: Social Science for the Twenty-first Century*. Minneapolis: University of Minnesota Press.

Watkins, J.W.N. 1968. "Methodological Individualism and Social Tendencies." In M. Brodbeck (ed.), *Readings in the Philosophy of the Social Sciences*. New York: Macmillan.

Weber, M. 1949. *The Methodology of the Social Sciences*. New York: Free Press.

Weber, M. and E. Shils (eds.). 1949. *Max Weber on the Methodology of the Social Sciences*. Glencoe, IL: Free Press.

Woodward, J. 2000. "Explanation and Invariance in the Special Sciences." *British Journal of the Philosophy of Science* 51: 197–254.

Woodward, J. 2003. *Making Things Happen: A Theory of Causal Explanation.* Oxford Studies in Philosophy of Science. New York: Oxford University Press.

Zahle, J. 2003. "The Individualism-Holism Debate on Intertheoretic Reduction and the Argument from Multiple Realization." *Philosophy of the Social Sciences* 33.1: 77–99.

Zahle, J. 2007. "Holism and Supervenience." In S.P. Turner and M.W. Risjord (eds.), *Philosophy of Anthropology and Sociology.* Amsterdam: Elsevier.

Zald, M. 1992. "Sociology as a Discipline: Quasi-Science and Quasi Humanities." *American Sociologist* 22: 5–27.

Zald, M. 1995. "Progress and Cumulation in the Human Sciences after the Fall." *Sociological Forum* 10.3: 455–81.

13   Philosophy of Economics

Daniel M. Hausman

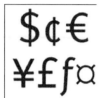 At the frontiers between economics and philosophy lie many intriguing questions concerned with methodology, rationality, ethics, and normative social and political philosophy. These questions are diverse, as are the responses that have been given to them. Although different works in philosophy of economics may be related to one another in many ways, philosophy of economics is not a single unified enterprise. It is a collection of separate inquiries linked to one another by connections among the questions and by the dominating influence of mainstream economic models and techniques. Economics is of particular interest to those interested in epistemology and philosophy of science because of its detailed peculiarities and because it resembles the natural sciences, while its object consists of social phenomena. This essay will address only epistemological and ontological issues concerning economics.

1.   Economics and Philosophy of Economics

Philosophy of economics consists of inquiries concerning: rational choice; the appraisal of economic outcomes, institutions and processes; and the ontology of economic phenomena and the possibilities of acquiring knowledge of them. Although these inquiries overlap in many ways, it is useful to divide philosophy of economics in this way into three subject matters which can be regarded respectively as branches of action theory, ethics (or normative social and political philosophy), and philosophy of science. Economic theories of rationality, welfare, and social choice defend substantive philosophical theses often informed by relevant philosophical literature and of evident interest to those interested in action theory, philosophical psychology, and social and political philosophy. For an overview of economics and ethics with some discussion of rationality, see Hausman and McPherson (2006). Questions concerning rationality and welfare will not be discussed in this essay, which will focus on epistemological and ontological issues concerning economics.

Contemporary economics consists of many schools and many branches. Even so-called "orthodox" or "mainstream" economics has many variants. Most mainstream economics is applied and relies on only rather rudimentary theory, but there is also a good deal of theoretical inquiry. Both theoretical and applied work can be distinguished as microeconomics or macroeconomics.

Microeconomics focuses on relations among individuals (though firms and households often count as honorary individuals and consumer demand is in practice often treated as an aggregate). Individuals have complete and transitive preferences that govern their choices. Consumers prefer more commodities to fewer and have "diminishing marginal rates of substitution" – i.e., they will pay less for units of a commodity when they already have lots of it than when they have little of it. Firms attempt to maximize profits in the face of diminishing returns: holding fixed other inputs, with more units of one input output increases, but at a diminishing rate. Economists idealize and suppose that in competitive markets, firms and individuals cannot influence prices, but economists are also interested in strategic interactions, in which the rational choices of separate individuals are interdependent. Game theory is devoted to the study of strategic interactions. Economists model the outcome of the profit-maximizing activities of firms and the attempts of consumers to best satisfy their preferences as an *equilibrium* in which there is no excess demand on any market. What this means is that anyone who wants to buy anything at the going market price is able to do so. There is no excess demand, and unless a good is free, there is no excess supply.

Macroeconomics grapples with the relations among economic aggregates, focusing especially on problems concerning the business cycle and the influence of monetary and fiscal policy on economic outcomes. Most mainstream economists would like to unify macroeconomics and microeconomics, but fewer are satisfied with the attempts that have been made to do so. Econometrics is a third main branch of economics, devoted to the empirical estimation, elaboration, and to some extent testing of specific microeconomic and macroeconomic models. Among macroeconomists, disagreement is much sharper than among microeconomists or econometricians. In addition to Keynesians and monetarists, "new classical economics" (rational expectations theory) has spawned several approaches, such as so-called "real business cycle" theories (Begg 1982, Minford and Peel 1983, Carter and Maddock 1984, Hoover 1988, Sent 1998). Branches of mainstream economics are also devoted to specific questions concerning growth, finance, employment, agriculture, natural resources, international trade, and so forth. Within orthodox economics, there are also many different approaches, such as *agency theory* (Jensen and Meckling 1976, Fama 1980), the *Chicago School* (Becker 1976), and *public choice theory* (Buchanan 1975, Brennan and Buchanan 1985).

In addition to mainstream economics, there are many other schools, but none is nearly so influential. *Austrian economists* accept orthodox views of choices and constraints, but they emphasize uncertainty and question whether one should regard outcomes as equilibria, and they are skeptical about the value of mathematical modeling (Mises 1949, Rothbard 1957, Dolan 1976, Kirzner 1976, Mises 1978, Buchanan and Vanberg 1979, Mises 1981, Wiseman 1983). Traditional *institutionalist economists* question the value of abstract general theorizing (Veblen 1898, Dugger 1979, Wisman and Rozansky 1991, Hodgson 2000). They emphasize the importance of generalizations concerning norms and behavior within particular institutions. More recent work in economics, which is also called institutionalist, attempts to explain features of institutions by emphasizing the costs of transactions, the inevitable incompleteness of contracts, and the problems "principals" face in monitoring and directing their agents (Williamson 1985, Mäki, Gustafsson, and Knudsen 1993). *Marxian economists* traditionally articulated and developed Karl Marx's economic theories, but recently many Marxian economists have revised traditional Marxian concepts and themes with tools borrowed from orthodox economic theory (Morishima 1973, Roemer 1981, 1982). There are also *socio-economists* (Etzioni 1988), *behavioral economists* (Winter 1962, Ben-Ner and Putterman 1998, Rabin 1998, Camerer 2003, Camerer, Loewenstein, and Rabin 2003), *post-Keynesians* (Kregel 1976, Dow 1985), and *neo-Ricardians* (Sraffa 1960, Roncaglia 1978, Pasinetti 1981). Economics is not one homogeneous enterprise.

2.   Six Central Methodological Problems

Although the different branches and schools of economics raise a wide variety of methodological issues, six problems have been central to methodological reflection concerning economics.

2.1   Positive versus normative economics

Policy makers look to economics to guide policy, and it seems inevitable that even the most esoteric issues in theoretical economics may bear on some people's material interests. The extent to which economics bears on and may be influenced by normative concerns raises methodological questions about the relationships between a *positive* science concerning "facts" and a *normative* inquiry into what ought to be. Most economists and methodologists believe that there is a reasonably clear distinction between facts and values, between what is and what ought to be, and they believe that most of economics should be regarded as a positive science that helps policy makers choose means to accomplish their ends, though it does not bear on the choice of ends itself.

This view is questionable for several reasons (Mongin 2006). First, economists have to interpret and articulate the incomplete specifications of goals and constraints provided by policy makers (Machlup 1969b). Second, economic "science" is a human activity, and like all human activities it is governed by values. Those values need not be the same as the values that influence economic policy, but it is questionable whether the values that govern the activity of economists can be sharply distinguished from the values that govern policy makers. Third, much of economics is built around a normative theory of rationality. One can question whether the values implicit in such theories are sharply distinguishable from the values that govern policies. For example, it may be difficult to hold a maximizing view of individual rationality, while at the same time insisting that social policy should resist maximizing growth, wealth, or welfare in the name of freedom, rights, or equality. Fourth, people's views of what is right and wrong are, as a matter of fact, influenced by their beliefs about how people in fact behave. There is evidence that studying theories that depict individuals as self-interested leads people to regard self-interested behavior more favorably and to become more self-interested (Marwell and Ames 1981, Frank, Gilovich, and Regan 1993). Finally, people's judgments are clouded by their interests, which are influenced by economic policies and theories (Marx 1867, preface).

2.2   Reasons versus causes

Orthodox theoretical microeconomics is as much a theory of rational choices as it is a theory that explains and predicts economic outcomes. Since virtually all economic theories that discuss individual choices take individuals as acting for reasons, and thus in some way rational, questions about the role that views of rationality and reasons should play in economics are of general importance. Economists are typically concerned with the aggregate results of individual choices rather than with particular individuals, but their theories in fact offer both causal explanations for why individuals choose as they do and accounts of the *reasons* for their choices.

Explanations in terms of reasons have several features that distinguish them from explanations in terms of causes. Reasons justify the actions they explain. Reasons can be evaluated, and they are responsive to criticism. Reasons, unlike causes, must be intelligible to those for whom they are reasons. On grounds such as these, some philosophers have questioned whether explanations of human action can be causal explanations (Winch 1958, von Wright 1971). Yet merely giving a reason – even an extremely good reason – fails to explain an agent's action, if the reason was not in fact "effective." A young man might, for example, start attending church regularly and give as his reason a concern with salvation. But others might suspect that this agent is deceiving himself and that the minister's attractive daughter is in fact responsible for his renewed interest in religion. Donald Davidson (1963) argued that what distinguishes the reasons that explain an action from the reasons that fail to explain it are that the former are also causes of the action. Although the account of rationality within economics differs in some ways from the "folk psychology" people tacitly invoke in everyday explanations of actions, many of the same questions carry over (Rosenberg 1976, ch. 5, 1980).

An additional difference between explanations in terms of reasons and explanations in terms of causes, which some economists have emphasized, is that the beliefs and preferences that explain actions may depend on mistakes and ignorance (Knight 1935). As a first approximation, economists can abstract from such difficulties. They thus often assume that people have perfect information about all the relevant facts. In that way theorists need not worry about what people's beliefs are. By assumption people believe and expect whatever the facts are. But once one goes beyond this first approximation, difficulties arise which have no parallel in the natural sciences. Choice depends on how things look "from the inside," which may be very different from the actual state of affairs. Consider, for example, the stock market. The "true" value of a stock depends on the future profits of the company, which are, of course, uncertain. In 1999 and 2000, stock prices of "dot-com" companies were far above any plausible estimate of their true value. But what matters, at least in the short run, is what people believe.

No matter how overpriced the shares might have been, they were excellent investments if tomorrow or next month somebody would be willing to pay even more for them. Economists disagree about how significant this subjectivity is. Members of the Austrian school argue that these differences are of great importance and sharply distinguish theorizing about economics from theorizing about any of the natural sciences (Buchanan and Vanberg 1979, Mises 1981).

2.3 Social scientific naturalism

Of all the social sciences, economics most closely resembles the natural sciences. Economic theories have been axiomatized, and articles and books of economics are full of theorems. Of all the social sciences, only economics boasts a Nobel Prize. Economics is thus a test case for those concerned with the comparison between the natural and social sciences. Those who have wondered whether social sciences must differ fundamentally from the natural sciences seem to have been concerned mainly with three questions:

1. Are there fundamental differences between the structure or concepts of theories and explanations in the natural and social sciences? Some of these issues have already been mentioned in §2.2.
2. Are there fundamental differences in goals? Philosophers and economists have argued that in addition to or instead of the predictive and explanatory goals of the natural sciences, the social sciences should aim at providing us with *understanding*. Weber and others (Weber 1904, Knight 1935, Machlup 1969a) have argued that the social sciences should provide us with an understanding "from the inside," that we should be able to empathize with the reactions of the agents and to find what happens "understandable." This (and the closely related recognition that explanations cite reasons rather than only causes) seems to introduce an element of subjectivity into the social sciences that is not found in the natural sciences.
3. Owing to the importance of human choices (or perhaps free will), are social phenomena too "irregular" to be captured within a framework of laws and theories? Given human free will, perhaps human behavior is intrinsically unpredictable and not subject to any laws. But there are, in fact, many regularities in human action, and given the enormous causal complexity characterizing some natural systems, the natural sciences must cope with many irregularities, too.

2.4   *Abstraction, idealization, and* ceteris paribus *clauses in economics*

Economics raises questions concerning the legitimacy of severe abstraction and idealization. For example, mainstream economic models often stipulate that everyone is perfectly rational and has perfect information, or that commodities are infinitely divisible. Such claims are exaggerations, and they are clearly false. Other schools of economics may not employ idealizations that are this extreme, but there is no way to do economics if one is not willing to simplify drastically and abstract from many complications. How much simplification, idealization, and abstraction is legitimate?

In addition, because economists attempt to study economic phenomena as constituting a separate domain, influenced only by a small number of causal factors, the claims of economics are true only *ceteris paribus* – that is, they are true only if there are no interferences or disturbing causes. What are *ceteris paribus* clauses, and when if ever are they legitimate in science? Questions concerning *ceteris paribus* clauses are closely related to questions concerning simplifications and idealizations, since one way to simplify is to suppose that the various disturbing causes or interferences are inactive and to explore the consequences of some small number of causal factors. These issues and the related question of how well supported economics is by the evidence have been *the* central questions in economic methodology. They will be discussed further below in §3 and elsewhere.

2.5   *Causation in economics and econometrics*

Many important generalizations in economics are causal claims. For example, the law of demand asserts that a price increase will (*ceteris paribus*) diminish the quantity demanded. Econometricians have also been deeply concerned with the possibilities of determining causal relations from statistical evidence and with the relevance of causal relations to the possibility of consistent estimation of parameter values. Since concerns about the consequences of alternative policies are so central to economics, causal inquiry is unavoidable.

Before the 1930s, economists were generally willing to use causal language explicitly and literally, despite some concerns that there might be a conflict between causal analysis of economic changes and "comparative statics" treatments of equilibrium states. Some economists were also worried that thinking in terms of causes was not compatible with recognizing the multiplicity and mutuality of determination in economic equilibrium. In the anti-metaphysical intellectual environment of the 1930s and 1940s (of which logical positivism was at least symptomatic), any mention of causation became highly suspicious, and economists commonly pretended to avoid causal concepts. The consequence was that they ceased to reflect carefully on the causal concepts that they continued implicitly to invoke (Hausman 1983, Helm 1984, Hausman 1990, Runde 1998). For example, rather than formulating the law of demand in terms of the causal consequences of price changes for quantity demanded, economists tried to confine themselves to discussing the mathematical function relating price and quantity demanded. There were important exceptions (Haavelmo 1944, Simon 1953, Wold 1954), and during the past generation, this state of affairs has changed dramatically.

For example, in his *Causality in Macroeconomics* (2001), Kevin Hoover develops feasible methods for investigating large-scale causal questions, such as whether changes in the money supply (M) cause changes in the relate of inflation P or accommodate changes in P that are otherwise caused. If changes in M cause changes in P, then the conditional distribution of P on M should remain stable with exogenous changes in M, but should change with exogenous changes in P. Hoover argues that historical investigation, backed by statistical inquiry, can justify the conclusion that some particular changes in M or P have been exogenous. One can then determine the causal direction by examining the stability of the conditional distributions. Econometricians have made vital contributions to the contemporary revival of philosophical interest in the notion of causation. In addition to Hoover's work, see, for example, Geweke (1982), Granger (1969, 1980), Cartwright (1989), Sims (1977), and Zellner and Aigner (1988).

2.6 Structure and strategy of economics

In the wake of the work of Kuhn (1970) and Lakatos (1970), philosophers are more interested in the larger theoretical structures that unify and guide research within particular research traditions. Since many theoretical approaches in economics are systematically unified, they pose questions about what guides research, and many economists have applied the work of Kuhn or Lakatos to shed light on the overall structure of economics

(Coats 1969, Bronfenbrenner 1971, Kunin and Weaver 1971, Worland 1972, Stanfield 1974, Blaug 1976, Latsis 1976, Baumberger 1977, Dillard 1978, Hutchison 1978, Jalladeau 1978, Hands 1985b, Weintraub 1985, Blaug and de Marchi 1991, Hausman 1992b, ch. 6). Whether these applications have been successful is controversial, but the comparison of the structure of economics to Kuhn's and Lakatos' schema has highlighted distinctive features of economics. For example, asking what the "positive heuristic" of mainstream economics consists in permits one to see that mainstream models typically attempt to demonstrate that an economic equilibrium will obtain, and thus that mainstream models are unified in more than just their common assumptions. Since the success of research projects in economics is controversial, understanding their global structure and strategy may clarify their drawbacks as well as their advantages.

3.   Inexactness, *Ceteris Paribus* Clauses, and "Unrealistic Assumptions"

As mentioned in the previous section, the most important methodological issue concerning economics involves the simplification, idealization, and abstraction that characterize economic theory and resulting doubts about whether economics is well supported. If claims such as, "Agents prefer larger commodity bundles to smaller commodity bundles" are interpreted as universal generalizations, they are false. Can a science rest on false generalizations such as these? If these claims are not universal generalizations, then what is their logical form? And how can such claims be tested and confirmed or disconfirmed? These problems have bedeviled economists and economic methodologists from the first methodological reflections to the present day.

3.1   *Classical economics and the method* a priori

The first extended reflections on economic methodology appear in the work of Nassau Senior (1836) and John Stuart Mill (1836). Their essays must be understood against the background of the prevailing economic theory. Like Adam Smith's economics (1776), to which it owed a great deal, and modern economics, the "classical" economics of the middle decades of the nineteenth century traced economic regularities to the choices of individuals facing social and natural constraints. But, as compared to Smith, more reliance was placed on severely simplified models. In David Ricardo's *Principles of Political Economy* (1817), a portrait is drawn in which wages above the subsistence level lead to increases in the population, which in turn require more intensive agriculture or cultivation of inferior land. The extension of cultivation leads to lower profits and higher rents; and the whole tale of economic development leads to a gloomy stationary state in which profits are too low to command any net investment, population growth drives wages down to subsistence levels, and only the landlords are affluent.

Fortunately for the world, but unfortunately for classical economic theory, the data consistently contradicted these gloomy predictions (de Marchi 1970). Yet the theory continued to hold sway for more than half a century, and the consistently unfavorable data were explained away as due to various "disturbing causes." It is consequently not surprising that Senior's and Mill's accounts of the method of economics emphasize the relative autonomy of theory.

Mill distinguishes between two inductive methods. The method *a posteriori* is a method of direct experience. In his view, it is only suitable for phenomena in which few causal factors are operating or in which experimental controls are possible. Mill's famous methods of induction (1843, Book III) provide an articulation of the method *a posteriori*. In his method of difference, for example, one holds fixed every causal factor except one and checks to see whether the effect ceases to obtain when that one factor is removed.

Mill maintains that direct inductive methods cannot be used to study phenomena in which many causal factors are in play. For example, if one attempts to investigate whether tariffs enhance prosperity by comparing nations with high tariffs and nations without high tariffs, the results will be worthless because the prosperity of the countries studied depends on so many other causal factors. So, Mill argues, one needs instead to employ the method *a priori*. Despite its name, Mill emphasizes that this too is an inductive method. The difference between the method *a priori* and the method *a posteriori* is that the method *a priori* is an *indirect* inductive method. One first determines the laws governing individual causal factors in domains in which Mill's methods of induction are applicable. Having then determined the laws of the individual causes, one investigates their combined consequences deductively. Finally, there is a role for "verification" of the combined consequences, but owing to the causal complications, this testing has comparatively little weight. The testing of the conclusions serves only as a check on one's deductions and as an indicator of whether there are significant disturbing causes that one has not yet accounted for.

As an example of the method *a priori*, Mill discusses the science of the tides. Newton determined the law of gravitation by studying planetary motion, in which gravity is the only significant causal factor. Then he and others developed the theory of tides deductively from that law and information concerning the positions and motions of the moon and sun. The implications of the theory will be inexact and sometimes badly mistaken, because many subsidiary causal factors influence tides. By testing the theory one can uncover mistakes in one's deductions and evidence concerning the role of the subsidiary factors. Because of the causal complexity, such testing does little to confirm or disconfirm the law of gravitation, which has already been established. Although Mill does not often use the language of "*ceteris paribus*," his view that the principles or "laws" of economics hold in the absence of "interferences" or "disturbing causes" provides an account of how the principles of economics can be true *ceteris paribus* (Hausman 1992b, ch. 8, 12).

Because economic theory includes only the most important causes, its claims, like claims concerning tides, are inexact. Its predictions will be imprecise and sometimes completely wrong. Mill maintains that it is nevertheless possible to develop economic theory by studying in simpler domains the laws governing the major causal factors and then deducing their consequences in more complicated circumstances. For example, the statistical data are ambiguous concerning the relationship between minimum wages and unemployment of unskilled workers; and since the minimum wage has never been extremely high, there are no data about what unemployment would be in those circumstances. On the other hand, everyday experience teaches one that firms can choose among more or less labor-intensive processes and that a high minimum wage will make more labor-intensive processes more expensive. On the assumption that firms try to keep their costs down, one has good reason to believe that a high minimum wage will increase unemployment.

In defending a view of economics as in this way inexact and employing the method *a priori*, Mill was able to reconcile his empiricism and his commitment to Ricardo's economics. Although Mill's views on economic methodology were questioned later in the nineteenth century by some economists who believed that the theory was too remote from the contingencies of policy and history (Roscher 1874, Schmoller 1888, 1898), Mill's methodological views dominated the mainstream of economic theory for well over a century (for example, Cairnes 1875). Mill's vision survived the so-called neoclassical revolution in economics beginning in the 1870s, and it is clearly discernible in the most important methodological treatises concerning neoclassical economics, such as John Neville Keynes' *The Scope and Method of Political Economy* (1891) or Lionel Robbins' *An Essay on the Nature and Significance of Economic Science* (1932). Hausman (1992b) argues that current methodological practice closely resembles Mill's methodology, despite the fact that few economists would explicitly defend it.

3.2 Milton Friedman and the defense of "unrealistic assumptions"

Although some contemporary philosophers have argued that Mill's method *a priori* is defensible (Bhaskar 1978, Cartwright 1989, Hausman 1992b), by the middle of the twentieth century Mill's views appeared to many economists out of step with contemporary philosophy of science. Without studying Mill's text carefully, it was easy for economists to misunderstand his terminology and to believe that his method *a priori* is opposed to empiricism. Others took seriously Mill's view that the basic principles of economics should be empirically established and found evidence to cast doubt on some of the basic principles, particularly the view that firms attempt to maximize profits (Hall and Hitch 1939, Lester 1946, 1947). Methodologists who were well informed about contemporary developments in philosophy of science, such as Terence Hutchison (1938), denounced "pure theory" in economics as unscientific.

Philosophically reflective economists proposed several ways to replace the old-fashioned Millian view with a more up-to-date methodology that would continue to justify current practice (see particularly Machlup 1955, 1960 and Koopmans 1957). By far the most influential of these was Milton Friedman's contribution in his 1953 essay, "The Methodology of Positive Economics." This essay has had an enormous influence, far more than any other work on methodology.

Friedman begins his essay by distinguishing between positive and normative economics and conjecturing that policy disputes turn largely on disagreements about the consequences of alternatives and thus are capable of being resolved by progress in positive economics. Turning to positive economics, Friedman asserts (without argument) that the ultimate goal of *all* positive sciences is correct prediction concerning phenomena not yet observed. He holds a practical view of science and looks to science for predictions that will guide policy.

Since experimentation is often impractical and since the uncontrolled phenomena economists observe are difficult to interpret (owing to the same causal complexity that bothered Mill), it is hard to judge whether a particular theory is a good basis for predictions or not. Consequently, Friedman argues, economists have supposed that they could test theories by the "realism" of their "assumptions" rather than by the accuracy of their predictions. Friedman argues at length that this is a grave mistake. The realism of a theory's assumptions is, he maintains, irrelevant to its predictive value. Theories should be appraised exclusively in terms of the accuracy of their predictions. It does not matter whether the assumption that firms maximize profits is realistic. What matters is whether the theory of the firm makes correct and significant predictions.

As critics have pointed out (and almost all commentators have been critical), Friedman refers to several different things as "assumptions" of a theory and means several different things by speaking of assumptions as "unrealistic" (Brunner 1969). Since Friedman aims his criticism at those who investigate empirically whether firms in fact attempt to maximize profits, he must take "assumptions" to include central explanatory generalizations, such as "Firms attempt to maximize profits," and by "unrealistic," he must mean, among other things, "false." In arguing that it is a mistake to appraise theories in terms of the realism of assumptions, Friedman is arguing at least that it is a mistake to appraise theories by investigating whether their central explanatory generalizations are true or false.

It would seem that this interpretation makes Friedman's views inconsistent, because in testing whether the assumption that firms attempt to maximize profits is realistic, one is checking whether predictions of theory concerning the behavior of firms are true or false. An "assumption" such as "firms maximize profits" is itself a prediction. But there is a further wrinkle. Friedman is not concerned with every prediction of economic theories. In Friedman's view, "theory is to be judged by its predictive power *for the class of phenomena which it is intended to explain*" (1953, 8; emphasis added). Economists are interested in only some of the implications of economic theories. Other predictions, such as those concerning the results of surveys, are irrelevant to policy. What matters is whether economic theories are successful at predicting the phenomena that economists are interested in. In other words, Friedman believes that economic theories should be appraised in terms of their predictions concerning prices and quantities exchanged on markets. In his view, what matters is this "narrow predictive success" (Hausman 2008a), not overall predictive adequacy.

So Friedman permits economists to ignore the disquieting findings of surveys. They can ignore the fact that people do not always prefer larger bundles of commodities to smaller bundles of commodities. They need not be troubled that some of their models suppose that all agents know the prices of all present and future commodities in all markets. All that matters is whether their predictions concerning market phenomena turn out to be correct or not. And since anomalous market outcomes could be due to any number of uncontrolled causal factors, while experiments are difficult to carry out, it turns out that economists need not worry about ever encountering evidence that would disconfirm fundamental theory. Detailed models may be disconfirmed, but fundamental theory is safe. In this way one can understand how Friedman's methodology, which appears to justify the pragmatic view that economists should use any model that appears to "work" regardless of how unreasonable its assumptions might appear, has been put in service of theoretical orthodoxy. For other discussions of Friedman's essay, see Bear and Orr (1967), Boland (1979), Hammond (1992), Hirsch and de Marchi (1990), Mäki (1992), Melitz (1965), Rotwein (1959), and Samuelson (1963).

Over the last two decades there has been a surge of experimentation in economics, and Friedman's methodological views do not command the same near unanimity that they used to. But they remain enormously influential, and they still serve as a way of avoiding awkward questions concerning simplifications, idealizations, and abstraction in economics, rather than providing a response to them.

4. Contemporary Directions in Economic Methodology

The past half-century has witnessed the emergence of a large literature devoted to economic methodology. That literature explores many methodo-logical approaches and applies its conclusions to many schools and branches of economics. Much of the literature focuses on the fundamental theory of mainstream economics – the theory of the equilibria resulting from constrained rational individual choice. Since 1985, there has been a journal, *Economics and Philosophy*, devoted specifically to philosophy of economics, and since 1994 there has also been the *Journal of Economic Methodology*. This section will sample some of the methodological work that has been done during the past two decades.

4.1   Popperian approaches

Karl Popper's philosophy of science has been influential among economists, as among other scientists. Popper defends what he calls a falsificationist methodology (1968, 1969). Scientists should formulate theories that are "logically falsifiable" – that is, inconsistent with some possible observation reports. "All crows are black" is logically falsifiable, since it is inconsistent with (and would be falsified by) an observation report of a red crow. Second, Popper maintains that scientists should subject theories to harsh tests and should reject them when they fail the tests. Third, scientists should regard theories as conjectures. Passing a test does not confirm a theory or provide one with reason to believe it. It only justifies continuing to employ it (since it has not yet been falsified) and devoting increased efforts to attempting to falsify it (since it has thus far survived testing). Popper has also written in defense of what he calls "situational logic" (which is basically rational choice theory) as the correct method for the social sciences (1967, 1976). There appear to be serious tensions between Popper's falsificationism and his defense of situational logic, and his discussion of situational logic has not been as influential as his falsificationism. For discussion of how situational logic applies to economics, see Hands (1985a).

Given Popper's falsificationism, there seems little hope of understanding how extreme simplifications can be legitimate or how current economic practice could be scientifically reputable. Even if economic theories were logically falsifiable, the widespread acceptance of Friedman's methodological views insures that they are not subjected to serious test. When they apparently fail tests, they are rarely repudiated. Economic theories, which have not been well tested, are taken to be well established guides to policy, rather than merely conjectures. Some critics of neoclassical economics have made these criticisms (Eichner 1983). But most of those who have espoused Popper's philosophy of science have not repudiated mainstream economics.

Mark Blaug (1992) and Terence Hutchison (1938, 1977, 1978, 2000), who are the most prominent Popperian methodologists, criticize particular features of economics, and they both call for more testing and a more critical attitude. For example, Blaug (1980, ch. 14) praises Gary Becker (1976) for his refusal to explain differences in choices by differences in preferences, but criticizes him for failing to go on and test his theories severely. However, both Blaug and Hutchison understate the radicalism of Popper's views and take his message to be merely that scientists should be critical and concerned to test their theories.

Blaug's and Hutchison's criticisms have sometimes been challenged on the grounds that economic theories cannot be tested, because of their *ceteris paribus* clauses and the many subsidiary assumptions required to derive testable implications (Caldwell 1984). But this response ignores Popper's insistence that testing requires methodological *decisions* not to attribute failures of predictions to "interferences." For views of Popper's philosophy and its applicability to economics, see de Marchi (1988), Caldwell (1991), and Boland (1982, 1989, 1992).

Applying Popper's views on falsification literally would be destructive. Not only neoclassical economics, but all known economic theories would be condemned as unscientific, and there would be no way to discriminate among economic theories. One major problem is that one cannot derive testable implications from theories by themselves. To derive testable implications, one also needs subsidiary assumptions or hypotheses concerning distributions, measurement devices, proxies for unmeasured variables, the absence of various interferences, and so forth. This is the so-called "Duhem-Quine thesis" (Duhem 1906, Quine 1953, Cross 1982). These problems arise generally, and Popper proposes that they be solved by a methodological decision to regard a failure of the deduced testable implication to be a failure of the theory. But in economics the subsidiary assumptions are dubious and in many cases known to be false. Making the methodological decision that Popper requires is unreasonable and would lead one to reject all economic theories.

Imre Lakatos (1970), who was for most of his philosophical career a follower of Popper, offers a broadly Popperian solution to this problem. Lakatos insists that testing is always comparative. When theories face empirical difficulties, as they always do, one attempts to modify them. Scientifically acceptable (in Lakatos' terminology, "theoretically progressive") modifications must always have some additional testable implications and are thus not purely *ad hoc*. If some of the new predictions are confirmed, then the modification is "empirically progressive," and one has reason to reject the unmodified theory and to employ the new theory, regardless of how unsuccessful in general either theory may be. Though progress may be hard to come by, Lakatos' views do not have the same destructive implications as Popper's. Lakatos appears to solve the problem of how to appraise mainstream economic theory by arguing that what matters is empirical progress rather than empirical success. Lakatos' views have thus been more attractive to economic methodologists than Popper's.

Developing Thomas Kuhn's notion of a "paradigm" (Kuhn 1970) and some hints from Popper, Lakatos also developed a view of the global structure of what he called "scientific research programs." Lakatos emphasized that there is a "hard core" of basic theoretical propositions that define a research program and are not to be questioned within the research program. In addition members of a research program accept a common body of heuristics that guide them in the articulation and modification of specific theories. These views were also attractive to economic methodologists, since theory development in economics is so sharply constrained by prior commitments and specific heuristics. The fact that economists do not give up basic theoretical postulates that appear to be false might be justified by regarding them as part of the "hard core" of the neoclassical research program.

Yet Lakatos' views do not provide a satisfactory account of how economics can be a reputable science, despite its reliance on extreme simplifications. For it is questionable whether the development of neoclassical economic theory has demonstrated empirical progress. For example, the replacement of "cardinal" utility theory by "ordinal" utility theory in the 1930s, which is generally regarded as a major step forward, involved the replacement of one theory by another that was strictly weaker and which had no additional empirical content. Furthermore, despite his emphasis on heuristics as guiding theory modification, Lakatos still emphasizes testing. Science is for Lakatos more empirically driven than is contemporary economics (Hands 1992). It is also doubtful whether research enterprises in economics have "hard cores" (Hausman 1992b, ch. 6). For attempts to apply Lakatos' views to economics, see Latsis (1976), and Weintraub (1985). As is already apparent in Blaug and de Marchi (1991), writers on economic methodology have in recent years become increasingly disenchanted with Lakatos' philosophy.

There is a second major problem with Popper's philosophy of science, which plagues Lakatos' views as well. Both maintain that there is no such thing as empirical confirmation (for some late qualms of Lakatos, see Lakatos

1974). Popper and Lakatos maintain that evidence never provides reason to believe that scientific claims are true, and both deny that results of tests can justify relying on statements in practical endeavors or in theoretical inquiry. There is no better evidence for one unfalsified proposition than for another. Someone who questions whether there is enough evidence for some proposition to justify relying on it would be making the methodological "error" of supposing that there can be evidence in support of hypotheses. With the notable exception of Watkins (1984), few philosophers within the Popperian tradition have faced up to this implausible consequence.

4.2 The rhetoric of economics

One radical reaction to the difficulties of justifying the reliance on severe simplifications is to deny that economics passes methodological muster. Alexander Rosenberg (1992) maintains that economics can only make imprecise generic predictions, and it cannot make progress, because it is built around folk psychology, which is an inadequate theory of human behavior and which (owing to the irreducibility of intentional notions) cannot be improved. Complex economic theories are valuable only as applied mathematics, not as empirical theory. Since economics does not show the same consistent progress as the natural sciences, one cannot dismiss Rosenberg's suggestion that economics is an empirical dead end. But his view that it has made *no* progress and that it does not permit quantitative predictions is hard to accept. For example, contemporary economists are much better at pricing stock options than economists were even a generation ago.

An equally radical but opposite reaction is that of Deirdre McCloskey (1985, 1994), who denies that there are any non-trivial methodological standards that economics must meet. In her view, the only relevant criteria for assessing the practices and products of a discipline are those accepted by the practitioners. Apart from a few general standards such as honesty and a willingness to listen to criticisms, the only justifiable criteria for any conversation are those of the participants. Economists can thus dismiss the arrogant pretensions of philosophers to judge economic discourse. Whatever a group of economists takes to be good economics is good economics. Philosophical standards of empirical success are just so much hot air. Those who are interested in understanding the character of economics and in contributing to its improvement should eschew methodology and study instead the "rhetoric" of economics – that is, the means of argument and persuasion that succeed among economists.

McCloskey's studies of the rhetoric of economics have been valuable and influential (1985, esp. chs. 5–7), but much of her work consists not of such studies but of philosophical critiques of economic methodology. These are more problematic, because the position sketched in the previous paragraph is hard to defend and potentially self-defeating. It is hard to defend, because epistemological standards for good science have already infected the conversation of economists. The standards of predictive success which lead one to have qualms about economics are already standards that many economists accept. The only way to escape these doubts is to surrender the standards that gave rise to them. But McCloskey's position undermines any principled argument for a change in standards. Furthermore, as Rosenberg (1988) has argued, it seems that economists would doom themselves to irrelevance if they were to surrender standards of predictive success, for it is upon such standards that policy decisions are made.

McCloskey (1985, ch. 9) does not, in fact, want to insulate economists from all criticisms. For she herself criticizes the bad habit some economists have of conflating statistical significance with economic importance. Sometimes McCloskey (1985, ch. 2) characterizes rhetoric descriptively as the study of what in fact persuades, but sometimes she characterizes it normatively as the study of what ought to persuade. And if rhetoric is the study of what ought to persuade, then rhetoric is not an alternative to methodology: it is methodology. Questions about whether economics is a successful empirical science cannot be conjured away.

4.3   "Realism" in economic methodology

Economic methodologists have paid little attention to debates within philosophy of science between realists and anti-realists (van Fraassen 1980, Boyd 1984), because economic theories rarely postulate the existence of unobservable entities or properties, apart from variants of "everyday unobservables," such as beliefs and desires. Methodologists have, on the other hand, vigorously debated the goals of economics, but those who argue that the ultimate goals are predictive (such as Friedman) do so because of their interest in policy, not out of interest in epistemological and semantic puzzles concerning references to unobservables.

Nevertheless, there are two important recent realist programs in economic methodology. The first, developed mainly by Uskali Mäki, is devoted to exploring the varieties of realism implicit in the methodological statements and theoretical enterprises of economists (see Mäki 1990a, 1990b, 1992). The second, which is espoused by Tony Lawson and his co-workers, mainly at Cambridge University, derives from the work of Roy Bhaskar (1978) (see Lawson 1997 and Fleetwood 1999). In Lawson's view, one can trace many of the inadequacies of mainstream economics (of which he is a critic) to an insufficient concern with ontology. In attempting to identify regularities on the surface of the phenomena, mainstream economists are doomed to failure. Economic phenomena are, in fact, influenced by a large number of different causal factors, and one can achieve scientific knowledge only of the underlying mechanisms and tendencies, whose operation can be glimpsed obscurely in observable relations. Mäki's and Lawson's programs have little to do with one another, though Mäki (like Mill, Cartwright, and Hausman) shares Lawson's and Bhaskar's concern with underlying causal mechanisms.

4.4   *Economic methodology and social studies of science*

Throughout its history, economics has been the subject of sociological as well as methodological scrutiny. Many sociological discussions of economics, like Marx's critique of classical political economy, have been concerned to identify ideological distortions and thereby to criticize particular aspects of economic theory and economic policy. Since every political program finds economists who testify to its economic virtues, there is a never-ending source of material for such critiques.

The influence of contemporary sociology of science and social studies of science, coupled with the difficulties methodologists have had making sense of and rationalizing the conduct of economics, has led to a sociological turn within methodological reflection itself. Rather than showing that there is good evidence supporting developments in economic theory or that those developments have other broadly epistemic virtues, methodologists and historians such as D. Wade Hands (Hands and Mirowski 1998, Hands 2001), Philip Mirowski (2002), and E. Roy Weintraub (1991) have argued that these changes reflect a wide variety of non-rational factors, from changes in funding for theoretical economics, to political commitments, personal rivalries, attachments to metaphors, or mathematical interests.

Furthermore, many of the same methodologists and historians have argued that economics is not only an object of social inquiry, but also a tool of social inquiry. By studying the incentive structure of scientific disciplines and the market forces impinging on research (including of course research in economics), it should be possible to write the economics of science and the economics of economics itself (Hull 1988, Hands 1995, Leonard 2002).

Exactly how, if at all, this work is supposed to bear on questions concerning how well supported are the claims economists make is not clear. Though eschewing traditional methodology, Mirowski's monograph (1990) on the role of physical analogy in economics is often very critical of mainstream economics. In *Reflection without Rules* (2001), Hands maintains that general methodological rules are of little use. He defends a naturalistic view of methodology and is skeptical of prescriptions that are not based on detailed knowledge. But he does not argue that no rules apply.

4.5   *Detailed contemporary studies*

The above survey of approaches to the fundamental problems of appraising economic theory is far from complete. For example, there have been substantial efforts to apply structuralist views of scientific theories (Sneed 1971, Stegmüller 1976, 1979) to economics (Stegmüller, Balzer, and Spohn 1982, Hamminga 1983, Hands 1985c, Balzer and Hamminga 1989). The above discussion does at least document some of the disagreements concerning how to interpret and appraise economic theories. It is not surprising that there is no consensus among those writing on economic methodology concerning the overall empirical appraisal of specific approaches in economics, including mainstream microeconomics, macroeconomics, and econometrics. When practitioners cannot agree, it is questionable whether those who know more philosophy but less economics will be able to settle the matter. Since the debates continue, those who reflect on economic methodology should have a continuing part to play.

Meanwhile, there are many other more specific methodological questions to address, and it is a sign of the maturity of the sub-discipline that an increasing percentage of work on economic methodology addresses more specific questions. There is a plethora of work, as a perusal of any recent issue of the *Journal of Economic Methodology* or *Economics and Philosophy* will confirm. Some of the range of issues currently under discussion were mentioned above in §2. Other areas of current interest include:

1. *Feminist economics.* Although more concerned with the content of economics than with its methodology, the recent explosion of work on feminist economics is shot through with methodological (and sociological) self-reflection. The fact that a larger percentage of economists are men than is true of any of the other social sciences and indeed than several of the natural sciences raises methodological questions about whether there is something particularly masculine about the discipline (Nelson 1996). Since 1995, there has been a journal, *Feminist Economics*, which pulls together much of this work.

2. *Economic models.* A century ago economists talked of their work in terms of "principles," "laws," and "theories." Nowadays the standard intellectual tool is a "model." Is this just a change in terminological fashion, or does the concern with models signal a methodological shift? What are models? These questions have been discussed by Cartwright (1989, 1999), Hausman (1992b), Mäki (1991), Morgan (2001), Morgan and Morrison (1999), Rappaport (1998), and Sugden (2000).

3. *Experimental economics.* During the past generation, experimental work in economics has expanded rapidly. This work has many different objectives (see Roth 1988) and apparently holds out the prospect of bridging the gulf between economic theory and empirical evidence. Some of it casts light on the way in which methodological commitments influence the extent to which economists heed empirical evidence. For example, in the case of so-called preference reversals (where people claim to prefer A to B but are willing to pay more for B), economists devoted considerable attention to the experimental findings and conceded that they disconfirmed central principles of economics. But economists were generally unwilling to pay serious attention to the theories proposed by psychologists that predicted the phenomena before they were observed. The reason seems to be that these psychological theories do not have the same wide scope as the basic principles of mainstream economics (Hausman 1992b, ch. 13). The methodological commitments governing theoretical economics are much more complex and much more specific to economics than the general rules proposed by philosophers such as Popper and Lakatos.

The relevance of experimentation, however, remains controversial. There are many questions about whether experimental findings can be generalized to non-experimental contexts and, more generally, concerning the possibilities of learning from experiments. See Camerer (2003), Guala (2005),

Hey (1991), Kagel and Roth (1995), Plott (1991), V.L. Smith (1991), and Starmer (1999).

5. Conclusion

There is an enormous amount of activity at the frontiers between economics and philosophy concerned with methodology, rationality, ethics, and normative social and political philosophy. This work is diverse and addresses very different questions. Although many of these are related, philosophy of economics is not a single unified enterprise. It is a collection of separate inquiries linked to one another by connections among the questions and by the dominating influence of mainstream economic models and techniques. The methodological – that is, epistemological and ontological – questions surveyed here should be seen in the context of the wider range of relations between economics and philosophy.

Since economics has the subject matter of the social sciences and much of the structure of the natural sciences, it is a terrific object of inquiry for those interested in the relationship between the natural and social sciences. Furthermore, since its validity is contested, it provides an excellent case study for those concerned about the validity and reliability of methods that have apparently worked so well in the natural sciences. It is an unsettled area with more questions than answers. But could a philosopher ask for anything more enticing?

References

The following bibliography contains only works cited in this essay, which derives from my "Philosophy of Economics" entry in the *Stanford Encyclopedia of Philosophy*. Other overviews of the subject can be found in the *Journal of Economic Methodology*, vol. 8.1, March 2001 Millennium symposium on "The Past, Present and Future of Economic Methodology," in *The Handbook of Economic Methodology*, edited by Davis, Hands, and Mäki, and in *The Philosophy of Economics: An Anthology*, edited by Hausman. For a comprehensive bibliography of works on economic methodology up to 1988, see Redman 1989. The third edition of Hausman's *The Philosophy of Economics: An Anthology* contains an up-to-date and reasonably comprehensive bibliography of books on economic methodology. Essays from economics journals are indexed in the *Journal of Economic Literature*, and the *Index of Economic Articles in Journal and Collective Volumes* also indexes collections. Since 1991, works on methodology can be found under the number B4. Works on ethics and economics can be found under the numbers A13, D6, and I3. Discussions of rationality and game theory can be found under A1, C7, D00, D7, D8, and D9.

Balzer, W. and B. Hamminga (eds.). 1989. *Philosophy of Economics*. Dordrecht: Kluwer-Nijhoff.

Baumberger, J. 1977. "No Kuhnian Revolutions in Economics." *Journal of Economic Issues* 11: 1–20.

Bear, D. and D. Orr. 1967. "Logic and Expediency in Economic Theorizing." *Journal of Political Economy* 75: 188–96.

Becker, G. 1976. *The Economic Approach to Human Behavior*. Chicago: University of Chicago Press.

Begg, D. 1982. *The Rational Expectations Revolution in Macroeconomics: Theories and Evidence*. Baltimore: Johns Hopkins University Press.

Ben-Ner, A. and L. Putterman (eds.). 1998. *Economics, Values and Organization*. Cambridge: Cambridge University Press.

Bhaskar, R. 1978. *A Realist Theory of Science*. Brighton: Harvester Press.

Blaug, M. 1976. "Kuhn versus Lakatos *or* Paradigms versus Research Programmes in the History of Economics." In S. Latsis (ed.), *Method and Appraisal in Economics*. Cambridge: Cambridge University Press, 149–80.

Blaug, M. 1980. *The Methodology of Economics: Or How Economists Explain*. Cambridge: Cambridge University Press. 2nd edn, 1992.

Blaug, M. 1992. *The Methodology of Economics: Or How Economists Explain*. 2nd edn. Cambridge: Cambridge University Press.

Blaug, M. and N. de Marchi (eds.). 1991. *Appraising Modern Economics: Studies in the Methodology of Scientific Research Programs*. Aldershot: Edward Elgar.

Boland, L. 1979. "A Critique of Friedman's Critics." *Journal of Economic Literature* 17: 503–22.

Boland, L. 1982. *The Foundations of Economic Method*. London: George Allen & Unwin.

Boland, L. 1989. *The Methodology of Economic Model Building: Methodology after Samuelson*. London: Routledge.

Boland, L. 1992. *The Principles of Economics: Some Lies My Teachers Told Me*. London: Routledge.

Boyd, R. 1984. "The Current Status of Scientific Realism." In J. Leplin (ed.), *Scientific Realism*. Berkeley: University of California Press, 41–82.

Brennan, G. and J. Buchanan. 1985. *The Reason of Rules: Constitutional Political Economy*. New York: Cambridge University Press.

Bronfenbrenner, M. 1971. "The Structure of Revolutions in Economic Thought." *History of Political Economy* 3: 136–51.

Brunner, K. 1969. "'Assumptions' and the Cognitive Quality of Theories." *Synthese* 20: 501–25.

Buchanan, J. 1975. *The Limits of Liberty: Between Anarchy and the Leviathan.* Chicago: University of Chicago Press.

Buchanan, J. and V. Vanberg, 1979. "The Market as a Creative Process." *Economics and Philosophy* 7: 167–86.

Cairnes, J. 1875. *The Character and Logical Method of Political Economy.* 2nd edn. Reprinted. New York: A.M. Kelley, 1965.

Caldwell, B. (ed.). 1984. *Appraisal and Criticism in Economics.* London: Allen & Unwin.

Caldwell, B. 1991. "Clarifying Popper." *Journal of Economic Literature* 29: 1–33.

Camerer, C. 2003. *Behavioral Game Theory: Experiments in Strategic Interaction.* Princeton, NJ: Princeton University Press.

Camerer, C., G. Loewenstein, and M. Rabin (eds.). 2003. *Advances in Behavioral Economics.* Princeton, NJ: Princeton University Press.

Carter, M. and R. Maddock. 1984. *Rational Expectations: Macroeconomics for the 1980s?* London: Macmillan.

Cartwright, N. 1989. *Nature's Capacities and their Measurement.* Oxford: Clarendon Press.

Cartwright, N. 1999. *The Dappled World: A Study of the Boundaries of Science.* Cambridge: Cambridge University Press.

Coats, A. 1969. "Is There a 'Structure of Scientific Revolutions' in Economics?" *Kyklos* 22: 289–94.

Cross, R. 1982. "The Duhem-Quine Thesis, Lakatos and the Appraisal of Theories in Macroeconomics." *Economic Journal* 92: 320–40.

Davidson, D. 1963. "Actions, Reasons and Causes." *Journal of Philosophy* 60: 685–700.

Davis, J., D.W. Hands, and U. Mäki (eds.). 1998. *The Handbook of Economic Methodology.* Cheltenham: Edward Elgar.

de Marchi, N. 1970. "The Empirical Content and Longevity of Ricardian Economics." *Economica* 37: 257–76.

de Marchi, N. (ed.). 1988. *The Popperian Legacy in Economics.* Cambridge: Cambridge University Press.

Dillard, D. 1978. "Revolutions in Economic Theory." *Southern Economic Journal* 44: 705–24.

Dolan, E. (ed.). 1976. *The Foundations of Modern Austrian Economics.* Kansas City: Sheed & Ward.

Dow, S. 1985. *Macroeconomic Thought: A Methodology Approach.* Oxford: Blackwell.

Dugger, W. 1979. "Methodological Differences between Institutional and Neoclassical Economics." *Journal of Economic Issues* 13: 899–909.

Duhem, P. 1906. *The Aim and Structure of Scientific Theories,* trans. P. Wiener. Princeton, NJ: Princeton University Press, 1954.

Eichner, A. 1983. "Why Economics Is Not Yet a Science." In A. Eichner (ed.), *Why Economics Is Not Yet a Science.* Armonk, NY: M.E. Sharpe, 205–41.

Etzioni, A. 1988. *The Moral Dimension: Toward a New Economics*. New York: Macmillan.

Fama, E. 1980. "Agency Problems and the Theory of the Firm." *Journal of Political Economy* 88: 288–307.

Fleetwood, S. (ed.). 1999. *Critical Realism in Economics: Development and Debate*. London: Routledge.

Frank, R., T. Gilovich, and D. Regan. 1993. "Does Studying Economics Inhibit Cooperation?" *Journal of Economic Perspectives* 7: 159–72.

Friedman, M. 1953. "The Methodology of Positive Economics." In *Essays in Positive Economics*. Chicago: University of Chicago Press, 3–43.

Geweke, J. 1982. "Causality, Exogeneity and Inference." In W. Hildenbrand (ed.), *Advances in Econometrics*. Cambridge: Cambridge University Press.

Gordon, D. 1955. "Operational Propositions in Economic Theory." *Journal of Political Economy* 63: 150–61.

Granger, C. 1969. "Investigating Causal Relations by Econometric Models and Cross-Spectral Methods." *Econometrica* 37: 424–38.

Granger, C. 1980. "Testing for Causality: A Personal Viewpoint." *Journal of Economic Dynamics and Control* 2: 329–52.

Guala, F. 2005. *The Methodology of Experimental Economics*. Cambridge: Cambridge University Press.

Haavelmo, T. 1944. "The Probability Approach in Econometrics." *Econometrica* 12 (suppl.): 1–118.

Hall, R. and C. Hitch. 1939. "Price Theory and Business Behaviour." *Oxford Economic Papers* 2: 12–45.

Hamminga, B. 1983. Neoclassical Theory Structure and Theory Development: An Empirical-Philosophical Case Study Concerning the Theory of International Trade. Boston, MA: Springer.

Hammond, J.D. 1992. "An Interview with Milton Friedman on Methodology." In W. Samuels (ed.), *Research in the History of Economic Thought and Methodology*, vol. 10. Greenwich, CT: JAI Press, 91–118.

Hands, D.W. 1985a. "Karl Popper and Economic Methodology." *Economics and Philosophy* 1: 83–100.

Hands, D.W. 1985b. "Second Thoughts on Lakatos." *History of Political Economy* 17: 1–16.

Hands, D.W. 1985c. "The Structuralist View of Economic Theories: The Case of General Equilibrium in Particular." *Economics and Philosophy* 1: 303–36.

Hands, D.W. 1992. *Testing, Rationality and Progress*. Totowa, NJ: Rowman and Littlefield.

Hands, D.W. 1995. "Social Epistemology Meets the Invisible Hand: Kitcher on the Advancement of Science." *Dialogue* 34: 605–21.

Hands, D.W. 2001. *Reflection without Rules: Economic Methodology and Contemporary Science Theory*. Cambridge: Cambridge University Press.

Hands, D.W. and P. Mirowski. 1998. "Harold Hotelling and the Neoclassical Dream." In R. Backhouse, D. Hausman, U. Mäki, and A. Salanti (eds.), *Economics and Methodology: Crossing Boundaries*. London: Macmillan, 322–97.

Hausman, D. 1983. "Are There Causal Relations Among Dependent Variables?" *Philosophy of Science* 50: 58–81.

Hausman, D. 1990. "Supply and Demand Explanations and Their *Ceteris Paribus* Clauses." *Review of Political Economy* 2: 168–86.

Hausman, D. 1992a. *Essays on Philosophy and Economic Methodology*. Cambridge: Cambridge University Press.

Hausman, D. 1992b. *The Inexact and Separate Science of Economics*. Cambridge: Cambridge University Press.

Hausman, D. 2008a. "Why Look Under the Hood." In D. Hausman (ed.), 2008. *The Philosophy of Economics: An Anthology*. 3rd edn. Cambridge: Cambridge University Press, 183–7.

Hausman, D. (ed.). 2008b. *The Philosophy of Economics: An Anthology*. 3rd edn. Cambridge: Cambridge University Press.

Hausman, D. and M. McPherson. 2006. *Economic Analysis, Moral Philosophy, and Public Policy*. Cambridge: Cambridge University Press.

Helm, D. 1984. "Predictions and Causes: A Comparison of Friedman and Hicks on Method." *Oxford Economic Papers* 36 (suppl.): 118–34.

Hey, J.D. 1991. *Experiments in Economics*. Oxford: Blackwell.

Hirsch, A. and N. de Marchi. 1990. *Milton Friedman: Economics in Theory and Practice*. Ann Arbor: University of Michigan Press.

Hodgson, G. 2000. "What Is the Essence of Institutional Economics?" *Journal of Economic Issues* 34: 317–29. Reprinted in D. Hausman (ed.), *The Philosophy of Economics: An Anthology*. 3rd edn. Cambridge: Cambridge University Press, 2008, 397–410.

Hoover, K. 1988. *The New Classical Macroeconomics: A Sceptical Inquiry*. Oxford: Basil Blackwell.

Hoover, K. 2001. *Causality in Macroeconomics*. Cambridge: Cambridge University Press.

Hull, D. 1988. *Science as a Process: An Evolutionary Account of the Social and Conceptual Development of Science*. Chicago: University of Chicago Press.

Hutchison, T. 1938. *The Significance and Basic Postulates of Economic Theory*. Reprinted with a new Preface. New York: A.M. Kelley, 1960.

Hutchison, T. 1977. *Knowledge and Ignorance in Economics*. Chicago: University of Chicago Press.

Hutchison, T. 1978. *On Revolutions and Progress in Economic Knowledge*. Cambridge: Cambridge University Press.

Hutchison, T. 2000. *On the Methodology of Economics and the Formalist Revolution*. Cheltenham: Edward Elgar.

Jalladeau, J. 1978. "Research Program versus Paradigm in the Development of Economics." *Journal of Economic Issues* 12: 583–608.

Jensen, M. and W. Meckling. 1976. "Theory of the Firm: Managerial Behavior, Agency Costs and Ownership Structure." *Journal of Financial Economics* 3: 305–60.

Kagel, J.H. and A.E. Roth (eds.). 1995. *The Handbook of Experimental Economics*, Princeton, NJ: Princeton University Press.

Kaufmann, F. 1933. "On the Subject-Matter and Method of Economic Science." *Economica* 13: 381–401.

Keynes, J.N. 1917. *The Scope and Method of Political Economy*. 4th edn. Reprinted. New York: A.M. Kelley, 1955 [1891].

Kirzner, I. 1976. *The Economic Point of View*. 2nd edn. Kansas City: Sheed & Ward.

Knight, F. 1935. "Economics and Human Action." Reprinted in D. Hausman (ed.), *The Philosophy of Economics: An Anthology*. 3rd edn. Cambridge: Cambridge University Press, 2008, 100–7.

Koopmans, T. 1957. *Three Essays on the State of Economic Science*. New York: McGraw-Hill.

Kregel, J. 1976. "Economic Methodology in the Face of Uncertainty: The Modeling Methods of Keynes and the Post-Keynesians." *Economic Journal* 86: 209–25.

Kuhn, T. 1970. *The Structure of Scientific Revolutions*. 2nd edn. Chicago: University of Chicago Press.

Kunin, L. and F. Weaver. 1971. "On the Structure of Scientific Revolutions in Economics." *History of Political Economy* 3: 391–7.

Lakatos, I. 1970. "Falsification and the Methodology of Scientific Research Programmes." In I. Lakatos and A. Musgrave (eds.), *Criticism and the Growth of Knowledge*. Cambridge: Cambridge University Press, 91–196.

Lakatos, I. 1974. "Popper on Demarcation and Induction." In P. Schlipp (ed.), *The Philosophy of Karl Popper*. LaSalle, IL: Open Court, 241–73.

Lakatos, I. and A. Musgrave (eds.). 1970. *Criticism and the Growth of Knowledge*. Cambridge: Cambridge University Press.

Lange, O. 1945. "The Scope and Method of Economics." *Review of Economic Studies* 13: 19–32.

Latsis, S. (ed.). 1976. *Method and Appraisal in Economics*. Cambridge: Cambridge University Press.

Lawson, T. 1997. *Economics and Reality*. London: Routledge.

Leonard, T.C. 2002. "Reflection on Rules in Science: An Invisible-Hand Perspective." *Journal of Economic Methodology* 9: 141–68.

Lester, R.A. 1946. "Shortcomings of Marginal Analysis for Wage-Employment Problems." *American Economic Review* 36: 62–82.

Lester, R.A. 1947. "Marginal Costs, Minimum Wages, and Labor Markets." *American Economic Review* 37: 135–48.

Machlup, F. 1955. "The Problem of Verification in Economics." *Southern Economic Journal* 22: 1–21.

Machlup, F. 1960. "Operational Concepts and Mental Constructs in Model and Theory Formation." *Giornale Degli Economisti* 19: 553–82.

Machlup, F. 1969a. "If Matter Could Talk." Reprinted in Machlup 1978, 309–32.

Machlup, F. 1969b. "Positive and Normative Economics." Reprinted in F. Machlup, *Methodology of Economics and Other Social Sciences.* New York: Academic Press, 1978, 425–50.

Machlup, F. 1978. *Methodology of Economics and Other Social Sciences.* New York: Academic Press.

Mäki, U. 1990a. "Mengerian Economics in Realist Perspective." *History of Political Economy* 22: 289–310.

Mäki, U. 1990b. "Scientific Realism and Austrian Explanation." *Review of Political Economy* 2: 310–44.

Mäki, U. (ed.). 1991. *Fact and Fiction in Economics: Models, Realism and Social Construction.* Cambridge: Cambridge University Press.

Mäki, U. 1992. "Friedman and Realism." *Research in the History of Economic Thought and Methodology* 10: 171–95.

Mäki, U., B. Gustafsson, and C. Knudsen. 1993. *Rationality, Institutions and Economic Methodology.* London: Routledge.

Marwell, G. and R. Ames. 1981. "Economists Free Ride. Does Anyone Else? Experiments on the Provision of Public Goods. IV." *Journal of Public Economics* 15: 295–310.

Marx, K. 1867. *Capital,* vol. 1, trans. S. Moore and E. Aveling. New York: International Publishers, 1967.

McCloskey, D. 1985. *The Rhetoric of Economics.* Madison: University of Wisconsin Press.

McCloskey, D. 1994. *Truth and Persuasion in Economics.* Cambridge: Cambridge University Press.

Melitz, J. 1965. "Friedman and Machlup on the Significance of Testing Economic Assumptions." *Journal of Political Economy* 73: 37–60.

Mill, J.S. 1836. "On the Definition of Political Economy and the Method of Investigation Proper to it." Reprinted in *Collected Works of John Stuart Mill,* vol. 4. Toronto: University of Toronto Press, 1967.

Mill, J.S. 1843. *A System of Logic.* London: Longmans, Green & Co., 1949.

Minford, P. and D. Peel. 1983. *Rational Expectations and the New Macroeconomics.* Oxford: Martin Robertson & Co.

Mirowski, P. 1990. *More Heat Than Light.* Cambridge: Cambridge University Press.

Mirowski, P. 2002. *Machine Dreams: Economics Becomes a Cyborg Science.* Cambridge: Cambridge University Press.

Mises, L. von. 1949. *Human Action. A Treatise on Economics.* New Haven: Yale University Press.

Mises, L. von. 1978. *The Ultimate Foundation of Economic Science: An Essay on Method.* 2nd edn. Kansas City: Sheed Andrews.

Mises, L. von. 1981. *Epistemological Problems of Economics*, trans. G. Reisman. New York: New York University Press.

Mongin, P. 2006. "Value Judgments and Value Neutrality in Economics." *Economica* 73: 257–86.

Morgan, M. 2001. "Models, Stories, and the Economic World." *Journal of Economic Methodology* 8: 361–84.

Morgan, M. and M. Morrison (eds.). 1999. *Models as Mediators: Perspectives on Natural and Social Science.* Cambridge: Cambridge University Press.

Morishima, M. 1973. *Marx's Economics: A Dual Theory of Value and Growth.* Cambridge: Cambridge University Press.

Nelson, J. 1996. *Feminism, Objectivity and Economics.* London: Routledge.

Pasinetti, L. 1981. *Structural Change and Economic Growth: A Theoretical Essay on the Dynamics of the Wealth of Nations.* Cambridge: Cambridge University Press.

Plott, C.R. 1991. "Will Economics Become an Experimental Science?" *Southern Economic Journal* 57: 901–19.

Popper, K. 1967. "La Rationalité et le Statut du Principe de Rationalité." In E. Classen (ed.), *Les Fondements Philosophiques des Systèmes Économiques.* Paris: Paypot, 142–50.

Popper, K. 1968. *The Logic of Scientific Discovery.* Rev. edn. London: Hutchinson & Co.

Popper, K. 1969. *Conjectures and Refutations: The Growth of Scientific Knowledge.* 3rd edn. London: Routledge & Kegan Paul.

Popper, K. 1976. "The Logic of the Social Sciences." In T. Adorno et al. (eds.), *The Positivist Dispute in German Sociology*, trans. G. Adey and D. Frisby. New York: Harper, 87–104.

Quine, W.V. 1953. "Two Dogmas of Empiricism." In *From a Logical Point of View.* Cambridge, MA: Harvard University Press, 20–46.

Rabin, M. 1998. "Psychology and Economics." *Journal of Economic Literature* 36: 11–46.

Rappaport, S. 1998. *Models and Reality in Economics.* Cheltenham: Edward Elgar.

Redman, D. 1989. *Economic Methodology: A Bibliography with References to Works in the Philosophy of Science, 1860–1988.* New York: Greenwood Press.

Ricardo, D. 1817. *On the Principles of Political Economy and Taxation*, vol. 1 of *The Collected Works of David Ricardo*, P. Sraffa and M. Dobb (eds.). Cambridge: Cambridge University Press, 1951.

Robbins, L. 1932. *An Essay on the Nature and Significance of Economic Science.* 2nd edn. 1935. 3rd edn. 1983. London: Macmillan.

Roemer, J. 1981. *Analytical Foundations of Marxian Economic Theory.* Cambridge: Cambridge University Press.

Roemer, J. 1982. *A General Theory of Exploitation and Class.* Cambridge, MA: Harvard University Press.

Roncaglia, A. 1978. *Sraffa and the Theory of Prices.* Chichester: John Wiley.

Roscher, W. 1874. *Geschichte der National-oekonomik in Deutschland.* Munich: R. Oldenbourg.

Rosenberg, A. 1976. *Microeconomic Laws: A Philosophical Analysis.* Pittsburgh: University of Pittsburgh Press.

Rosenberg, A. 1980. *Sociobiology and the Preemption of Social Science.* Baltimore: Johns Hopkins University Press.

Rosenberg, A. 1988. "Economics Is Too Important to Be Left to the Rhetoricians." *Economics and Philosophy* 4: 129–49.

Rosenberg, A. 1992. *Economics – Mathematical Politics or Science of Diminishing Returns.* Chicago: University of Chicago Press.

Roth, A. 1988. "Laboratory Experimentation in Economics: A Methodological Overview." *Economic Journal* 98: 974–1031.

Rothbard, M. 1957. "In Defense of 'Extreme Apriorism.'" *Southern Economic Journal* 23: 314–20.

Rotwein, E. 1959. "On 'The Methodology of Positive Economics.'" *Quarterly Journal of Economics* 73: 554–75.

Runde, J. 1998. "Assessing Causal Economic Explanations." *Oxford Economic Papers* 50: 151–72.

Samuelson, P. 1963. "Problems of Methodology – Discussion." *American Economic Review Papers and Proceedings* 53: 232–6.

Schmoller, G. 1888. *Zur Literatur-geschichte der Staats- und Sozialwissenschaften.* Leipzig: Duncker & Humblot.

Schmoller, G. 1898. *Über einige Grundfragen der Sozialpolitik und der Volkswirtshaftslehre.* Leipzig: Duncker & Humblot.

Senior, N. 1836. *Outline of the Science of Political Economy.* Reprinted. New York: A.M. Kelley, 1965.

Sent, E.-M. 1998. *The Evolving Rationality of Rational Expectations.* Cambridge: Cambridge University Press.

Simon, H. 1953. "Causal Ordering and Identifiability." In W. Hood and T. Koopmans (eds.), *Studies in Econometric Method.* New York: John Wiley & Sons, 49–74.

Sims, C. 1977. "Exogeneity and Causal Orderings in Macroeconomic Models." In C. Sims (ed.), *New Methods in Business Cycle Research.* Minneapolis, MN: Federal Reserve Bank, 23–43.

Smith, A. 1776. *An Inquiry into the Nature and Causes of the Wealth of Nations.* Reprinted. New York: Random House, 1937.

Smith, V.L. 1991. *Papers in Experimental Economics.* Cambridge: Cambridge University Press.

Sneed, J. 1971. *The Logical Structure of Mathematical Physics.* Dordrecht: Reidel.

Sraffa, P. 1960. *Production of Commodities by Means of Commodities: Prelude to a Critique of Economic Theory.* Cambridge: Cambridge University Press.

Stanfield, R. 1974. "Kuhnian Revolutions and the Keynesian Revolution." *Journal of Economic Issues* 8: 97–109.

Starmer, C. 1999. "Experiments in Economics: Should We Trust the Dismal Scientists in White Coats?" *Journal of Economic Methodology* 6: 1–30.

Stegmüller, W. 1976. *The Structure and Dynamics of Theories*, trans. William Wohlhueter. New York: Springer-Verlag.

Stegmüller, W. 1979. *The Structuralist View of Theories*. New York: Springer-Verlag.

Stegmüller, W., W. Balzer, and W. Spohn (eds.). 1982. *Philosophy of Economics: Proceedings, Munich, July 1981*. New York: Springer-Verlag.

Sugden, R. 2000. "Credible Worlds: The Status of Theoretical Models in Economics." *Journal of Economic Methodology* 7: 1–31.

Veblen, T. 1898. "Why Is Economics Not an Evolutionary Science?" *Quarterly Journal of Economics* 12: 373–97.

van Fraassen, B. 1980. *The Scientific Image*. Oxford: Oxford University Press.

von Wright, G.H. 1971. *Explanation and Understanding*. Ithaca, NY: Cornell University Press.

Watkins, J. 1984. *Science and Scepticism*. Princeton: Princeton University Press.

Weber, M. 1904. "'Objectivity' in Social Science and Social Policy." In E. Shils and H. Finch (eds.), *The Methodology of the Social Sciences*. New York: Free Press, 1949, 49–112.

Weintraub, E.R. 1985. *General Equilibrium Analysis: Studies in Appraisal*. Cambridge: Cambridge University Press.

Weintraub, E.R. 1991. *Stabilizing Dynamics: Constructing Economic Knowledge*. Cambridge: Cambridge University Press.

Williamson, O. 1985. *The Economic Institutions of Capitalism: Firms, Markets, Relational Contracting*. New York: Free Press.

Winch, P. 1958. *The Idea of a Social Science*. London: Routledge.

Winter, S. 1962. "Economic 'Natural Selection' and the Theory of the Firm." *Yale Economic Essays* 4: 255–72.

Wiseman, J. (ed.). 1983. *Beyond Positive Economics?* London: British Association for the Advancement of Science.

Wisman, J. and J. Rozansky. 1991. "The Methodology of Institutionalism Revisited." *Journal of Economic Issues* 25: 709–37.

Wold, H. 1954. "Causality and Econometrics." *Econometrica* 22: 162–77.

Worland, S. 1972. "Radical Political Economy as a 'Scientific Revolution.'" *Southern Economic Journal* 39: 274–84.

Zellner, A. and D. Aigner (eds.). 1988. *Causality. Annals, Journal of Econometrics*.

Index